Ecological Studies, Vol. 141

Analysis and Synthesis

Ecological Studies

Volumes published since 1993 are listed at the end of this book.

Springer

Berlin
Heidelberg
New York
Barcelona
Hong Kong
London
Milan
Paris
Singapore
Tokyo

S. Halle N. C. Stenseth (Eds.)

Activity Patterns in Small Mammals

An Ecological Approach

With 59 Figures and 11 Tables

Prof. Dr. Stefan Halle
Institute of Ecology
Friedrich-Schiller-University Jena
Dornburger Str. 159
07743 Jena
Germany

Prof. Dr. Nils Christian Stenseth
Department of Biology
Division of Zoology
University of Oslo
P.O. Box 1050, Blindern
03160 Oslo 3
Norway

ISSN 0070-8356
ISBN 3-540-59244-X Springer-Verlag Berlin Heidelberg New York

Library of Congress Cataloging-in-Publication Data

Activity patterns in small mammals: an ecological approach / S. Halle, N. C. Stenseth (eds.) p. cm. – (Ecological studies, ISSN 0070-8356 ; v. 141) Includes bibliographical references. ISBN 354059244X (alk. paper) 1. Mammals – Behaviour – Evolution. 2. Mammals – Ecology. I. Halle, S. (Stefan), 1956 – II. Stenseth, Nils Chr. III. Series.
QL739.3 .A28 2000; 599.15–dc21; 99-058890

Springer-Verlag is a company in the BertelsmannSpringer publishing group.

Printed in Germany

Production: PRO EDIT GmbH, Heidelberg
Cover Design: design & production GmbH, Heidelberg
Typesetting: Ulrich Kunkel Textservice, Reichartshausen

Printed on acid free paper SPIN 10428072 31/3136/Di 5 4 3 2 1 0

Preface

Stefan Halle and Nils Chr. Stenseth

The study of animal activity behaviour represents a small but active field of research, partly relating to ecology, but somewhat isolated from the main ecological developments. Modern ecology, focusing on the performances of individuals, assumes, nonetheless, some understanding of individual activity patterns, but pays little implicit attention to the study of activity behaviour and the evolutionary ecology of temporal strategies. Our purpose with this book is to help to bridge this gap – or, to put it differently: we aim at seeing the activity patterns of individuals in an ecological context.

The idea for this book was born during a 4-year postdoctoral stay of SH in Oslo with the group of NCS, and particularly so while working together on a theoretical paper where we re-evaluated hypotheses on the triggering mechanism of microtine short-term activity rhythms (Halle and Stenseth 1994). During this work we realised that current modelling and conceptual approaches were very poorly developed in the field of activity behaviour, although the question of optimal activity timing clearly has all the ingredients for being an exciting issue in behavioural ecology. Both of us felt uneasy about this situation, and wanted, therefore, to draw together what is known today and to point out possible connections of this field with the mainstream of modern ecology. We focus on small mammals partly because of our own experience, but also – and primarily so – because of much relevant information being available for this group. However, throughout the book we take the approach of seeing small mammals as a model of some general value.

We as editors have very different backgrounds. SH is a field ecologist with a focus on empirical studies of small rodent activity in their natural habitats, but having recently become more and more interested in theoretical and conceptual considerations on activity behaviour. NCS is a theoretically oriented population ecologist who in the past has primarily worked on questions related to dynamic consequences of given assumptions about how individuals interact, but recently has become more interested in linking theoretical considerations (which easily may become sterile) with empirical findings of different kinds (often being rather muddled). During our time of close cooperation, the empiricist got to appreciate the value of translating rather loose verbal arguments into more precise mathematically formulated models, and the theorist got to appreciate the many exciting issues relating to the wealth of observations on activity patterns of small mammals. This book embodies the merging of the two perspectives we represent. Rather than giving a set of final 'solutions' we have aimed at providing – within the one book – theoretical considerations pertinent to activity patterns of animals as well as observations on small mammal activity patterns under various environmental conditions.

It is our hope that this book will serve as a reference book for established researchers in the field, and as a starting point – and an entry to the relevant literature – for newcomers to the field, be it already established scientists wanting to approach these issues or new research students. To help newcomers in the field, we have included an extensive Appendix providing a guided tour into both empirical aspects of how to monitor activity behaviour as well as how to analyse the obtained data.

We have benefited from collaboration with a variety of people; too many to mention all of them. In particular, however, we would like to thank William Z. Lidicker Jr., who shaped the view of NCS for the extensive benefits being gained from joining theoretical and empirical approaches during the work on a book on animal dispersal (Stenseth and Lidicker 1992). The late Ulrich Lehmann and Hermann Remmert were of crucial importance for developing SH's view on activity behaviour. We would also like to thank the authors and the publisher for good co-operation and their patience when we had, for various reasons, to modify the time schedules for publication again and again. In particular we would like to thank Andrea Schlitzberger for her sympathetic handling of our book project, and Theodora Krammer for her careful copy editing of the book manuscript. The postdoc stay of SH, which was essential for the development of the idea for this book, was funded by the German Science Foundation DFG (grant number HA 1513 / 2-1). NCS would like to thank the Norwegian Science Council and the University of Oslo for their continuous and generous support. Finally, we would like to thank Barbara Halle, the staff of the Institute of Ecology at Jena and the Friedrich-Schiller-University Jena for supporting our project in a variety of ways.

Jena and Oslo, March 2000 Stefan Halle and Nils Chr. Stenseth

References

Halle S, Stenseth NC (1994) Microtine ultradian rhythm of activity: an evaluation of different hypotheses on the triggering mechanism. Mammal Rev 24:17–39

Stenseth NC, Lidicker WZ (1992) Animal dispersal – small mammals as a model. Chapman and Hall, London

Foreword

Richard M. Sibly

How animals spend their days and nights is a topic of compelling interest to every student of animal behaviour. The discovery of what animals do, however, is not the end of the story. At once one wants to know the answers to all of Tinbergen's (1963) classic questions about how and why they do it. Here, after a respectful nod at causal explanations which answer the question 'how?', I consider the necessity for and difficulty of establishing survival-value explanations ('why?').

Much is now known about the relationships between animal activity patterns and the four major environmental cycles, of day and night, of the tides, of the moon and of the seasons (Aschoff 1984). The causal mechanisms underlying such entrainments between activity patterns and environmental cycles have been studied formally using mathematical models, and physiologically with the aim of identifying the cellular and molecular components of the entrainment mechanisms (Rusak 1989). Complementary to these causal considerations are analyses of survival value that aim to identify the "good and bad times for a particular activity" (Enright 1970) and, going further, to quantify fitness costs and benefits in relation to environmental cycles (Daan and Aschoff 1982). These studies of survival value nowadays lie within the purview of behavioural ecology, and, as the editors of this volume point out, it is surprising that they have not been more vigorously pursued in recent years. There have, however, been some notable exceptions. In a useful and comprehensive review Daan (1981) treated the daily routines in the behaviour of individual animals as strategies to cope with the time structure of the environment. Many illuminating examples were given of ways in which selection pressures act or may act, either directly, on mortality or mating success, or indirectly, by long-term accumulation of physiological factors which affect the long-term expectation of progeny.

In a stimulating discussion of the reasons for nocturnal activity in small nocturnal mammals, Daan and Aschoff (1982) suggested that fitness costs of spontaneous activity decline rapidly around dusk, when diurnal predation fades out (Fig. 1). Simultaneously, the costs of being inactive rise because of intraspecific competition for space and mates. The intersection of the two cost curves (Fig. 1) represents the optimum time for the onset of activity. This survival-value approach was different from the traditional causal chronobiological analyses, and opened up new lines of research. In particular, it provides the starting point for the present book. As Daan and Aschoff em-

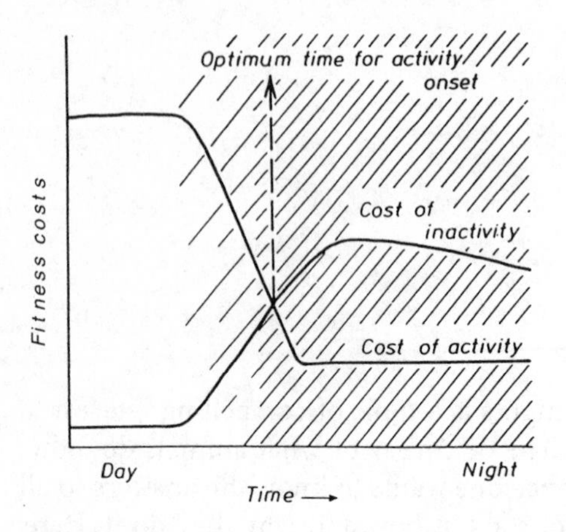

Fig. 1. Hypothetical fitness costs for a small nocturnal mammal of being active or inactive as a function of time around sunset. See text for discussion. (From Daan and Aschoff 1982)

phasised, however, the fitness costs and benefits represented in Fig. 1 are not easily measured. To see the reason for these measurement and observability problems we need to go back to the subject's foundations, to see which variables need to be measured, and why.

The theoretical foundations of behavioural ecology lie in the genetical theory of life-history evolution. These foundations are now firmly secured. We know that in general genes spread in populations if they decrease the mortality rates or increase the production rates of their carriers, allowing them to breed earlier and/or to produce more offspring (see e.g. Sibly and Antonovics 1992). It follows from this that there are, in general terms, two main goals of the daily cycle, to maximise production rate, and to minimise mortality rate. When these goals oppose each other, for instance when animals can only secure more or better food by taking risks, then it may be necessary to calculate the importance of each goal relative to the other. To see how this is done, suppose the feeding opportunity would result in increased production and that this would have a large effect on fitness, whereas increased mortality rate only has as small effect. Clearly, feeding would be favoured by natural selection in these circumstances. The fitness effects can be calculated by taking into account the sensitivity of Darwinian fitness to performance relative to each goal. Both sensitivity and Darwinian fitness are defined using established formulae from life-history theory expressed in terms of life-history parameters, i.e., ages at breeding, fecundities and survivorships (see e.g. Sibly and Calow 1986; Caswell 1989). It turns out that the fitness calculations not only indicate the selection pressures on the animal, but also reveal the priorities that should be given to different activities (here feeding and vigilance). Decision-making in herring gulls has been analysed using this sort of approach by Sibly and McCleery (1985).

In the social context a further dimension is added, when animals compete or co-operate, so affecting each others' production and mortality rates. Here the relatedness of the animals is crucially important, but difficult to establish in the field. However if this is known then it can be taken into account to calculate whether genes make net fitness profits from particular behaviours they may induce in their carriers (see e.g. Sibly 1994).

Now consider how the twin goals of production maximisation and mortality minimisation can be achieved in practice. In an environment of unchanging opportunity, there is likely to be a simple alternation of production-enhancing and mortality-reducing behaviours, and we would not expect any temporal variation or patterning on any larger time scale. An unchanging opportunity, however, is a contradiction in terms, since opportunities occur by definition when environments offer advantages for limited periods only.

The most important examples of opportunities are related to the day/night cycle. This is because, for most organisms, the day/night cycle is the dominant feature affecting the fitness costs and benefits of many behaviours, and as a result most organisms are primarily adapted in their sensory and empowering mechanisms to either day or night. Necessarily therefore, in these cases either day or night provides opportunity, the other does not.

Once a predictable, limited window of opportunity is established, however, there are immediately further consequences for the patterning of behaviour. Firstly, since conspecifics will also be unusually active in these periods, intraspecific competition must then be increased. This may reduce the production rates of individual animals. The same holds for interspecific competitors occupying a similar niche. Secondly, the increased activity of primary foragers may create opportunities for their predators and parasites. This is expected to increase prey mortality rates, since if more predators are hunting, more prey may be killed. On the other hand this could be offset in part by the 'dilution' effect that arises because there is greater activity on the part of conspecifics/competitors.

Further complexity may be added if the processing of resources is to some extent independent of their acquisition, as is sometimes true of the digestion of food. Then resource acquisition may occur in the window of opportunity, but processing may occur both inside and outside the window. There may then be an advantage in storing resources at the end of the window for processing later, as in gut-filling strategies (Daan 1981; Sibly 1981), and when the next window opens, there may be a premium on early collection of resources so that processing facilities can be used immediately that would otherwise have lain idle. These processes of collection at the start of the window and storage at its end suggest that the ends of the window will be unusually busy. If this happens, however, the character of the opportunity changes during the window and, as a result, the activities of conspecifics and competitors, predators, prey and parasites may change further.

It follows from the above that we need to measure, for each opportunity and for each resource, the animal's need for the resource, the rate at which it can be acquired, and the mortality risks that would accompany its collection. Since all of these may change during the window of opportunity, an assessment of time changes is also necessary.

The easiest of these variables to measure are the rates at which animals can feed and drink. At least in this case measured rates of resource acquisition may, with luck, be obtainable for analysis. There is the possibility that the animal uses different methods at different times – it may put more effort into foraging when it is hungrier – but by careful observation of behaviour this can be checked. There remains the problem of ascertaining what might have been achieved when the animal did not attempt to acquire resources. However, circumstantial evidence may allow some kind of assessment to be made in this case.

The animal's need to collect resources is much harder to ascertain. This is because internal variables – state of energy reserves, amount of digesta in the gut – are not readily measured without catching the animal, but if the animal is caught normal behaviour may not be resumed after release at least for some time. This is a classic observability problem – the act of measurement precludes further observation. The hardest variables to measure are mortality risks. Mortality risks are always crucially important in behavioural ecology but are generally extremely difficult to measure precisely. The difficulties arise partly because the sample sizes needed are experimentally very difficult to arrange, and partly because of ethical worries about the suffering caused to the animals involved.

The challenge of quantifying opportunities in terms of fitness costs and benefits is taken up in the following chapters, allowing development of the research agenda implicit in Daan and Aschoff (1982). Those ideas in turn, however, owe something to a seminal essay by J.T. Enright (1970), and it seems appropriate to conclude this foreword with the words with which he began:

"No description of where an animal lives and what it does can be complete without considering when the activity takes place, because animals are obviously adapted to perform given activities at given environmental times: certain seasons, times of day, or phases of the tides. These environmental cycles result in a temporal subdivision of an organism's life cycle into good and bad times for a given activity, and the adaptations which permit an organism to function most efficiently at a particular time of day, stage of the tide, or season cover the entire gamut: morphological, physiological, and behavioural."

References

Aschoff J (1984) Circadian timing. Ann NY Acad Sci 423:442–468

Caswell H (1989) Matrix population models. Sinauer, Sunderland, Massachusetts

Daan S (1981) Adaptive daily strategies in behaviour. In: Aschoff J (ed) Biological rhythms. Handbook of behavioural neurobiology, vol 4. Plenum Press, New York, pp 275–298

Daan S and Aschoff J (1982) Circadian contributions to survival. In: Aschoff J, Daan S, Groos GA (eds) Vertebrate circadian systems. Springer, Berlin Heidelberg New York, pp 305–321

Enright JT (1970) Ecological aspects of endogenous rhythmicity. Annu Rev Ecol Syst 1:221–238

Rusak B (1989) The mammalian circadian system: models and physiology. J Biol Rhythms 4: 121–134
Sibly RM (1981) Strategies of digestion and defecation. In: Townsend CR, Calow P (eds) Physiological ecology: an evolutionary approach to resource use. Blackwell, Oxford, pp 109–139
Sibly RM (1994) An allelocentric analysis of Hamilton's Rule for overlapping generations. J Theor Biol 167:301–305
Sibly RM, Antonovics J (1992) Life-history evolution. In: Berry RJ, Crawford TJ, Hewitt GM (eds) Genes in ecology. Symposium of the British Ecological Society, Blackwell, Oxford, pp 87–122
Sibly RM, Calow P (1986) Why breeding earlier is always worthwhile. J Theor Biol 123:311–319
Sibly RM, McCleery R (1985) Optimal decision rules for herring gulls. Anim Behav 33:449–465
Tinbergen N (1963) On aims and methods in ethology. Z Tierpsychol 20:410–433

Contents

Section IV Conclusion

Appendix

Subject Index [illegible]

Contributors

ALBERS, H. E., Departments of Psychology and of Biology, Georgia State University, Atlanta, GA 30303-3083, USA

BARTNESS, T. J., Departments of Psychology and of Biology, Georgia State University, Atlanta, GA 30303-3083, USA

BEHRENDS, P. R., Department of Psychology, McMaster University, Hamilton, Ontario, L8S 4K1, Canada

DALY, M., Department of Psychology, McMaster University, Hamilton, Ontario, L8S 4K1, Canada

DELIBES, M., Estación Biológica de Doñana, Consejo Superior de Investigaciones Científicas, Avda. de Maria Luisa s/n, Pabellón del Perú, 41013 Sevilla, Spain

ERKERT, H. G., Zoological Institute/Animal Physiology, Eberhard-Karls-University Tübingen, Auf der Morgenstelle 28, 72076 Tübingen, Germany

FLOWERDEW, J. R., Department of Zoology, University of Cambridge, Downing Street, Cambridge CB2 3EJ, UK

HALLE, S., Institute of Ecology, Friedrich-Schiller-University Jena, Dornburger Str. 159, 07743 Jena, Germany

HELDMAIER, G., Department of Biology/Zoology, Philipps-University Marburg, 35032 Marburg, Germany

MERRITT, J. F., Powdermill Biological Station, Carnegie Museum of Natural History, Rector, PA 15677-9605, USA

PALOMARES, F., Estación Biológica de Doñana, Consejo Superior de Investigaciones Científicas, Avda. de Maria Luisa s/n, Pabellón del Perú, 41013 Sevilla, Spain

RUF, T., Research Institute of Wildlife Ecology, University of Veterinary Medicine at Vienna, 1160 Vienna, Austria

SIBLY, R. M., School of Animal and Microbial Sciences, University of Reading, Reading RG6 6AJ, UK

SMALLWOOD, J. A., Department of Integrative Biology, University of California, Los Angeles, CA 90024-1606, USA

STENSETH, N. C., Division of Zoology, Department of Biology, University of Oslo, P.O. Box 1050, Blindern, 0316 Oslo 3, Norway

VESSEY, S. H., Department of Biological Sciences, Bowling Green State University, Bowling Green, OH 43403-0212, USA

WAUTERS, L. A., Department of Biology, University of Antwerp, U.I.A., 2610 Wilrijk, Belgium. *Present address:* Instituto Oikos, Viale Borri 148, 21100 Varese, Italy

WEINER, J., Institute of Environmental Sciences, Jagiellonian University, ul. Ingardena 6, 30-060 Krakow, Poland

WEINERT, D., Institute of Zoology, Martin-Luther-University Halle-Wittenberg, 06099 Halle, Germany

WILSON, M. I., Department of Psychology, McMaster University, Hamilton, Ontario, L8S 4K1, Canada

ZIELINSKI, W. J., Redwood Sciences Laboratory, Pacific Southwest Experiment Station, USDA Forest Service, Arcata, CA 95521, USA

ZIV, Y., Department of Life Sciences, Ben Gurion University, Beer Sheva 84105, Israel

Section I Introduction

1 Introduction

Stefan Halle and Nils Chr. Stenseth

1.1 Two States of Life

The life of almost every animal is divided into two basically different behavioural states: activity and rest. During activity a variety of vital tasks essential to ensure the maintenance of life are performed. The most prominent is foraging, but exploration, search for mates, and, in territorial species, patrolling and defence of the home range are also indispensable enterprises. On the other hand, activity is a state of increased mortality risk due to predators, harsh weather conditions, or any other source of disaster. Also, energetic expenditures are considerably increased during activity, caused by locomotion, stress, and thermoregulation.

Resting may be considered the antipode of activity, since conditions are just the opposite. Small mammals – the focal group of species in this book – normally rest in dens, burrows, or nests. These facilities provide insulation and often serve as relatively safe refuges from predation. Energy expenditures during rest are low, but on the other hand the animal has to live on the stock of energy reserves gathered during activity. These reserves may be presumptive energy in the form of freshly ingested food, which first has to be

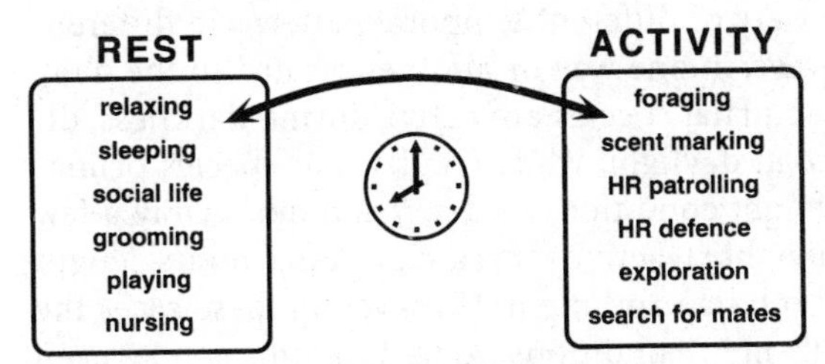

Fig. 1.1. Essentials of diel activity patterns as temporally structured alternations between 'rest' and 'activity', in which both behavioural states are generic terms for a variety of different events: 'activity' comprises everything which is indispensable for keeping alive, for protecting the home range (*HR*), and for mating. 'Rest', on the other hand, is more than simply 'non-activity' because important social interactions occur besides recovery. 'Activity *patterns*' arise because the two states and the transitions from one state to the other are systematically related to specific times

Ecological Studies, Vol. 141
Halle/Stenseth (eds.) Activity Patterns in Small Mammals

processed by digestion before being available. The reserves may also consist of externally stored food, allowing for prolonged stays in the safe refuges when unfavourable conditions prevail outside. During rest periods, animals primarily perform what is often called 'comfort behaviour', such as sleeping, grooming, playing, and intense social contacts with other group members in communal nesting species. Also, lactation and care of young is commonly performed in the safer nests or burrows.

Activity may be regarded as the working part of life, necessary to afford the economic base for the luxury of resting. Alternatively resting may be considered a phase of recovery, necessary to gather new strength for the challenges of activity. Irrespective of the point of view, activity and rest are alternating, since neither one or the other behavioural state can be retained permanently. Repeated alternations between the two states result in an explicit temporal pattern of one or several activity cycles (i.e., a complete sequence of one activity and one rest phase) during the 24-h daily cycle, commonly called the 'diel activity pattern' or, in short, the 'activity pattern' of a species. One should always be aware, however, that activity patterns in fact reflect systematic transitions between two types of behaviour, rather than a switch between 'activity' and 'non-activity' (Rusak 1989). Furthermore, both 'activity' and 'rest' are in fact generic terms for a variety of very different events (Fig. 1.1).

1.2 Diel Patterns and the Biological Clock

When observing animals in captivity or in the field, one readily realises that the diel activity pattern is characteristic of each species. Although we are confronted with a puzzling variety of different temporal patterns in different species, these patterns are always, in one way or another, related to the daily oscillation in illumination. Nocturnal species are active during darkness, diurnal species are active in broad daylight, while crepuscular species primarily roam about under the twilight conditions of dawn and dusk. Only a few species exhibit extraordinarily short activity cycles of some hours length and, as a result, are active at both day and night. However, in these cases the activity level often differs with light conditions. Activity patterns may even vary in one and the same species, e.g. in the course of seasons or in relation to the animal's sex, age or reproductive state. The striking diversity of activity patterns between species suggests that timing of activity and rest is a crucial trait for survival, and that the different temporal strategies reflect different ecological constraints. Temporal patterns are particularly diverse in small mammals, which is one of the reasons why we focus on this group in our book.

The existence of characteristic and distinctive temporal patterns raises two inherently different questions (Fig. 1.2). The first is *why* a particular species or individual restricts activity to a particular time of the day, and the second is *how* the species or individual manages to time its activity accurately (cf. Sibly in the Foreword). The first question is related to behavioural ecology and the evolution of temporal activity patterns, while the latter is related to time assessment. Largely due to historic reasons and methodological constraints, the second question was the first to be taken up - soon after Szymanski published the earliest systematic study on animal activity patterns in 1920.

The resulting biased development in this field of research may partly be explained by the early discovery of circa-rhythms (see Bünning 1977). When animals are exposed to constant light or constant darkness, the characteristic alternation between activity and rest persists and the activity phases follow the schedule exhibited under natural light conditions for an extended period. Indeed, this was a most exciting and sensational finding, since it implied that activity is not simply switched on and off by exogenous environmental signals, but that animals are instead capable of assessing time to keep to the rhythm by some kind of an intrinsic biological clock. This clock was assumed to give the individual a feeling, if not knowledge, for the flow of time, such that a constant continuum without any entraining signal is nevertheless divided up into subjective days and subjective nights. The excitement increased even further by the observation that such clocks were found in almost every living organism, ranging from unicellular algae to higher plants, from worms to insects, fish and birds up to mammals, including humans (Aschoff 1965; Bünning 1977; Winfree 1988).

The search for the physiological processes driving the biological clock clearly dominated the research in this area for most of this century, and clock mechanisms are still under lively debate today. Chronobiology, as this field of research was termed, rapidly developed to an efficient and productive branch of life sciences, with its own academic societies, journals, and jargon. Chronobiology was also one of the pioneering fields where modelling was employed as a new tool for the analysis and interpretation of data and as a concept for developing and evaluating hypotheses (e.g. Aschoff and Wever 1962; Wever 1965). Until now, the literature on biological rhythms has grown to an almost endless shelf of research papers and books, which can hardly be digested by a single researcher. This furious development was also catalysed by the practical importance of chronobiology for applied problems in medical sciences, psychology, and even labour research (Aschoff 1963, 1966, 1984; Winfree 1988), which will probably be appreciated by everyone who has experienced jet-lag or the doubtful pleasures of shift-work.

Besides insects and birds, small mammals - with small rodents like rats and hamsters in particular - were favourite model-systems for experimental studies in chronobiology. Practical handiness was probably one of the most important reasons, since small rodents are easy to keep and to breed in the laboratory and do not need much space. Small rodents are also most 'cooperative' in experimental work since they eagerly use the exercise wheel, as experienced by anyone who has ever had a golden hamster as a pet (those people will probably also subscribe to the statement that temporal conflicts may occur, since the activity patterns of kids and pets are not well synchronised in the case of golden hamsters). The rotating running wheel facilitates the technique of activity recording considerably and gives an easy and superficially straightforward activity measure (see, however, the Appendix for a critical evaluation of wheel running and a review on the methodology of activity recording). Finally, the activity rhythm of small rodents can easily be manipulated by changing the light regime, which allows one to explore the features of the rhythm. As a consequence, a lot is known today about the activity triggering mechanism in this group of animals, although the actual system of time assessment still remains somewhat mysterious (see Chap. 2 for a review on biological clocks in small mammals).

1.3 Doing the Right Thing at the Right Time

Unfortunately, the broad interest in biological clocks distracted attention from the first, and somewhat more basic question: *why* is a particular time of the day chosen for activity? In his important 1970 review, Enright presented an argument of disarming logic: the extensive research on biological clocks revealed that time assessment mechanisms are most complex and sophisticated devices. Such complex mechanisms would only have a chance to develop and establish in the course of evolution if considerable fitness benefits are related to the ability to assess time. This means that choosing the right time for activity and rest and performing specific activities at the most appropriate time of the day have to result in better survival and/or increased reproductive success. Since the adaptive value of a specific trait is determined by how it fits to the environment the individual is living in, one undoubtedly has to deal with the ecological implications of temporal behaviour in order to understand the diel activity patterns we observe in nature, and to unravel the perplexing diversity of daily time schedules.

Following Chiba's view expressed in a volume on circadian clocks and ecology (Chiba 1989), the ecological approach implies a shift in the concerned level of the problem: "An ecological aspect of circadian rhythm study is to aim at investigating the causal relationship between levels from community down to the individual or even to some physiological level." There are complex interactions between physiological processes, individual behaviour, and the ever changing environment, and selection will favour individuals that manage best to keep all these components in pace. However, whereas the physiological approach is mainly related to the proximate reasons for activity timing, the ecological approach aims at an ultimate and evolutionary understanding of temporal behaviour (DeCoursey 1990). In the ecological approach it is not of primary importance how the animal assesses time and how the clock functions. We simply assume – and accept the fact – that animals have clocks, and that these clocks are in one way or another working with impressive precision. Instead, we ask how time assessment helps to improve survival, or in other words, we ask for the *purpose* of biological clocks.

Environmental conditions vary in many respects in the course of the daily 24-h period. First of all, there is the ubiquitous daily light/dark (LD) cycle, but temperature and humidity follow along. The chance, or danger, of encountering someone else, be it of the same or a different species, is likely to be higher during certain periods compared with others, depending on the activity pattern of the counterpart. As a result, the costs of activity in terms of energetic expenses and mortality risk will vary. An individual is obviously better off when it is able to pick just the hours when activity is relatively cheap and to rest at those hours when the net gain of activity is likely to be negative. Or, as Sibly framed it in the Foreword, there are "limited windows of opportunity" for different activities. If we accept this, and there is hardly any reason not to do so, selective pressures for the development of mechanisms that restrict activity to the most advantageous time windows are easily to imagine.

The ultimate reasons for a specific diel activity pattern may be as different as the ways of life of different species (see Chap. 4 for a more detailed discussion). For predators, the activity pattern of the prey may be the most relevant feature because it results in predictable temporal variations of food availability. Prey, on the other hand, will profit most by choosing the time of the day for activity when predation risk is relatively low. Temporal variation in food availability over the 24-h day is presumably not pertinent in herbivores, whose activity cycles may instead be largely governed by the bulkiness of food and digestive abilities. Activity behaviour may also be ruled by the social environment, since activity timing determines whether potential mates are met or whether competitors are avoided. Similarly, competitive species may be separated by utilising different temporal niches. However, reflecting on possible ultimate reasons for activity behaviour we should not forget that activity timing may also simply be a direct response to unfavourable conditions like cold temperatures or poor light at some hours, and a more favourable environment at other times of the day.

Seen in this light, the enormous diversity of temporal strategies observed in different species is no longer astounding, but in fact expected. The meaning of biotic and abiotic features of the environment is obviously different in different species, and will change considerably in the course of seasons or the life stages of an individual in some cases. Even more so, the activity pattern of a given species or individual may be seen as a reflection of the ecological situation it is living under, and may thus reveal relevant information about the general characteristics of the ecological niche.

Accepting that it is advantageous to restrict activity and rest to specific time windows, we have to face the subsequent question, which is why *that* particular time of the day is chosen by a given species. Obviously, this is the most challenging question to answer. To approach it, we have to differentiate between various types of activity rather than pooling highly contrasting enterprises such as foraging, mating, home range defence, and exploration under just one heading, 'activity'. Presumably we also have to break down the activity range into different sections, because behavioural routines like patrolling, for instance, will probably differ between the interior and the edges of the home range. We also have to consider habitat structure and vegetation cover, because the possibilities of hiding and escape will probably affect activity timing. Finally, population density and inter-individual variation in temporal strategies must be looked at since individual fitness as a consequence of the activity pattern will probably be subject to frequency-dependent selection. So, detailed knowledge about the everyday life of single individuals and field data of high complexity are required.

1.4 The History of an Idea

The idea of an ecological dimension of diel temporal patterns is by no means new. In fact, most of the arguments can already be found in a series of significant and aged papers. The first in this row is the paper by Park (1940), who discussed the possible ultimate reasons for the evolutionary establishment of nocturnal habits. In this paper he complained about the lack of data from the natural environment when writing that "much of this (available) information concerns common laboratory animals, under more or less unnatural surroundings, or which have their routine activities controlled or affected by man." This point was in fact already made three years earlier by Hamilton (1937), who carried out one of the first quantitative studies on vole activity in the field. Five years after Park's contribution, Calhoun (1945) claimed that his results from sophisticated cage experiments with voles and cotton rats "should be viewed from the standpoint that an animal's behavior gives some indication as to its relationship to the environment and more particularly to its adjustments to the community of which it is a part." Crowcroft in his well-known study on shrew activity stressed the importance of temporal behaviour for interspecific strife in writing that "the investigation of the activity patterns of associated species is of particular interest as it is a necessary prelude to the study of competition between them" (Crowcroft 1953). Kavanau (1969) distinguished between a 'visual activity type' that can be determined in the laboratory and primarily reflects adaptations of the visual system, and the much more flexible 'ecological activity type' as an expression of complex interactions of the organism with habitat factors like predator-prey interactions and disturbances, for instance.

In 1970, Enright put forward his still appealing, but somewhat disturbing, view that biological clocks were presumably not invented for the pleasure of chronobiologists, but for the good of their possessors: "The vast and ever-increasing body of literature on these endogenous rhythmicities is apt to give the ecologist the uncomfortable impression that he is confronted by an aspect of behavioural physiology which he understands only slightly and yet which, in some way not readily apparent, should have a direct bearing on ecology. … The suspicion, therefore, is unavoidable that the rhythms have a direct ecological significance and have arisen through evolution as a result of appreciable selective advantages." In this view the use of time assessment mechanisms is to assure "that the appropriate activity be performed at the right time in the right place." Consequently, the perspective of Enright's view was "to examine the ecological consequences of internal timing mechanisms under field conditions and the possible selective advantages and disadvantages of endogenous rather than exogenous timing." Interestingly, he considered his contribution a "personal, exploratory essay" rather than an ordinary review, which would have been impossible "because too little attention has been given in the previous literature to an evaluation of ecological significance."

Most of the later papers about the ecological consequences of and the constraints for activity timing, and this book in particular, can be considered as intellectual successors to Enright. His 1970 review clearly framed questions as well as conceptual approaches, and was published in a well-respected and widelyspread journal, so it could have marked the birth of a new branch in behavioural ecology research. However, only a few original papers and field studies picked up Enright's arguments in the following years, while the conceptual issues were developed further by several reviews. Two years after Enright, Ashby (1972) compared diurnal and nocturnal mammals and asked how the specific conditions during the two halves of the 24-h day shaped the design of species. With respect to future research he concluded that "a combination of much more extensive field observation and of laboratory experiments designed to replicate natural conditions as closely as possible seems to be the main essential for further progress in the study of the activity patterns of mammals." He also supposed that small mammals with body weights below about 500 g may be of particular interest, because in this size class the energetic and metabolic constraints are expected to be especially burdening. Again 2 years later Schoener (1974), in his review on resource partitioning, considered temporal behaviour as a possible dimension for niche separation, although he regarded this dimension to be the least important, except for some special cases (see Chap. 4).

The next milestone is represented by two outstanding reviews, one by Daan (1981) and the other by Daan and Aschoff (1982). These two papers are still the most distinctive and concise surveys of selective advantages of activity timing that specifically accentuated the evolutionary ecology view: "Daily routines in the behavior of individual animals can be viewed as strategies to cope with the time structure of the environment", and "ultimate factors in the environment, such as predation or swarming conditions (air temperatures), favor the genetic material of those animals responding appropriately to the proximate cues" (both quotations from Daan 1981). Again, the considerable bias in activity data to ecologically incomprehensible cage experiments was clearly denounced: "A major impediment is that our knowledge of circadian systems is largely based on spontaneous locomotor activity, especially wheel running, which in itself is poorly understood" (Daan and Aschoff 1982), and "the phenomenon has not been consequently studied in animals in natural circumstances" (Daan 1981). And finally, with still not much to add today after almost 20 years of research, "individual daily habits have hardly been the subject of experimental analysis. Yet, they form a central theme in the discussion of the adaptiveness of circadian rhythms" (Daan 1981).

The most recent attempt to establish an evolutionary ecology approach in chronobiology stems from DeCoursey when she reviewed her 35 years research on mammal biorhythms in two papers (DeCoursey 1989, 1990). The line of arguments cover the two decades after Enright's "personal essay" without any discernible disruption or transformation: "In terms of time-organized behavioral ecology of many species, a number of highly developed behavior patterns imply increased fitness shaped by selective environmental factors", and "animals presumably profit from doing the appropriate act at the right place and at the right time, paced by physiological endogenous timers." Also, pungent objection against the kind of hitherto collected data turn up again with almost predictable regularity: "The majority of research on circadian rhythms has been conducted on highly domesticated laboratory organisms", hence "little is known of circadian systems of free living or wild caught species including in a proximate sense the environmental parameters to which the circadian system of these species is responding or in an ultimate evolutionary sense the degree of selection on circadian pacemaker systems by factors of an organism's niche" (quotations from DeCoursey 1990).

1.5 Why Do We Know That Little?

From the previous section it is obvious that the ecological relevance of diel activity patterns was obvious to at least some early students in chronobiology and ecology. The idea was raised about every 10 years (Enright 1970; Daan 1981; DeCoursey 1990) by well-respected scientists and in much cited journals or books. Like the spatial organisation of animal populations, which has been intensely scrutinised during recent decades, the problem of temporal behaviour has all the ingredients of a challenging issue in behavioural ecology, because evolutionary aspects and fitness maximisation as well as mutual inter- and intraspecific relations are addressed. Now, at the turn of the millennium, this book is coming forward as a repeated attempt to emphasise these points, and again the same old themes sound like a moderately refreshed cover version of a long-known evergreen.

So the reader is left to the riddle of why the daily timing of activity has not yet become a central topic in behavioural ecology. The almost comprehensive lack in alertness about temporal behaviour may be demonstrated by the fact that activity patterns are not even mentioned by most textbooks on behavioural ecology. One exception, however, is time budgets (i.e., the question of how much time is spent on different kinds of activity) which has received some ecological attention. Time budgets are indeed an important aspect of behaviour since they are directly related to metabolism and energetic constraints that will often change in the course of seasons, in the course of an individual's life span, or in relation to the reproductive state. Hence, comparisons of time budgets under different ecological conditions allows us to draw conclusions about behavioural strategies and the animal's priorities for decision making. Furthermore, time budgeting was fruitfully related to energy budgets and the theory of risk-sensitive foraging (e.g. Lima and Dill 1990; McNamara 1992; Kacelnik and Bateson 1996; Smallwood 1996), which indeed is a state-of-the-art conceptual framework in evolutionary and behavioural ecology for approaching temporal variations in the environment. However, the environment also changes considerably in a 24-h cycle; hence just the same is also true of the diel distribution of activity. The relevant questions and the subdisciplines addressed for the examination of activity pat-

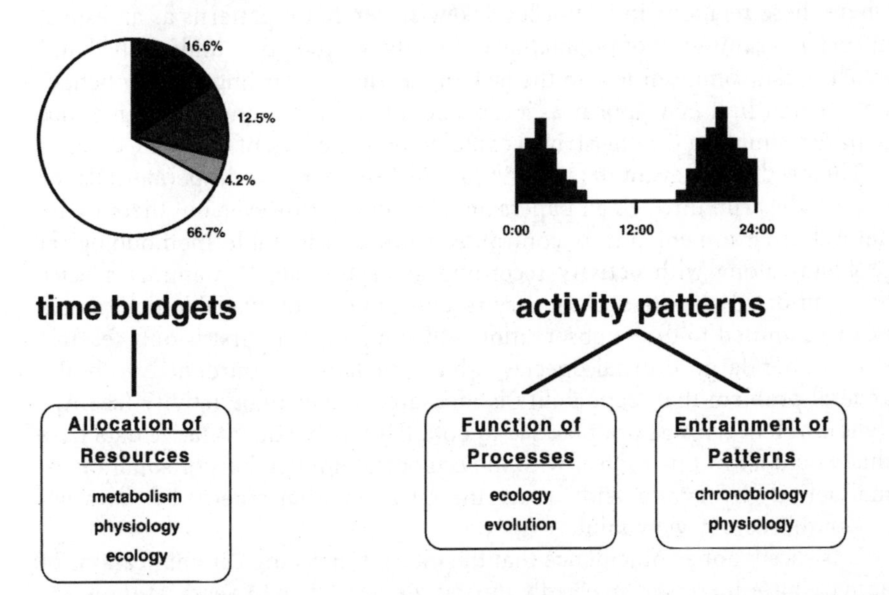

Fig. 1.2. Time budgets and activity patterns as two different aspects of the temporal behaviour of animals. Time budgets represent the allocation of time and other resources to different kinds of activity. Activity patterns describe the temporal distribution of activity over the 24-h day, which can be examined under two different aspects, i.e., the function of the underlying processes, and the mechanisms that entrain the pattern with the environment. Depending on the approach, the subdisciplines addressed are different

terns are quite different from the aspect of time budgeting (Fig. 1.2) as they aim at the function of the processes that are represented by the pattern and on the mechanisms that entrain behaviour and physiology with the changing environment.

According to Daan (1981) the fact that diel patterns were not enthusiastically taken up in research is, however, not surprising: "The detection of daily habits requires that animals be followed individually many days in a row, while their movements and behaviors are continuously monitored." With time budgets this is far easier because one only has to distinguish between different kinds of activity, and data can be collected cumulatively with rather crude temporal accuracy. An important aspect in recording activity patterns is to grasp short-term transitions from one type of activity to another, and to correlate this with various biotic and abiotic parameters in order to find hints at the proximate as well as ultimate reasons for the observed behaviour. Cage experiments, that were the major research tool in chronobiology for decades, do not provide the kind of data needed for this purpose because "the problem of interpreting laboratory results in terms of their ecological implications is nowhere more prominent than when studying activity patterns of small mammals" (Pearson 1962). Obviously, the relevance of activity patterns for predator-prey interactions and interspecific competition, for instance, can only be reasonably investigated in the natural environment where these relations in fact occur. Likewise, temporal patterns as an aspect of social organisation of populations can only be studied in undisturbed and intact social communities. So the lack of alertness regarding activity behaviour, that at first may appear as a cruel act of scientific ignorance, turns out to have a simple and rather trivial explanation – the lack of adequate data.

The need for relevant data was in fact realised early, as desperate calls for field studies run through all papers on this subject. However, out there in the natural environment one is confronted with considerable methodological problems along with activity recording of cryptic small mammal species living in untidy habitats. Until very recently, the set of available field methods was limited to direct observation and trapping. The first is only feasible with rather large diurnal species, while the latter is burdened with the general problem that trappability is measured rather than activity (see Appendix). Hence, it was not possible to obtain a reliable quantitative data base that would allow one to deal with the ecological implications of small mammal activity behaviour and to face the questions that emerged from a behavioural ecology viewpoint.

It is surely not a coincidence that the interest in ecological implications of activity have increased markedly during the last 10 or 15 years, at the same time as new and advanced techniques of animal surveillance in the field were established. The improvement of radiotelemetry and the fast progressing miniaturisation of transmitters and batteries were indeed most important innovations, but other encouraging techniques were developed as well, and the technical equipment is constantly elaborating further. The crucial dependence of approachable ecological questions on recent advances in field techniques is well demonstrated by several chapters of this book within the section on empirical findings, and by the Appendix on methodological aspects of activity recording in particular. We feel that an exciting and promising new phase in this field of research has just begun. Hence, it seems timely to document the present state of affairs in a comprehensive review.

1.6 Why Small Mammals?

In the face of severe methodological problems with studies on activity behaviour in the field, a fault-finding reader may argue that it is unwise to focus on small mammals in which the practical difficulties are especially challenging due to their secretive habit. There are, nonetheless, some good reasons to choose small mammals as target organisms. Already mentioned is the great variety of temporal strategies in this group, sometimes even in closely related species, which indeed is a most precious feature for the comparative approach that we take in this book. Since the technical problems related to monitoring small animals can generally be solved, one has the additional pragmatic advantage of relatively small activity ranges, which considerably facilitates data sampling with high temporal and spatial resolution. Furthermore, small species are relatively easy to trap and to handle in the field, which helps in obtaining data on the biological state of individuals.

Besides such practical considerations there are additional reasons for our focus on small mammals which are related to the ecological peculiarities of this group. 'Small mammals' are, by definition, rather small (corresponding to body weights of something between 3 kg and 10 g). Small endothermic species have a high metabolic rate, with the consequence that the maintenance of a balanced energy household is a particularly difficult task (see Chap. 3 for a more detailed contemplation on this aspect). Because of fundamental physiological constraints, the timing of activity has rather limited degrees of freedom. Nevertheless, in some cases small mammal species of comparable body sizes reveal very different types of activity patterns, as in murid and microtine rodents for instance (see Chaps. 10 and 11). This allows us to ask rather specifically what may be the causes of the differences in temporal organisation, taking feeding habits, social organisation, and any other ecological factors into account.

Weather and other environmental conditions have a more severe impact on small species, since large surface-volume ratios and confined energy reserves result in limited buffering capacities. Direct responses to extrinsic factors are, therefore, more likely to occur in small rather than in large species. This allows us to draw conclusions on how the species concerned experience the environment (i.e., what conditions are considered to be preferable and adverse, respectively). Only if it has been established what is good and bad from the animal's perspective – and not from the researchers point of view – it is possible to search for behavioural responses to different conditions.

As another consequence of body size, small mammals commonly have short generation times and high reproductive rates. This allows many small mammals to reach high population densities fast, but on the other hand, following an outbreak densities often decline again in rather short times. In some groups, outbreaks and crashes are almost proverbial, with the Norwegian lemming as the most popular representative (Stenseth and Ims 1993). Such density fluctuations may allow us to relate observed changes of activity behaviour to population dynamics (Halle and Lehmann 1992). Rapid changes in densities also means that the social environment changes considerably in relatively short time. Hence, one may be able to correlate variations in the activity pattern with the demographic and/or social structure of the population. Finally, many small mammal species are important prey for raptors, owls, and carnivorous mammals, which for their part are heavily affected by changes in the abundance of their staple food. Even better, some of these predators, like weasels and stouts for instance, are themselves small mammals. Thus, it is possible to deal with predator-prey relations with both interacting species belonging to the same group. Thorough analyses of such systems may also allow us to value different temporal patterns with respect to predation risk and optimal foraging strategies.

From the collection of pros and cons for small mammals as study organisms in activity surveys it is obvious that there are many methodological difficulties. However, trying to solve these difficulties is definitely worth the effort due to a reasonable chance of finding a rather convenient experimental systems for approaching the ecology of activity patterns.

1.7 About This Book

The ecological approach taken in this book is illustrated by the figure on the cover. The individual – in this case a vole – is placed in the centre together with some plants representing its food resource. On the opposite side there are other individuals, in this case offspring, symbolising social interactions. The individual is surrounded by a circle showing sun and moon, which represents the steady periodical change of environmental conditions in the course of the day/night cycle. The thermometer indicates that not only light conditions are changing, but other abiotic environmental factors are following along. The raptor and the owl in the circle depict temporal changes in the relation to other species, which in the case of predation results in different levels of mortality risk at different times of the day. The challenging task for the individual is to find an optimal strategy of activity timing by a trade-off in which all the conflicting aspects are considered and reconciled. The solution to this temporal optimisation problem is, in our view, an important part of adaptation, or better adaptedness, to the specific ecological niche and environment, respectively. The optimum will also depend on what other individuals – both of the same and of other species – do. Hence, the establishment of characteristic activity patterns can only reasonably be analysed within the conceptual framework of evolutionarily stable strategies, ESS (Maynard-Smith 1982).

Questions about the relevance of ecological constraints for activity timing can certainly not be answered in terms of universally valid statements about small mammal behaviour. Priorities for decision making will vary depending on species and life style, and it cannot be excluded that optimal temporal strategies may even differ within species and populations, between males and females or between juveniles and adults for instance. However, in comparing an increasing number of case studies within the group of small mammals, some general principles of temporal organisation will hopefully emerge. Indeed, this is one of the highlighted objectives of the comparative approach taken in this book.

Following this Introduction, the book is divided into two major sections. In the first section, the theoretical background is evaluated in three chapters, providing reviews of what is known about the biological clock in small mammals, about the crucial relationship between activity and metabolism, and about the ecological relevance of temporal activity patterns. The second and largest section of the book is devoted to empirical data with a distinct focus on field studies. In each of the ten chapters, specialists in their area review the literature and give their views on the critical ecological factors governing activity timing in a certain species or species group. We start off with relatively large species among the small mammals (i.e., weasels and mongooses) and go stepwise down in size to shrews and bats at the other end of the scale. Groups of species were chosen such that as broad as possible a spectrum of life styles and ecological situations could be covered. In the subsequent Conclusion, we draw together the empirical findings and try to give a synthesis with respect to general characteristics within the whole group of small mammals. In the Conclusion we also reflect on the future developments in this field of research, asking what we currently know and which questions remain to be addressed. The final Appendix provides a methodological overview and literature review of techniques for measuring small mammal activity as well as of methods of time series analyses.

This book is primarily intended to provide an ecologically biased overview of recent studies on small mammal activity. As argued in this chapter, the ideas that are treated on the following pages are not ingeniously new. However, tools are available today not only to frame such questions, but to obtain the kind of data needed to approach them experimentally. Hence, we are convinced that something substantial has changed in comparison to previous contributions to this issue. It is our hope that this book may be a starting point for a more thorough discussion of the ecological implications of activity timing and to establish diel activity patterns as a topic in behavioural ecology. We also hope that this book will make students of ecology more aware of diel activity pattern as a feature worth bothering about, because in one point Enright (1970) was apparently wrong: "It is self-evident to most ecologists that the time domain is an important aspect of adaptation."

References

Aschoff J (1963) Gesetzmäßigkeiten der biologischen Tagesperiodik. Dtsch Med Wochenzeitschr 88:1-21
Aschoff J (ed) (1965) Circadian clocks. North-Holland Publishing Company, Amsterdam
Aschoff J (1966) Physiologie biologischer Rhythmen. Ärtzliche Praxis 18:1569, 1593-1597
Aschoff J (1984) Circadian timing. Ann NY Acad Sci 423:442-468
Aschoff J, Wever R (1962) Beginn und Ende der täglichen Aktivität freilebender Vögel. J Ornithol 103:1-27
Ashby KR (1972) Patterns of daily activity in mammals. Mamm Rev 1:171-185
Bünning E (1977) Die physiologische Uhr. Circadiane Rhythmik und Biochronometrie. Springer, Berlin Heidelberg New York
Calhoun JB (1945) Diel activity rhythms of the rodents, *Microtus ochrogaster* and *Sigmodon hispidus hispidus*. Ecology 26:251-273
Chiba Y (1989) Circadian rhythms versus circadian clocks in ecology. In: Hiroshige T, Honma K (eds) Circadian clocks and ecology. Hokkaido University Press, Sapporo, pp 157-159
Crowcroft P (1953) The daily cycle of activity in British shrews. Proc Zool Soc Lond 123: 715-720
Daan S (1981) Adaptive daily strategies in behavior. In: Aschoff J (ed) Biological rhythms. Handbook of behavioural neurobiology, vol 4. Plenum Press, New York, pp 275-298
Daan S, Aschoff J (1982) Circadian contributions to survival. In: Aschoff J, Daan S, Groos GA (eds) Vertebrate circadian systems. Springer, Berlin Heidelberg New York, pp 305-321
DeCoursey PJ (1989) Photoentrainment of circadian rhythms: an ecologist's viewpoint. In: Hiroshiga T, Honma K (eds) Circadian clocks and ecology. Hokkaido University Press, Sapporo, pp 187-206
DeCoursey PJ (1990) Circadian photoentrainment in nocturnal mammals: ecological overtones. Biol Behav 15:213-238
Enright JT (1970) Ecological aspects of endogenous rhythmicity. Annu Rev Ecol Syst 1:221-238
Halle S, Lehmann U (1992) Cycle-correlated changes in the activity behaviour of field voles, *Microtus agrestis*. Oikos 64:489-497
Hamilton WJ (1937) Activity and home range of the field mouse, *Microtus pennsylvanicus pennsylvanicus* (Ord.). Ecology 18:255-263
Kacelnik A, Bateson M (1996) Risky theories - the effects of variance on foraging decisions. Am Zool 36:402-434
Kavanau JL (1969) Influences of light on activity of small mammals. Ecology 50:548-557
Lima SL, Dill LM (1990) Behavioral decisions made under the risk of predation: a review and prospectus. Can J Zool 68:619-640
Maynard-Smith J (1982) Evolution and the theory of games. Cambridge University Press, Cambridge
McNamara JM (1992) Risk-sensitive foraging: a review of the theory. J Math Biol 54:355-378
Park O (1940) Nocturnalism - The development of a problem. Ecol Monogr 10:485-536
Pearson AM (1962) Activity patterns, energy metabolism, and growth rate of the voles *Clethrionomys rufocanus* and *C. glareolus* in Finland. Ann Zool Soc Vanamo 24:1-58
Rusak B (1989) The mammalian circadian system: models and physiology. J Biol Rhythms 4:121-134
Schoener TW (1974) Resource partitioning in ecological communities. Science 185:27-39

Smallwood PD (1996) An introduction to risk sensitivity: the use of Jensen's inequality to clarify evolutionary arguments of adaptation and constraint. Am Zool 36:392–401
Stenseth NC, Ims RA (1993) The biology of lemmings. Academic Press, London
Szymanski JS (1920) Aktivität und Ruhe bei Tieren und Menschen. Z Allg Physiol 18:105–162
Wever R (1965) A mathematical model for circadian rhythms. In: Aschoff J (ed) Circadian clocks. North-Holland Publishing Company, Amsterdam, pp 47–63
Winfree AT (1988) Biologische Uhren: Zeitstrukturen des Lebendigen. Spektrum der Wissenschaft Verlagsgesellschaft, Heidelberg

Section II Theoretical Considerations

Theoretical Considerations – Introduction

Stefan Halle and Nils Chr. Stenseth

Entrainment of individuals with their fluctuating environment is, in very general terms, achieved by resetting the internal clock by some zeitgeber. When the same signal acts on several individuals in a similar manner, synchronisation is the inevitable consequence. Synchronisation may be the result of strong external signals from the environment, but it may also be caused by the interactions among individuals. The latter case of social synchronisation is of particular theoretical interest because such a system may show peculiarities comparable to the dynamics of two coupled pendulum clocks with self-organising features.

A second interesting aspect is the phase relation between the internal rhythm and the zeitgeber signal, which in the case of environmental synchronisation will most often reveal a 24-h periodicity. If 24 h is not an exact multiple of the internal rhythm, terminated cycles are the consequence, as common in circa-rhythms. In this book we will neither look at phase response curves nor at the physiological machinery behind it, which is the domain of chronobiology. Instead we will treat the processes that are driven by the resultant behavioural pattern in detail and – in particular – the consequences for individual survival.

In the first section of this volume, three chapters will provide an update about the theoretical background of activity behaviour with a focus on small mammals. Before looking at different species living under specific ecological constraints in the subsequent section, it seems reasonable to consider the common features first. How is 'time measuring', which is the precondition for any temporal strategy, achieved physiologically in this group, what are the general constraints common in all species, and in what frame have temporal strategies evolved? This section of the book is intended to open the field by setting some landmarks, in which the following reviews of empirical studies may be seen as example cases of what may come out from such a general approach.

In Chap. 2 Bartness and Albers lay the foundations by reviewing the information available on biological clocks. Since observable behavioural rhythms will always depend on some kind of a pacemaker, the subject of this chapter is indeed the base for all following considerations and empirical

Ecological Studies, Vol. 141
Halle/Stenseth (eds.) Activity Patterns in Small Mammals

findings. The authors show that the diel melatonin rhythm is a common feature of all small mammals, but that a great variety of habits arise from different translations of the hormonal signal into behaviour. They also state that biological clocks are such powerful tools because of a unique combination of precision and flexibility.

Weiner, in Chap. 3, scales down from activity to physiology by dealing with the metabolic constraints of activity behaviour, which is one of the two sides of the essential trade-off in small mammals between eating and been eaten. He develops a theoretical framework to see the activity pattern of a specific small mammal species as the particular solution to a general optimisation problem in which time has to be minimised and energy intake has to be maximised.

In Chap. 4 Halle scales up from activity to ecology by outlining the constraints under which activity patterns have evolved. Synchronisation with a seasonally fluctuating environment, predation, interspecific competition and spatio-temporal organisation of populations are rather obvious aspects where temporal strategies have direct fitness consequences. However, other relations, about which little is yet known, may also be of importance.

2 Activity Patterns and the Biological Clock in Mammals

Timothy J. Bartness and H. Elliott Albers

2.1 Introduction

One of the most consistent features of the physical and social environment is change. Many changes occur in an unpredictable manner, whereas others occur in precise rhythmic patterns. It is thought that internal clocks evolved as a way of adapting behaviour and physiology to these regularly occurring changes in the environment. Although biological clocks can be set to a number of different cycle lengths (see Aschoff 1981), clocks that exhibit daily cycles are prevalent in mammals. Internal clocks that time daily rhythms are called 'circadian clocks'. Their name is derived from the Latin 'circa' meaning about or around and 'dies' or 'dian' meaning day. It sounds as if these clocks might be fairly imprecise because their name suggests that they time cycles that only approximate a day. In fact, circadian clocks are extraordinarily precise and are found frequently to be inaccurate by less than a minute in each cycle (Pittendrigh and Daan 1976a). These clocks received their name because they measure out a genetically determined daily cycle that deviates slightly, but significantly from 24 h. In animals, the 'period' (i.e., time required to complete one full cycle) of circadian cycles usually ranges between 23–26 h. Although the period of circadian clocks varies among species and individuals of the same species, within an individual the period remains quite consistent.

Chronobiologists who study the biological rhythms of small rodents in the laboratory frequently use wheel running to assess the functioning of the biological clock. This approach was chosen because: (1) it is avidly exhibited by most small rodents with 5 000 to 10 000 wheel revolutions or more recorded in a single night in laboratory rats, mice and hamsters, (2) it is easily measured and quantified (Fig. 2.1), (3) it is a very precise rhythm in many species, and (4) early in the study of biological rhythms, and until recently, it did not appear that this behaviour affected the biological clock (see below for evidence to the contrary).

Ecological Studies, Vol. 141
Halle/Stenseth (eds.) Activity Patterns in Small Mammals

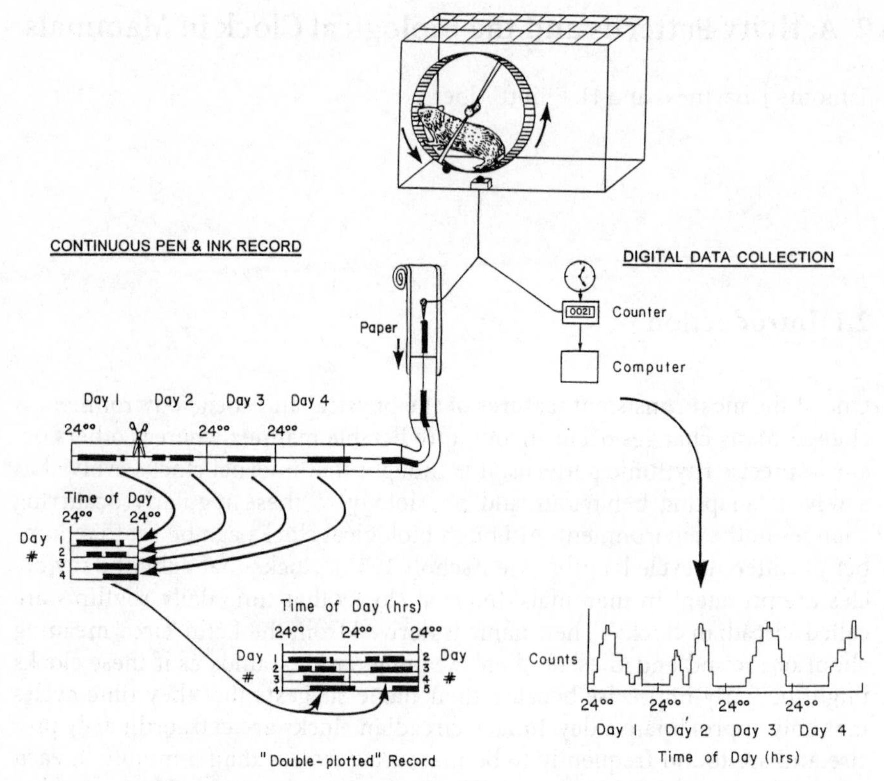

Fig. 2.1. Methods of recording wheel-running activity of small rodents. In the method on the *left* each wheel rotation activates a sideways movement of a pen on a continuously moving sheet of paper. When the 24-h strips are arranged one under the other, the rhythmic behaviour can be visualised. The 24-h strips are often double-plotted so that the continuity of the rhythmic pattern is more salient (*lower left*): on the first line, day 1 plus day 2 are plotted, and on the next line day 2 plus day 3 are plotted and so on. For more objective analyses, a digital record can be made. The amount of activity can then be plotted against time (*lower right*) and analysed statistically. (Adapted from Moore-Ede et al. 1982)

2.2 The Circadian Clock

Because it is thought that circadian clocks evolved to allow animals to adapt more easily to the 24-h rhythms in their environment, it might seem odd that they do not have a natural period of 24 h. There are significant potential problems with the strategy of attempting to match the 24-h rhythms in the environment with an internal clock genetically set to 24 h, however. One problem is that the adaptive value of the internal clock would be lost if the period of the circadian clock deviated, even a very small amount, from 24 h because this inaccuracy would be compounded each day. It would not be long before the internal clock was not synchronised with the environment. Even if the circadian clock had a precise 24-h period, however, but was not appropriately set to local time, its usefulness would be significantly compromised. It appears that an alternative strategy has evolved to ensure that the internal clock has a 24-h rhythm that is synchronised with the environment. The clock is reset to 24 h each day in a manner that ensures that it is appropriately in phase with the environment. The most important environmental stimulus for resetting the clock is light. The light/dark (LD) cycle is ideal for resetting the clock, not only because it is an extremely precise and consistent environmental rhythm, but also because it reflects many dramatic changes in the environment such as alterations in temperature, food availability and predation.

Two basic strategies of adaptation to the LD cycle have evolved, although exceptions to these patterns exist (see below). 'Diurnal' animals confine the majority of their locomotor activity to the day, whereas 'nocturnal' animals are primarily active at night. In addition to exhibiting opposite patterns of locomotor activity, the circadian patterns of all physiological and behavioural variables in diurnal and nocturnal species are opposite, except for one hormone rhythm – the rhythm of melatonin released from the pineal gland. The pineal gland transduces light/dark information into an internal endocrine signal in the form of the rhythmic pattern of secretion of this hormone. Regardless of whether an animal has a diurnal or nocturnal activity pattern, pineal and plasma melatonin concentrations are at their daily nadir during the light phase of the photocycle and at their peak during portions of the dark phase. In many species, especially those found in temperate zones, melatonin co-ordinates a constellation of responses that help confine physiological and behavioural responses to certain times of the year (i.e., reproduction, torpor, changes in pelage colour and body mass/fat – for a review see Bartness and Goldman 1989). The duration of nocturnal melatonin stimulation is the critical feature of the melatonin signal for both inhibitory and stimulatory effects of the hormone on these responses (for a review see Bartness et al. 1993). This pattern of melatonin synthesis and release results from the regulation of the pineal gland by an endogenous circadian oscillator (i.e., timer) that is entrained to (i.e., in synchronicity with) the LD cycle (Elliott and Goldman 1981). Furthermore, studies in Syrian hamsters (Elliott and Tamarkin 1994) and Siberian hamsters and laboratory rats (Illnerova 1991) suggest that a two-oscillator model of melatonin secretion can explain the persisting rhythm of melatonin secretion under conditions of constant darkness. One oscillator is thought to control the onset and another the offset of the secretion of melatonin by the pineal gland.

2.3 Circadian Clocks in Mammals

2.3.1 General Features

Some of the most important features of circadian clocks are observed when the natural rhythmic changes in the environment are eliminated. When animals are placed in controlled laboratory conditions where environmental time cues are absent, the circadian clock 'free-runs' and expresses its own endogenous, genetically determined natural periodicity (Fig. 2.2). For most species, elimination of the LD cycle by placing the animal in constant light or constant darkness is sufficient to establish a free-running circadian rhythm. For a rhythm to fulfill the criteria to be a 'true' circadian rhythm it must: (1) persist under constant conditions and (2) have a free-running rhythm with a period different from 24 h. Rhythms of responses that satisfy these criteria are called circadian and, by fulfilling these criteria, are considered to have an endogenous clock as their origin (cf. Brown et al. 1970). In contrast, if a rhythm had a free-running period (commonly denoted as 'tau' in chronobiology) of exactly 24 h in length, it is possible that the rhythm might be induced by some 24-h rhythm in the environment instead of an internal clock.

Although the circadian period is programmed by an internal clock, that does not mean that environmental factors do not influence the timing of the clock. When placed in continuous illumination, the intensity of that illumination produces consistent and predictable changes in a number of variables controlled by the clock. The effects of light on circadian clocks have been studied extensively and formalised into Aschoff's rules (Aschoff 1959). In nocturnal animals, the free-running period (tau) becomes longer and 'activity-time' or 'α', which is the duration of locomotor activity each day, becomes

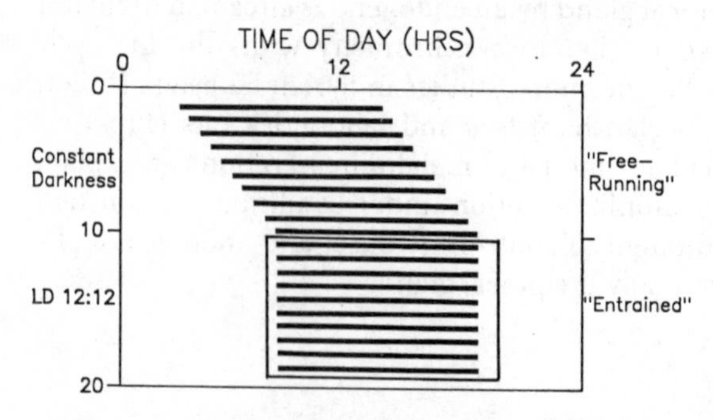

Fig. 2.2. Illustration of the free-running and entrained circadian locomotor activity rhythms of a nocturnal rodent. The 24-h day is divided into an active phase indicated by solid black lines and a rest phase that occurs during the remainder of each day (white). Days are indicated on the vertical axis. During the first 10 days of this record the rodent is housed in constant darkness. Under this condition, the circadian clock expresses its own intrinsic 'free-running' cycle. In this example the 'free-running' circadian cycle is slightly longer than 24 h. On day 10 a light/dark cycle consisting of 12 h light and 12 h darkness (LD 12 : 12) is imposed. The circadian clock is synchronised or 'entrained' to 24 h by the LD cycle. Locomotor activity occurs during the 12-h dark phase (indicated by box) in this nocturnally-active rodent. (Adapted from Albers et al. 1992)

shorter as light intensity increases. In diurnal animals, increasing the intensity of light shortens tau and lengthens α. The circadian clocks of most mammals obey the basic features of Aschoff's rules except for primates that are diurnally active (Tokura and Aschoff 1978; Aschoff 1979; Sulzman et al. 1979).

How circadian clocks communicate timing information to the behavioural and physiological rhythms they are driving is not well understood. These output mechanisms are probably complex because the rhythms under circadian control exhibit vastly different phases and patterns. One possibility is that the clock provides a brief timing pulse or a 'trigger' as an output signal. Another possibility is that the output of the clock is more like that produced by self-sustained oscillator equations (e.g. van der Pol oscillators) in that the clock provides a continuous output to control rhythmicity. Understanding the temporal regulation of the locomotor activity rhythms is particularly complicated because animals exhibit so many different patterns of locomotor activity. In many species activity is initiated and terminated multiple times during the active phase. A major challenge will be to define how the mechanisms controlling the temporal patterning of activity signal the initiation and termination of activity.

2.3.2 Temperature Effects on the Circadian Clock

Another important feature of circadian clocks is that their day-to-day precision is not altered dramatically by changes in the ambient temperature. This is an amazing property when one considers that the speed of most biochemical reactions is altered by two to three times for every 10 °C change in temperature (Q_{10}), but changes in the speed of circadian clocks resulting from a 10 °C change in temperature are considerably smaller (Pittendrigh and Caldarola 1973). The functional significance of the ability of the clock to compensate for changes in ambient temperature are most obvious in heterothermic animals, where tissue temperatures can vary as much as ambient temperatures. Even in mammals where a narrow range of body temperatures is defended, however, there is evidence that the clock can at least partially compensate for changes in ambient temperature. In laboratory rats and Syrian hamsters (*Mesocricetus auratus*), change as great as 25 °C in core body temperature produces relatively small changes in the speed of the circadian clock (Gibbs 1981, 1983). Similar results have been obtained in a hibernating species, golden-mantled ground squirrels (*Spermophilus lateralis*), during the phase of the year when they are homeothermic (Lee et al. 1990). It is interesting that temperature does produce rather large changes in the free-running period of squirrels that exhibit hibernation during the heterothermic phase, however. Although the circadian clocks of at least some small mammals can be influenced by temperature, and under some circumstances to a relatively large extent, the functional consequences of loss or reduction in temperature compensation under these conditions remains to be identified clearly. Thus, temperature compensation may be an important, but not ubiquitous, trait of circadian clocks in mammals.

2.3.3 Physiological Influences on the Circadian Clock

In addition to environmental factors, several physiological variables can influence the speed at which the clock runs. Gonadal hormones can produce changes in the free-running period in both male and female rodents. Reductions in circulating concentrations of testosterone produced by castration shortens the free-running period and administration of testosterone to castrated mice lengthens the period (Daan et al. 1975). In female laboratory rats and Syrian hamsters, the changing levels of estradiol and progesterone during the ovulatory cycle produce corresponding changes in the circadian period (Morin et al. 1977; Albers 1981; Albers et al. 1981). Specifically, estradiol shortens the circadian period in these species, but progesterone, administered in combination with estradiol, blocks the ability of estradiol to shorten the free-running period. When progesterone is administered alone, however, it has no effect on the circadian clock (Takahashi and Menaker 1980; Axelson et al. 1981). The functional significance of these hormonal effects remains to be determined.

Another phenomenon that appears to influence the circadian clock is age. The period of the clock appears to shorten with advancing age (Pittendrigh and Daan 1976b) and there is recent evidence that the response of the clock to light changes with age in Syrian hamsters (Zee et al. 1992).

2.4 Entrainment of Rhythms by Light and Other Signals

2.4.1 Light

One of the most important properties of circadian clocks is their ability to become synchronised with the 24-h LD cycle. The synchronisation process is called 'entrainment' and is the result of the ability of the clock to be reset by light. The mechanisms that reset the circadian clock to 24 h can be illustrated by observing how brief pulses of light reset circadian clocks that are free-running in constant darkness. If a rodent housed in constant darkness is exposed to a 10-min pulse of light around the time that its daily activity begins (i.e., the subjective phase of activity), then its circadian clock will be reset such that a delay in the normal onset of activity will occur. If a 10-min pulse of light is provided 6 h after the onset of activity, however, then its circadian clock will be reset such that an advance in the onset of activity will occur. If a 10-min pulse of light is provided at other times of the circadian cycle, then the timing of the clock is not altered. The phase-dependent sensitivity of the clock to light is summarised in a 'phase response curve' (PRC, Moore-Ede et al. 1982; Fig. 2.3). If the clock is reset in the advance (or delay) direction each day, then it can be forced to adopt the 24-h period of the LD cycle. In this same fashion, circadian clocks are reset daily to 24 h and thereby serve as relatively accurate indicators of the time of day. Because light can reset clocks in the advance or delay direction, it is possible for light to reset circadian clocks to exactly 24 h. Stimuli such as light that synchronise or entrain circadian rhythms are termed 'zeitgeber' (from the German meaning 'time-giver').

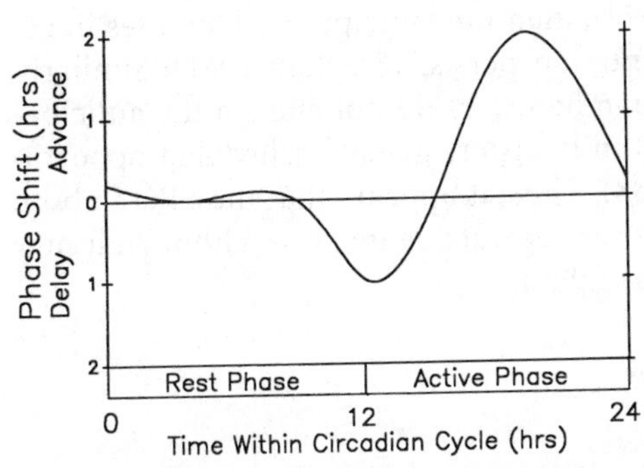

Fig. 2.3. A phase response curve illustrating the ability of 10-min pulses of light to reset the timing of circadian rhythms in a nocturnal rodent housed in continuous darkness. The circadian cycle is represented on the horizontal axis and divided into rest (i.e., 0–12) and activity (i.e., 12–24) phases. The resetting effects of light on the circadian clock are indicated on the vertical axis. Light provided around the beginning of the daily active phase resets the clock in the delay direction. Light provided approximately 6 h after the onset of the active phase resets the clock in the advance direction. Light provided at other times in the circadian cycle has little or no effect on the clock. (Adapted from Albers et al. 1992)

2.4.2 Food Availability

The first evidence that scheduled availability of food influences circadian control came from studies showing that locomotor activity was increased markedly before the time that food was presented each day in rats (Richter 1922). There is now clear evidence that restricted food availability is capable of entraining a circadian clock; that is, limited food access can serve as a zeitgeber and it has many of the same effects on clocks as other entraining stimuli (for a review see Rusak 1990). For example, following entrainment of laboratory rats to a restricted food schedule, circadian rhythms begin to free-run if the food schedule is replaced by total food deprivation (Boulos et al. 1980; Stephan 1981). The majority of research on the entraining effects of food availability have been conducted in laboratory rats and have found that the scheduling of food can influence the circadian control of a number of circadian rhythms including wheel running and drinking. It is not feasible to conduct similar experiments in Syrian hamsters because they do not survive under conditions where food availability is scheduled only for short intervals (Silverman and Zucker 1976). If freely-feeding Syrian hamsters are given restricted opportunities to hoard food, however, then this time cue is capable of serving as an entraining stimulus (Rusak et al. 1988). Specifically, most hamsters housed in constant light and allowed to leave their home cage for 30 min to retrieve seeds entrain to the 24-h cycle of food availability. Although the efficacy of food availability as a zeitgeber has only been examined in a few species, it seems easy to envision that food availability may serve as a potent zeitgeber in many small mammals. The mechanism underlying the ability of scheduled food presentations to generate the anticipatory locomotor activities has been termed a 'food-entrainable oscillator' (FEO). When the hypothesised circadian clock, the suprachiasmatic nucleus of the hypothalamus (SCN; see below for a discussion of the role of the SCN in circadian rhythm generation) is destroyed completely in laboratory rats (a treatment that abolishes circadian rhythms of locomotor activity in these and other species, e.g. Stephan and Zucker 1972), then the anticipatory increases in activity before scheduled food presentation persist (Stephan 1981). Similarly, complete lesions of the SCN in Syrian hamsters do not abolish the anticipatory wheel-running behaviour seen in hamsters given a scheduled opportunity to hoard food (Rusak et al. 1988). Thus, it appears that the FEO in both laboratory rats and Syrian hamsters resides outside the SCN, although it may be linked to it (for a review see Rusak 1990).

2.4.3 Non-photic Zeitgebers

Recent data indicate that the opportunity to engage in locomotor activity can also entrain the circadian clock and thus serve as a non-photic zeitgeber. This discovery came in the course of circadian rhythm experiments when it was noticed that when Syrian hamsters are provided with clean activity wheel cages, phase shifts (i.e., advances or delays in the normal onset of wheel running) in their circadian wheel-running rhythms often occur (Mrosovsky 1988). When placed in a clean cage hamsters often engage in a number of very active behaviours including running in the activity wheel. In subsequent studies, hamsters were confined in novel running wheels for specified intervals to characterise more fully the phase shifts produced by this form of locomotor activity (Reebs and Mrosovsky 1989a, b; Janik and Mrosovsky 1993). Results from these studies have shown that maximal phase shifts are stimulated by novelty-induced activity that lasts for approximately 3 h, that the number of wheel revolutions predicts the occurrence of large phase shifts (i.e., there is a high probability that a phase shift will occur if a hamster engages in > 5 000 wheel revolutions), and that wheel running can induce phase shifts only at certain phases of the circadian cycle. That is, novelty-induced activity produces phase advances during the middle of the subjective day and phase delays during the later half of the subjective night. Although novelty-induced wheel running is a robust zeitgeber in Syrian hamsters, it is not known if activity can entrain the circadian rhythms of other small mammals or whether the effects of activity on circadian rhythms has any adaptive value.

2.4.4 Social Factors

Comparatively little is known about the influence of social factors on circadian clocks; however, examples where social cues act as zeitgebers have been reported in small mammals. Studies in colonies of microchiropteran bats that live in caves where environmental stimuli such as temperature, humidity and darkness appear to remain constant, indicate that social cues are potent zeitgebers (Marimuthu et al. 1978, 1981). These animals, like other animals that spend long periods of time without exposure to the LD cycle such as burrow-dwelling rodent species (Pratt and Goldman 1986), periodically expose themselves to the photoperiod to sample the current photic information (i.e., whether it is light or dark outside) – a behaviour termed 'light-sampling'. If individual bats are held captive in caves in the absence of conspecifics, then their circadian rhythm of flight activity free-runs. In caves containing a colony of bats entrained to the LD cycle as a result of their normal light-sampling behaviour, however, individual bats held captive in the cave and only exposed to the social cues of the colony have 24-h activity rhythms that are in phase with the colony.

Another example of entrainment by social cues comes from studies in deer mice (*Peromyscus maniculatus*; Crowley and Bovet 1980). Mice housed in constant dim illumination and exhibiting free-running periods of different lengths develop similar free-running periods when housed in the same enclosure. When subsequently separated, the length of free-running period becomes different again. One particularly interesting feature of this experiment was that the length of the free-running period when the two mice were housed together tended to be closer to the circadian period of the dominant mouse. In contrast, other experiments have shown that reproductive state, another potent social cue, is not capable of acting as a zeitgeber in either laboratory rats or Syrian hamsters (Richter 1970; Davis et al. 1987).

A final example of a social cue serving as a zeitgeber comes from the interaction of mother mice with their pups (Viswanathan and Chandrashekaran 1985; Viswanathan 1989). If single field mice (*Mus booduga*) pups are housed in cages equipped with activity wheels and their tethered mothers (tethered to keep them from running in the wheels) are presented on a 12 : 12 h schedule, then the activity rhythms of the pups entrain to their mother's presence/absence. Specifically, pups only run in the wheel when their mothers are absent and do so whether housed in constant light or darkness. The ability of the mother to serve as a zeitgeber seems adaptive in that under natural conditions, pups of this age most likely are never directly exposed to the prevailing photocycle because they are burrow-dwelling animals. The mother stays with the pups during the day, and leaves the burrow to forage at night. Therefore, the pups may use the presence/absence of the mother to remain in phase with the photocycle in preparation for their future entry into the world above their burrows as they mature (Viswanathan and Chandrashekaran 1985).

In summary, although there are clear cases of entrainment of circadian rhythms by social factors, they are limited in number and little is known about the nature of the social stimuli involved. Further research on this subject will be important in defining how social factors may contribute to the temporal organisation of behaviour through effects on the circadian clock.

2.5 Localisation of the Circadian Clock in Mammals

2.5.1 The Suprachiasmatic Nucleus

Following an exhaustive series of experiments where brain lesions were made in rats, Richter identified an area in the ventral hypothalamus as necessary for the expression of circadian rhythms called the suprachiasmatic nucleus or SCN (Richter 1965). It was not until 1972 when a direct projection from the retina to the SCN was demonstrated conclusively in laboratory rats (Hendrickson et al. 1972; Moore and Lenn 1972), however, that the SCN was identified as a putative circadian clock. The first data indicating that the SCN might contain a circadian clock came from the independent findings of Moore and Eichler (1972) and Stephan and Zucker (1972). In these experiments, destruction of the SCN eliminated the circadian rhythms of locomotor activity and adrenal corticosterone content. Subsequently, a large number of studies have found destruction of the SCN to eliminate the circadian rhythmicity of a wide range of behavioural and physiological variables in a variety of mammalian species (Meijer and Rietveld 1989).

The hypothesis that the SCN contains a circadian clock has been supported by studies employing a broad spectrum of experimental approaches. Some of the most compelling evidence showing that the SCN functions as a circadian clock comes from data indicating that the SCN is capable of generating circadian rhythmicity. For example, circadian rhythms in electrical activity persist in the SCN in vivo even though it is isolated from other hypothalamic tissue as a result of microknife cuts (Inouye and Kawamura 1979) or when it is isolated as a small hypothalamic explant in vitro (Groos and Hendriks 1979). More recently, evidence supporting the hypothesis that the SCN functions as an endogenous clock has been provided by fetal brain transplantation studies. Circadian rhythmicity can be restored in rodents with SCN lesions by transplantation of fetal tissue containing the SCN, but not other neural tissues (Drucker-Colin et al. 1984). In fact, some of the characteristics of the circadian clock of the donor can be observed in the host following SCN transplantation (Ralph et al. 1990); transplantation of the SCN obtained from mutant Syrian hamsters with unusually short free-running periods produces unusually short circadian rhythms in the host.

2.5.2 Anatomy and Neurochemistry of the Suprachiasmatic Nucleus

Based on a variety of anatomical criteria, the SCN has been divided into dorsomedial and ventrolateral subdivisions (Moore and Card 1985; van den Pol 1985; Fig. 2.4). Neurons within the two subdivisions produce different neurotransmitters and receive different patterns of afferent input (Fig. 2.4). The ventrolateral subdivision is the site of termination for three of the major afferent projections to the SCN. The retinohypothalamic tract (RHT), which is the direct projection from the retina to the SCN, appears to be necessary for the entrainment of circadian rhythms to light (Johnson et al. 1988). The two other major pathways to the SCN are the geniculohypothalamic tract (GHT), that originates in the intergeniculate leaflet of the thalamus (Ribak and Peters 1975) and a projection from the raphe nuclei (Fuxe 1965). There is evidence that these pathways may influence the response of the SCN to light/dark information (see Meijer and Rietveld 1989; Albers et al. 1991), although neither pathway appears to be necessary for entrainment to light (Block and Zucker 1976; Pickard et al. 1987). It also has been hypothesised that these pathways are involved in mediating at least some of the phase shifting effects of non-photic cues (Biello et al.1991; Edgar et al. 1993).

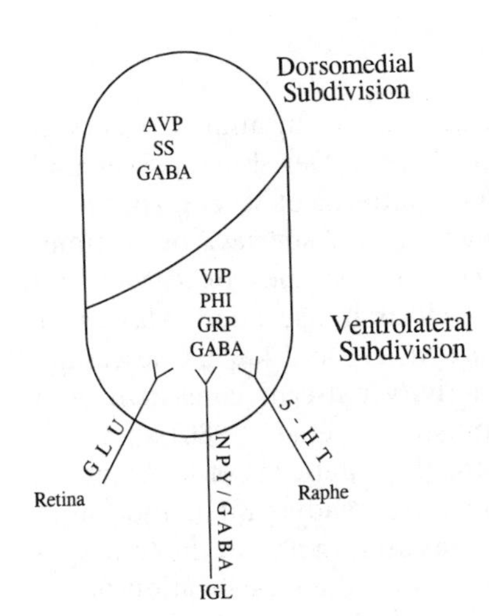

Fig. 2.4. Illustration of the major features of the suprachiasmatic nucleus. Based on differences in a variety of morphological and neurochemical criteria, the nucleus has been divided into the ventrolateral subdivision and dorsomedial subdivision. The ventrolateral subdivision receives the majority of afferent input to the nucleus including projections from the retina, intergeniculate leaflet of the thalamus (*IGL*) and the raphe. The primary neurotransmitter in the projection from the retina appears to be glutamate (*GLU*), the projection from the *IGL* contains both neuropeptide Y (*NPY*) and gamma-aminobutyric acid (*GABA*), and serotonin (*5-HT*) is found in the projection from the raphe. Neurons within the ventrolateral subdivision produce a large number of neurochemical signals including vasoactive intestinal peptide (*VIP*), peptide histidine isoleucine (*PHI*), gastrin releasing peptide (*GRP*) and *GABA*. Neurochemical signals produced within the dorsomedial subdivision include vasopressin (*AVP*), somatostatin (*SS*) and *GABA*. For more details see Albers et al. (1992)

2.6 Overview of the Major Activity Pattern Types in Mammals

2.6.1 Terminology

In addition to the nomenclature associated with circadian rhythms (see above), several terms are used frequently to describe the activity patterns of animals that are exposed to LD cycles. As stated above, it is not appropriate to refer to these activity patterns as circadian unless they fulfill the criteria for true circadian rhythms. Several more general terms can be used to describe these activity patterns, however. For example, the terms '24 h', 'diel' or 'daily' rhythms simply describe the length or frequency of the cycle of activity, but do not indicate whether the rhythms are induced by internal or external factors. Another term used to describe the day/night rhythm is 'nycthemeral' (from the Greek 'nycthemeron', night and 'hemera', day), and means a full day/night cycle. Although this term is used infrequently, it is appropriate in cases where conditions are not constant or if the response has not been verified as a true circadian rhythm. Other terms used to describe daily rhythms are more descriptive of the activity pattern. The most common activity patterns are 'unimodal' (i.e., having one peak) and 'monophasic', occurring almost exclusively during the day or at night (see below). Designations and examples of species that display such rhythms, and other rhythms ('bimodal' and/or 'polyphasic'), are found below.

2.6.2 Nocturnal

As stated above, when the majority of the activity of an animal occurs at night it is said to have a nocturnal activity pattern. Dewsbury (1980) and Baumgardner et al. (1980) surveyed the activity patterns of several species of muroid rodents using wheel-running behaviour (12 species) or a time-sampling technique (14 species), respectively. Five species of *Peromyscus* (mice), four species of *Microtus* (voles) as well as house mice (*Mus mus*), northern grasshopper mice (*Onychomys leucogaster*) and laucha de campo (*Calomys callosus*) all had wheel-running activity patterns consistent with the 'nocturnal' classification (Fig. 2.5). Baumgardner et al. (1980) surveyed these and two other muroid rodent species for their patterns of locomotion, stereotypy, grooming, eating, drinking and postural readjustment and found that the nocturnality of the activity patterns was seen in all but the four species of voles. These disparate results emphasise that the classification of activity patterns can depend on the method used to measure the locomotor activity. Dewsbury (1980), in an attempt to resolve the difference in the classification of these activity patterns of voles, hypothesised that voles may forage close to their burrow during the day and extend their foraging range during the cover of darkness. This latter foraging pattern may be reflected in the laboratory by the nocturnal pattern of wheel running found by Dewsbury (1980) for these vole species.

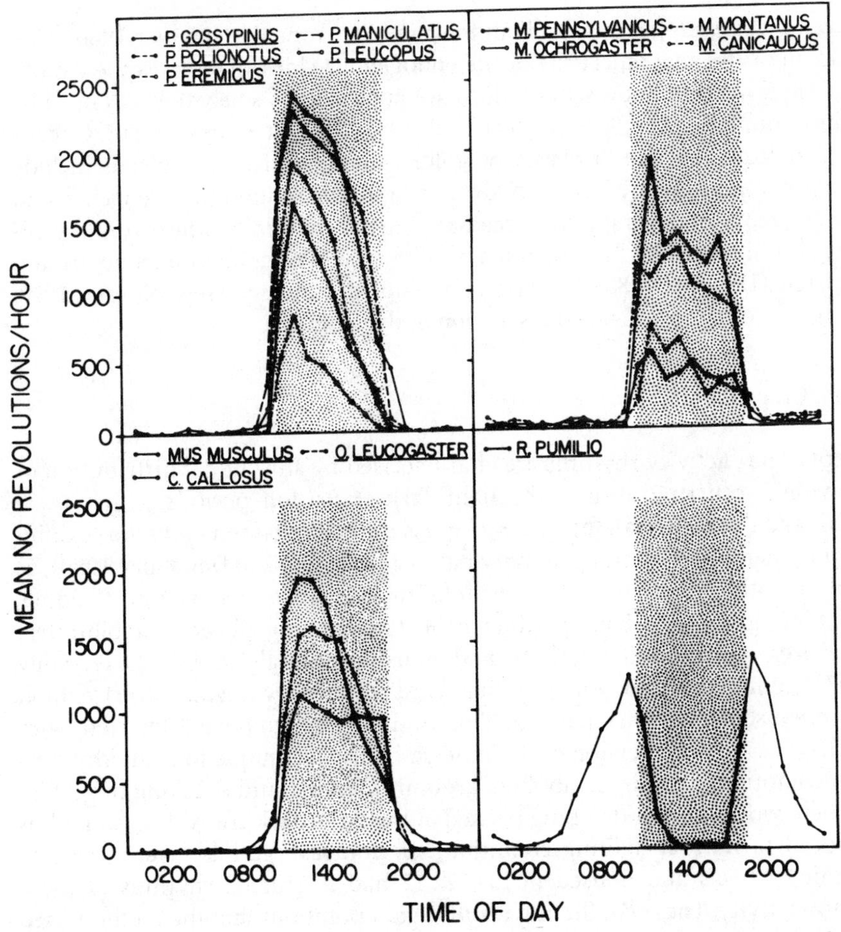

Fig. 2.5. Mean number of running wheel revolutions/h across the LD 12 : 12 for 13 species of muroid rodents (adapted from Dewsbury 1980). The *shaded period* represents the dark phase of the photocycle and appears in the middle of what normally would be the day in the field as a convenience to the researchers who also used the time-sampling technique to survey behaviours other than wheel running that could not be automated

2.6.3 Diurnal

As stated above, animals that show a diurnal pattern of activity exhibit the majority of their locomotor activity during the day or light portion of the LD cycle. Notable examples of day-active small mammals are ground squirrels. For example, a diurnal activity pattern of wheel running is shown by golden-mantled ground squirrels (Lee et al. 1986) and by antelope ground squirrels (*Ammospermophilus leucurus*; Kenagy 1978). Ground squirrels also exhibit rhythmic changes in their free-running period that approximates 1 year, even when housed in constant environmental conditions. These rhythms are driven by an endogenous clock and termed 'circannual' (Gwinner 1986). The location of the anatomical site of the endogenous clock that drives the circannual rhythms is unknown, but it does not appear to be solely the SCN or to be tightly coupled to the SCN (Zucker et al. 1983; Dark et al. 1985, 1990). Circannual rhythms that are displayed by golden-mantled ground squirrels include changes in reproductive status, body mass and hibernation (e.g. Pengelley and Fisher 1963). Interestingly, the circadian diurnal rhythm of wheel-running activity is influenced by the circannual timing system of golden-mantled ground squirrels (Lee et al. 1986), but not in antelope ground squirrels (Kenagy 1978; see below for a more detailed discussion of the former).

2.6.4 Crepuscular

Crepuscular activity rhythms are characterised by a bi-modal distribution of activities occurring around the light/dark transition periods – that is, at dawn and dusk. African four-striped grass mice (*Rhabdomys pumilio*) exhibit crepuscular wheel-running activity patterns (see Fig. 2.5; Dewsbury 1980), as well as other activities that include maintenance behaviours (feeding, drinking, grooming; Baumgardner et al. 1980). Using tilt-box (stabilimiter) measures of locomotor activity and housing animals in LD 12 : 12, Stutz (1972) found that Mongolian gerbils (jirds; *Meriones unguiculatus*) exhibit the crepuscular activity pattern. This finding was supported by Pietrewicz et al. (1982) using a version of the time-sampling technique to study rhythms of locomotion, digging, scrabbling, grooming, eating and drinking in gerbils. All behaviours except drinking peaked at the light/dark transitions in gerbils exposed to natural lighting conditions. In contrast, gerbils given access to running wheels and housed in LD 12 : 12 had nocturnal rhythms of locomotor activity. These conflicting results again point out that the method used to assess activity can influence the activity pattern that is displayed.

2.6.5 Ultradian

Rhythms with a period length significantly less than 24 h are termed 'ultradian'. Even less is known about the generation of these polyphasic rhythms and their influence by environmental factors than for circadian rhythms (see above). Activity rhythms of this type are best exemplified by the feeding rhythm of voles (for a review see Madison 1985 and Chap. 11). In common voles (*Microtus arvalis*), 2–3-h rhythms of feeding have been observed in both field and laboratory studies (see Gerkema et al. 1993 for a brief review). The ultradian rhythm of feeding occurs and, at the same time, a circadian rhythm of wheel running activity also is expressed (Fig. 2.6; Gerkema et al. 1993). Interestingly, SCN lesions block the circadian rhythm of wheel running, but not the ultradian rhythms of feeding in common voles (Gerkema et al. 1990).

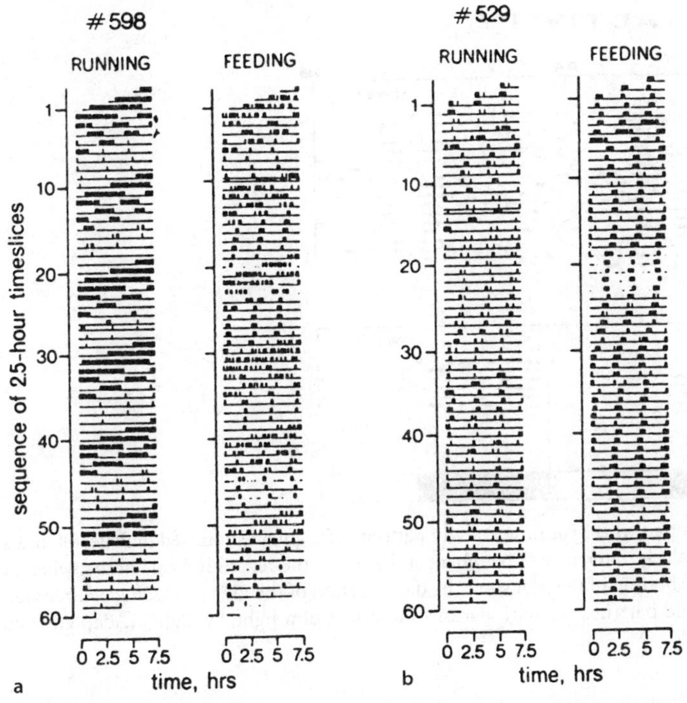

Fig. 2.6. Actograms for wheel running (*left record* of each pair) and feeding (*right record* of each pair) of voles during 6 days of constant darkness. Records are triple-plotted on a 2.5-h basis to facilitate visualisation of the activity patterns. a Intact vole showing circadian rhythmicity, and only the ultradian rhythm is prevalent. b Vole with a sham lesion of the SCN without circadian rhythmicity. (Adapted from Gerkema et al. 1993)

2.6.6 Acyclic

Some species of small mammals show no apparent pattern of activity when exposed to LD cycles. This arrhythmic pattern of activity is termed 'acyclic'. One of the clearest examples of acyclic wheel-running behaviour is seen in European ferrets (*Mustela furo*). No apparent temporal organisation to their wheel running is seen even when the LD cycle includes a twilight transition (Fig. 2.7, top; Stockman et al. 1985). This arrhythmic pattern of activity should be contrasted with that of Syrian hamsters that have a clear nocturnal activity pattern under identical conditions (see Fig. 2.7, bottom). Acyclic activity patterns are also seen in three other carnivorous species – bobcats (*Felix rufus*; Kavanau 1971) and domestic cats and dogs (Sternman et al. 1965; Hawking et al. 1971). Not all carnivores exhibit acyclic activity patterns, however. Least weasels (*Mustela rixosa*) show a nocturnal activity pattern under lighting conditions of simulated dusk/dawn and an acyclic activity pattern when changes in the LD cycle occur abruptly (Kavanau 1969).

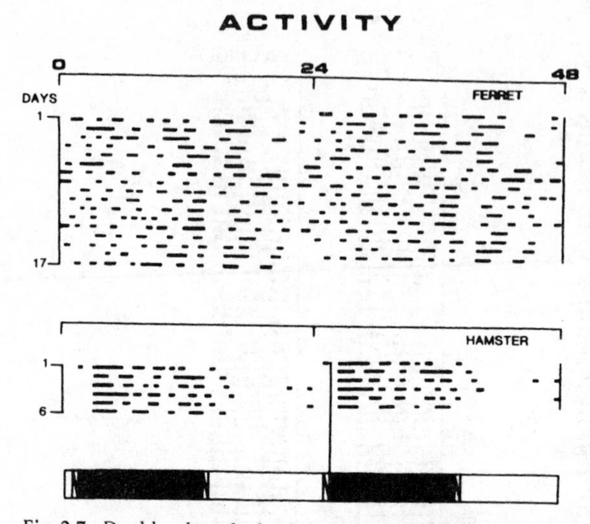

Fig. 2.7. Double-plotted wheel-running activity patterns of a gonadectomised male ferret and a gonadally-intact male Syrian hamster. Numbers at the *top* of the record indicate hours, whereas the *black bars* at the bottom of the plot show the dark portion of the LD 13 : 11. *Shaded areas* left and right of the solid bar (i.e., the dark period) indicate a dim light, 'twilight'. (Adapted from Stockman et al. 1985)

2.7 Effects of Semi-natural Environments in the Laboratory on Activity Patterns

Many of the species for which the activity patterns were described above have a semi-fossorial mode of living in the field. Thus, when they are exposed to LD cycles in standard laboratory cages they cannot escape from the light portion of the photoperiod. Obviously nocturnally-active animals in the field seek the shelter of their burrows during the day, and only sample the light periodically. Several researchers have designed semi-natural environments in the laboratory that give the animals the option of staying in their burrows and control their own exposure to the LD cycle as well as when they exhibit wheel-running behaviour. For example, DeCoursey (1986) studied light-sampling behaviour and photoentrainment in flying squirrels (*Glaucomys volans*). Squirrels only viewed light for several minutes a day, a finding similar to that in Syrian hamsters housed in a similar-type simulated burrow system (Pratt and Goldman 1986). Even with a simulated den/burrow and only when sampling light for a few minutes a day, however, the animals still remained entrained to the LD cycle, and therefore maintained their nocturnal pattern of locomotor activity. Thus, despite the artificial nature of the standard wheel-running cages, the use of abrupt transitions in the photoperiod and the forced exposure of the animals to both portions of the LD cycle, classification of locomotor activity patterns can correspond to the categorisation made in semi-natural environments.

2.8 Daily Activity Patterns: Flexibility, Variability and Interaction with Other Rhythms

2.8.1 Flexibility Between Species in Their Activity Patterns

It has been demonstrated repeatedly that alterations in gonadal steroids affect the amount of activity and its patterning in a variety of species including laboratory rats and Syrian hamsters (see above). Marked naturally-occurring changes in gonadal steroids and other hormones, such as prolactin, are seen in females during pregnancy and lactation. These changes in reproductive status can also affect activity patterns. For example, Perrigo (1987, 1990) explored the relationship between feeding effort (i.e., working for food – see below) on several responses related to growth and reproduction in house mice (*Mus domesticus*) and deer mice (*Peromyscus maniculatus*). Animals of both species were subjected to one of several increasing work requirements where delivery of a food pellet was contingent upon completion of a given number of revolutions made in a running wheel. Because wheel running was monitored continuously, the pattern of locomotor activity was also monitored coincidentally under these energetically demanding conditions.

These two species exhibited quite different degrees of flexibility in their activity patterns. Deer mice maintained a rigid pattern of nocturnal wheel running even with increasing work requirements (Fig. 2.8, bottom). House mice were also primarily nocturnal in their activity, although they occasionally displayed activity bouts that extended into the light following lights-out (Fig. 2.8, top). This tendency to show some diurnal activity was exaggerated as the work requirements were increased (Fig. 2.8, top). In contrast, the exclusive nocturnal activity pattern was only violated slightly at the highest work requirement in deer mice (Fig. 2.8, bottom). Perrigo (1990) hypothesised that this relatively inflexible pattern of nocturnal locomotor activity seen in deer mice might reflect an underlying circadian strategy strongly shaped by the risk of predation. In contrast, house mice may have developed a more flexible underlying circadian strategy that would permit them to decrease some of the energetic costs of foraging during the period of decreased ambient temperature at night. This increased flexibility in the foraging patterns of house mice was even more apparent during pregnancy and especially during lactation (Fig. 2.8, top). Indeed, the activity patterns of house mice appear to free-run during lactation despite the continued presence of an LD cycle. Deer mice, however, showed nocturnal foraging (wheel running) patterns that were nearly inflexible, albeit with some excursions into the light period occurring toward the end of the lactation period (Fig. 2.8, bottom). Thus, these two species are examples of the varying amounts of flexibility in the activity patterns that are exhibited by small rodents. Differences between these species may reflect species-specific evolutionary pressures of predation and thermoregulation (Perrigo 1990).

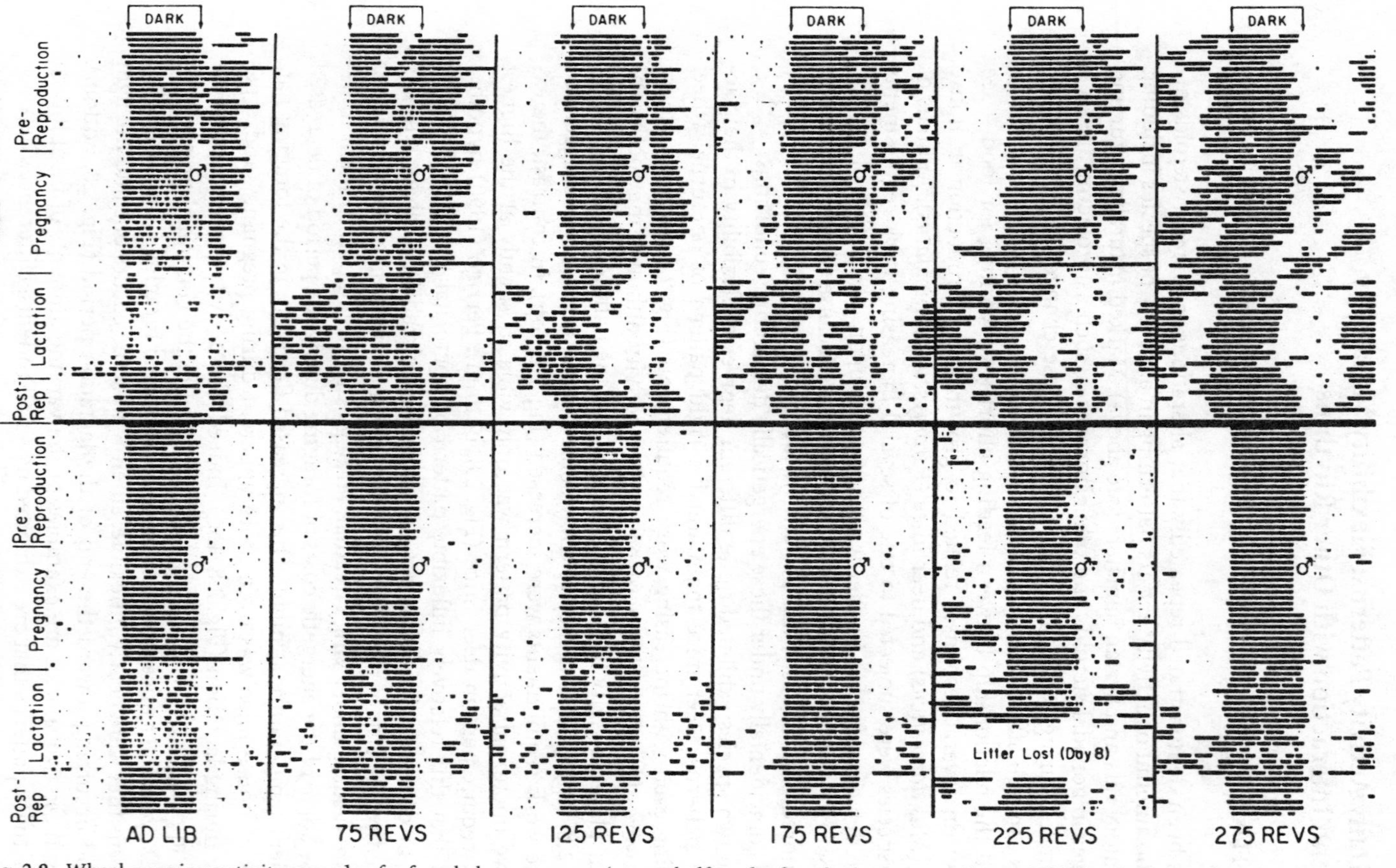

Fig. 2.8. Wheel-running activity records of a female house mouse (*upper half*) and a female deer mouse (*lower half*) that had to meet different work requirements (0, 75, 125, 175, 225, or 275 revolutions) to receive a food pellet. Daily records are aligned with the 8 h of darkness for the LD 16 : 8 photoperiod. *Male symbol* indicates a 2-h period of exposure of the mouse to a male during the pregnancy phase. Reproductive status is indicated on the *left side* of each record. *Post-Rep* post reproduction status. (Adapted from Perrigo 1987)

2.8.2 Within Species Variability in Activity Patterns

Within a species, considerable variability in activity patterns can be seen. Recently, variations in activity patterns have been shown to be due to an inheritable alteration in the organisation of the underlying circadian system in Siberian hamsters (*Phodopus sungorus sungorus*). The altered circadian system in these animals results in a unique pattern of entrainment to photoperiods of decreasing length (i.e., 'winter-like' short days, SD). Siberian hamsters typically show a collection of adaptive physiological and behavioural responses for overwintering when transferred from 'summer-like' long days to SDs (Hoffmann 1973). Specifically, SD exposure results in decreases in body mass and body fat, change to a white pelage and gonadal regression in most, but not all hamsters of this species (for a review see Bartness and Wade 1985).

The SD-insensitivity of some hamsters appears to be due to atypical entrainment to SD by their underlying circadian system (Puchalski and Lynch 1991). These SD non-responders have a longer free-running period and a unique phase-resetting response to light pulses (e.g. PRC) compared with their photoperiod-responsive counterparts. These two effects result in a delayed onset of activity in the dark and a shorter period of activity (α) than in photoresponsive hamsters (Puchalski and Lynch 1991). In a recent experiment, hamsters of both phenotypes were exposed to a brief pulse of light (30 min) at various circadian times. Just as there are times in the circadian cycle of locomotor activity that induce phase delays or advances (see above), there is also a circadian rhythm of photosensitivity to light that mediates photoperiodic time measurement (for a review see Elliott and Goldman 1981). Puchalski and Lynch (1994) found that photosensitive and photoinsensitive hamsters both show gonadal regression when illumination occurs just before and up to nearly 12 h before activity onset. Therefore, in this species, variability in activity patterns among two phenotypes has, as its basis, an inability of one phenotype (SD-insensitive hamsters) to entrain normally to full SD photoperiods.

2.8.3 Interactions of Circannual Rhythms with Circadian Rhythms

Ground squirrels have an annual 6-month period of inactivity during which time deep torpor will be exhibited with appropriate decreases in the ambient temperature. If housed at conventional vivarium temperatures, deep torpor does not occur, although the animals will decrease their body temperature to approximately 25 °C intermittently and are sluggish (Lee et al. 1986, 1990). These animals still exhibit wheel-running behaviour and so it is possible to investigate the interactions between circannual and circadian systems using the locomotor rhythm as an output of the circadian and circannual clock. Golden-mantled ground squirrels kept at 23 °C in constant dim light show marked seasonal alterations in the onset and length of their locomotor ac-

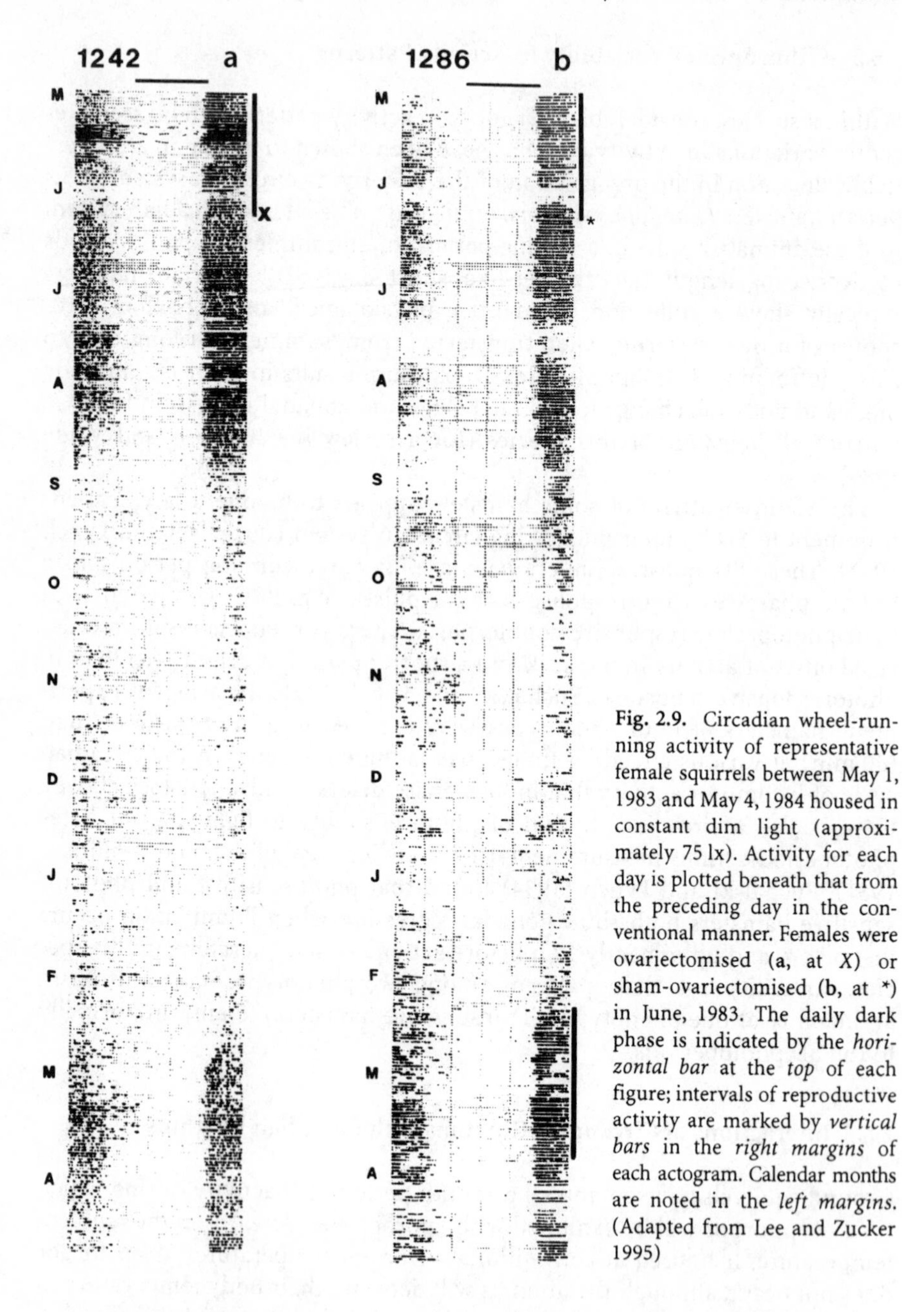

Fig. 2.9. Circadian wheel-running activity of representative female squirrels between May 1, 1983 and May 4, 1984 housed in constant dim light (approximately 75 lx). Activity for each day is plotted beneath that from the preceding day in the conventional manner. Females were ovariectomised (**a**, at *X*) or sham-ovariectomised (**b**, at *) in June, 1983. The daily dark phase is indicated by the *horizontal bar* at the *top* of each figure; intervals of reproductive activity are marked by *vertical bars* in the *right margins* of each actogram. Calendar months are noted in the *left margins*. (Adapted from Lee and Zucker 1995)

tivity rhythm (Zucker et al. 1983). Specifically, there is a delay in activity onset and a decrease in activity length (α) between February and September, a time of reproductive and general activity in the field, whereas the rest of the year the activity period is lengthened and the delay in activity onset corrected (Lee and Zucker 1995; Fig. 2.9). This 'cycle of cycles' (Mrosovsky 1976) also is present if golden-mantled ground squirrels are exposed to a LD cycle (LD 14 : 10) continuously (Lee et al. 1986).

As discussed above, the increased free-running period during the 'heterothermic' phase of the cycle and the decreased free-running period during the 'euthermic' phase of the annual cycle suggests that circadian rhythms in golden-mantled ground squirrels are temperature dependent (Lee et al. 1990; see above). The seasonal decrease in activity levels and the delay in activity onset relative to the LD cycle coincides with the normal hibernation season in these animals. The 4-h advance of the onset of activity coincides with the onset of their yearly estrous cycles and the time in the spring when golden-mantled ground squirrels emerge from their burrows in the field (Lee et al. 1986). Although linked to the estrous cycle, this cycle of cycles persists in long-term ovariectomised squirrels and thus may be independent of the seasonal reproductive cycle (Lee and Zucker, unpubl. obs.).

2.9 Concluding Remarks

The locomotor activity rhythm has proved to be invaluable as a measure of the output of the circadian clock. It will continue to be used by chronobiologists as a means to study the functions of circadian clocks and to investigate the physiological processes that mediate clock function. In addition, we now know that activity itself can phase shift circadian rhythms, a finding that should prove useful in determining how behaviour influences the clock.

Most of our information on the mechanisms underlying the biological clock in mammals comes from studies using Syrian hamsters and laboratory rats. This raises the frequently asked question: can these findings be generalised to other species? For example, the organisation of locomotor activity patterns in the Turkish hamster (*Mesocricetus brandti*), and the closely related Syrian hamster appear quite different. That is, although the entrainment of locomotor activity to LD cycles appears similar between the species (Albers et al. 1983), under constant light or following blinding and gonadectomy, the free-running activity rhythms were often characterised by frequent changes in the free-running period that appeared to occur spontaneously. Thus, a comparative approach to the study of activity patterns and the circadian clock may increase our understanding of species-specific differences in the organisation of locomotor and other activities as well as the neuroanatomical and neurochemical bases for the generation and modulation of activity rhythms.

Acknowledgements. The authors thank Drs. Donald Dewsbury, Teressa Lee and Martin Moore-Ede for sending us prints of their original figures, Dr. Irving Zucker for his comments on the manuscript and Ms. Cynthia Lane for her assistance in the preparation of this manuscript and for drawing the schematic of the SCN. Preparation of this chapter and original data were supported in part by NIMH Research Scientist Development Award K02 MH00841, NIMH RO1 MH48462 and NIH RO1 DK35254 to TJB and NIH RO1 NS30022 and NIH RO1 NS34586 to HEA.

References

Albers HE (1981) Gonadal hormones organize and modulate the circadian system of the rat. Am J Physiol 241:R62–R66

Albers HE, Gerall AA, Axelson JF (1981) Effect of reproductive state on circadian periodicity in the rat. Physiol Behav 26:21–25

Albers HE, Carter DS, Darrow JM, Goldman BD (1983) Circadian organization of locomotor activity in the Turkish hamster (*Mesocricetus brandti*). Behav Neural Biol 37:362–366

Albers HE, Liou SY, Ferris CF, Stopa EG, Zoeller RT (1991) Neurochemistry of circadian timing. In: Klein DC, Moore RY, Reppert SM (eds) The suprachiasmatic nucleus: the mind's clock. Oxford University Press, New York, pp 263–288

Albers HE, Liou SY, Stopa EG, Zoeller RT (1992) Neurotransmitter co-localization and circadian rhythms. In: Joss J, Buijs RM (eds) Tilders progress in brain research. Elsevier, Amsterdam, pp 289–307

Aschoff J (1959) Periodik licht- and dunkelaktiver Tiere unter konstanten Umgebungsbedingungen. Pflügers Arch Ges Physiol 270:9

Aschoff J (1979) Circadian rhythms: influences of internal and external factors on the period measured in constant conditions. Z Tierpsychol 49:225–249

Aschoff J (1981) A survey in biological rhythms. In: Aschoff J (ed) Biological rhythms. Handbook of behavioral neurobiology, vol 4 . Plenum Press, New York, pp 3–10

Axelson JF, Gerall AA, Albers HE (1981) The effect of progesterone on the estrous activity cycle of the rat. Physiol Behav 25:631–635

Bartness TJ, Goldman BD (1989) Mammalian pineal melatonin: a clock for all seasons. Experientia 45:939–945

Bartness TJ, Powers JB, Hastings MH, Bittman EL, Goldman BD (1993) The timed infusion paradigm for melatonin delivery: what has it taught us about the melatonin signal, its reception and the photoperiodic control of seasonal responses? J Pineal Res 15:161–190

Bartness TJ, Wade GN (1985) Photoperiodic control of seasonal body weight cycles in hamsters. Neurosci Biobehav Rev 9:599–611

Baumgardner DJ, Ward SE, Dewsbury DA (1980) Diurnal patterning of eight activities in 14 species of muroid rodents. Anim Learn and Behav 8:322–330

Biello SM, Harrington ME, Mason R (1991) Geniculo-hypothalamic tract lesions block chlordiazepoxide-induced phase advance in Syrian hamsters. Brain Res 552:47–52

Block M, Zucker I (1976) Circadian rhythms of rat locomotor activity after lesions of midbrain raphe nuclei. J Comp Physiol 109:235–247

Boulos Z, Rosenwasser AM, Terman M (1980) Feeding schedules and the circadian organization of behavior in the rat. Behav Brain Res 1:39–65

Brown FA, Hastings JW, Palmer JD (1970) The biological clock: two views. Academic Press, New York

Crowley M, Bovet J (1980) Social synchronization of circadian rhythms in deer mice (*Peromyscus maniculatus*). Behav Ecol Sociobiol 7:99–105

Daan S, Damassa D, Pittendrigh CS (1975) An effect of castration and testosterone replacement on a circadian pacemaker in mice (*Mus musculus*). Proc Natl Acad Sci USA 72:3744–3747

Dark J, Pickard GE, Zucker I (1985) Persistence of circannual rhythms in ground squirrels with lesions of the suprachiasmatic nuclei. Brain Res 332:201–207

Dark J, Kilduff TS, Heller HC, Licht P, Zucker I (1990) Suprachiasmatic nuclei influence hibernation rhythms of golden-mantled ground squirrels. Brain Res 509:111–118

Davis FC, Stice SB, Menaker M (1987) Activity and reproductive state in the hamster: independent control by social stimuli and a circadian pacemaker. Physiol Behav 40:583–590

DeCoursey PJ (1986) Light-sampling behavior in photoentrainment of a rodent circadian rhythm. J Comp Physiol A 159:161–169

Dewsbury DA (1980) Wheel-running behavior in 12 species of muroid rodents. Behav Proc 5: 271–280

Drucker-Colin R, Aguilar-Roblero R, Fernandez-Cancino F, Rattoni FB (1984) Fetal suprachiasmatic nucleus transplants: diurnal rhythm recovery of lesioned rats. Brain Res 311:353–357

Edgar DM, Miller JD, Prosser RA, Dean RR, Dement WC (1993) Serotonin and the mammalian circadian system: II. Phase-shifting rat behavioral rhythms with serotonergic agonists. J Biol Rhythms 8:17–31

Elliott JA, Goldman BD (1981) Seasonal reproduction: photoperiodism and biological clocks. In: Adler NT (ed) Neuroendocrinology of reproduction. Plenum Press, New York, pp 377–423

Elliott JA, Tamarkin L (1994) Complex circadian regulation of pineal melatonin and wheel-running in Syrian hamsters. J Comp Physiol A 174:469–484

Fuxe K (1965) Evidence of the existence of monoamine neurons in the central nervous system. IV. Distribution of monoamine nerve terminals in the central nervous system. Acta Physiol Scand 64:37–85

Gerkema MP, Groos GA, Daan S (1990) Differential elimination of circadian and ultradian rhythmicity by hypothalamic lesions in the common vole, *Microtus arvalis*. J Biol Rhythms 5:81–95

Gerkema MP, Daan S, Wilbrink M, Hop MW, van der Leest F (1993) Phase control of ultradian feeding rhythms in the common vole (*Microtus arvalis*): the roles of light and the circadian system. J Biol Rhythms 8:151–171

Gibbs FP (1981) Temperature dependence of rat circadian pacemaker. Am J Physiol 241: R17–R20

Gibbs FP (1983) Temperature dependence of the hamster circadian pacemaker. Am J Physiol 244:R607–R610

Groos GA, Hendriks J (1979) Regularly firing neurons in the rat suprachiasmatic nucleus. Experientia 35:1597–1598

Gwinner E (1986) Circannual rhythms. Springer, Berlin Heidelberg New York

Hawking F, Lobban MC, Gammage K, Worms MJ (1971) Circadian rhythms (activity, temperature, urine and microfilariae) in dog, cat, hen, duck, *Thamnomys* and *Gerbillus*. J Interdiscipl Cycle Res 2:455–473

Hendrickson AE, Wagoner N, Cowan WM (1972) An autoradiographic and electron microscopic study of retino-hypothalamic connections. Z Zell 135:1–26

Hoffmann K (1973) The influence of photoperiod and melatonin on testis size, body weight and pelage colour in the Djungarian hamster (*Phodopus sungorus*). J Comp Physiol 85:267–282

Illnerova H (1991) The suprachiasmatic nucleus and rhythmic pineal melatonin production. In: Klein DC, Moore RY, Reppert SM (eds) Suprachiasmatic nucleus: the mind's clock. Oxford University Press, New York, pp 197–216

Inouye ST, Kawamura H (1979) Persistence of circadian rhythmicity in mammalian hypothalamic "island" containing the suprachiasmatic nucleus. Proc Natl Acad Sci USA 76:5961–5966

Janik D, Mrosovsky N (1993) Nonphotically induced phase shifts of circadian rhythms in the golden hamster: activity-response curves at different ambient temperatures. Physiol Behav 53:431–436

Johnson RF, Moore RY, Morin LP (1988) Loss of entrainment and anatomical plasticity after lesions of the hamster retinohypothalamic tract. Brain Res 460:297–313

Kavanau JL (1969) Influences of light on activity of small mammals. Ecology 50:548–557

Kavanau JL (1971) Locomotion and activity phasing of some medium-sized mammals. J Mammal 52:386–403

Kenagy GJ (1978) Seasonality of endogenous circadian rhythms in a diurnal rodent *Ammospermophilus leucurus* and a nocturnal rodent *Dipodomys merriami*. J Comp Physiol A 128: 21–36
Lee TM, Zucker I (1995) Seasonal variations in circadian rhythms persist in gonadectomized golden-mantled ground squirrels. J Biol Rhythms 10:188–195
Lee TM, Carmichael MS, Zucker I (1986) Circannual variations in circadian rhythms of ground squirrels. Am J Physiol 250:R831–R836
Lee TM, Holmes WG, Zucker I (1990) Temperature dependence of circadian rhythms in golden-mantled ground squirrels. J Biol Rhythms 5:25–34
Madison DM (1985) Activity rhythm and spacing. In: Tamarin RH (ed) Biology of new world *Microtus*. Am Soc Mammal Spec Publ 8, Shippensburg, pp 373–419
Marimuthu G, Subbaraj R, Chandrashekaran MK (1978) Social synchronization of the activity rhythm in a cave-dwelling insectivorous bat. Naturwissenschaften 65:600
Marimuthu G, Rajan S, Chandrashekaran MK (1981) Social entrainment of the circadian rhythm in the flight activity of the microchiropteran bat *Hipposideros speoris*. Behav Ecol Sociobiol 8:147–150
Meijer JH, Rietveld WJ (1989) Neurophysiology of the suprachiasmatic circadian pacemaker in rodents. Physiol Rev 69:671–707
Moore RY, Card JP (1985) Visual pathways and the entrainment of circadian rhythms. Ann NY Acad Sci 453:123–133
Moore RY, Eichler VB (1972) Loss of a circadian adrenal corticosterone rhythm following suprachiasmatic nuclear lesions in the rat. Brain Res 42:201–206
Moore RY, Lenn NJ (1972) A retinohypothalamic projection in the rat. J Comp Neurol 146:1–14
Moore-Ede MC, Sulzman FM, Fuller CA (1982) The clocks that time us. Harvard University Press, Cambridge, MA
Morin LP, Fitzgerald KM, Zucker I (1977) Estradiol shortens the period of hamster circadian rhythms. Science 196:305–307
Mrosovsky N (1976) Lipid programmes and life strategies in hibernators. Am Zool 16:685–697
Mrosovsky N (1988) Phase response curves for social entrainment. J Comp Physiol A 162:35–46
Pengelley ET, Fisher KC (1963) The effect of temperature and photoperiod on the yearly hibernating behavior of the captive golden-mantled ground squirrels (*Citellus lateralis tescorum*). Can J Zool 41:1103–1120
Perrigo G (1987) Breeding and feeding strategies in deer mice and house mice when females are challenged to work for their food. Anim Behav 35:1298–1316
Perrigo G (1990) Food, sex, time and effort in a small mammal: energy allocation strategies for survival and reproduction. Behaviour 114:1–4
Pickard GE, Ralph MR, Menaker M (1987) The intergeniculate leaflet partially mediates effects of light on circadian rhythms. J Biol Rhythms 2:35–56
Pietrewicz AT, Hoff MP, Higgins SA (1982) Activity rhythms in the Mongolian gerbil under natural light conditions. Physiol Behav 29:377–380
Pittendrigh CS, Caldarola PC (1973) General homeostasis of the frequency of circadian oscillations. Proc Natl Acad Sci USA 70:2697–2701
Pittendrigh CS, Daan S (1976a) A functional analysis of circadian pacemakers in nocturnal rodents. I. The stability and lability of spontaneous frequency. J Comp Physiol 106:223–252
Pittendrigh CS, Daan S (1976b) A functional analysis of circadian pacemakers in nocturnal rodents: V. Pacemaker structure: a clock for all seasons. J Comp Physiol 106:333–355
Pratt BL, Goldman BD (1986) Activity rhythms and photoperiodism of Syrian hamsters in a simulated burrow system. Physiol Behav 36:83–89
Puchalski W, Lynch GR (1991) Circadian characteristics of Djungarian hamsters: effects of photoperiodic pretreatment and artificial selection. Am J Physiol 261:R670–R676
Puchalski W, Lynch GR (1994) Photoperiodic time measurement in Djungarian hamsters evaluated from T-cycle studies. Am J Physiol 267:R191–R201
Ralph MR, Foster RG, Davis FC, Menaker M (1990) Transplanted suprachiasmatic nucleus determnines circadian period. Science 247:975–978

Reebs SG, Mrosovsky N (1989a) Large phase-shifts of circadian rhythms caused by induced running in a re-entrainment paradigm: the role of pulse duration and light. J Comp Physiol A 165:819–825
Reebs SG, Mrosovsky N (1989b) Effects of induced wheel running on the circadian activity rhythms of Syrian hamsters: entrainment and phase response curve. J Biol Rhythms 4:39–48
Ribak CE, Peters A (1975) An autoradiographic study of the projections from the lateral geniculate body of the rat. Brain Res 92:261–294
Richter CP (1922) A behavioristic study of the activity of the rat. Comp Psychol Monogr 1:1–54
Richter CP (1965) Biological clocks in medicine and psychiatry. Thomas, Springfield, IL
Richter CP (1970) Dependence of successful mating in rats on functioning of the 24-hour clocks of the male and female. Commun Behav Biol 5:1–5
Rusak B (1990) Biological rhythms: from physiology to behavior. In: Montplaisir J, Godbout R (eds) Sleep and biological rhythms. Oxford University Press, New York, pp 11–24
Rusak B, Mistlerger RE, Losier B, Jones CH (1988) Daily hoarding opportunity entrains the pacemaker for hamster activity rhythms. J Comp Physiol A 164:165–171
Silverman HJ, Zucker I (1976) Absence of post-fast food compensation in the golden hamster (*Mesocricetus auratus*). Physiol Behav 17:271–285
Stephan FK (1981) Limits of entrainment to periodic feeding in rats with suprachiasmatic lesions. J Comp Physiol A 143:401–410
Stephan FK, Zucker I (1972) Circadian rhythms in drinking behavior and locomotor activity of rats are eliminated by hypothalamic lesions. Proc Natl Acad Sci USA 69:1583–1586
Sterman MB, Knauss T, Hehmann D, Clements CD (1965) Circadian sleep and waking patterns in the laboratory cat. Electroencephalogr Clin Neurophysiol 19:506–517
Stockman ER, Albers HE, Baum MJ (1985) Activity in the ferret: oestradiol effects and circadian rhythms. Anim Behav 33:150–154
Stutz AM (1972) Diurnal rhythms of spontaneous activity in the Mongolian gerbil. Physiol Zool 45:325–334
Sulzman FM, Fuller CA, Moore-Ede MC (1979) Tonic effects of light on the circadian system of the squirrel monkey. J Comp Physiol 129:43–50
Takahashi JS, Menaker M (1980) Interaction of estradiol and progesterone: effects on circadian locomotor rhythms of female golden hamsters. Am J Physiol 239:R497–R504
Tokura H, Aschoff J (1978) Circadian activity rhythms of the pig-tailed macaque, *Macaca nemestrina*, under constant illumination. Pflügers Arch Ges Physiol 376:241–243
van den Pol AN (1985) The hypothalamic suprachiasmatic nucleus of the rat: intrinsic anatomy. J Comp Neurol 15:1049–1086
Viswanathan N (1989) Presence-absence cycles of the mother and not light-darkness are the zeitgeber for the circadian rhythm of newborn mice. Experientia 45:383–385
Viswanathan N, Chandrashekaran MK (1985) Cycles of presence and absence of mother mouse entrain the circadian clock of pups. Nature 317:530–531
Zee PC, Rosenberg RS, Turek FW (1992) Effects of aging on entrainment and rate of resynchroniztion of circadian locomotor activity. Am J Physiol 263:R1099–R1103
Zucker I, Boshes M, Dark J (1983) Suprachiasmatic nuclei influence circannual and circadian rhythms of ground squirrels. Am J Physiol 244:R472–R480

3 Activity Patterns and Metabolism

January Weiner

3.1 Introduction

A living organism may be regarded as a converter of resources into copies of itself (Calow and Townsend 1981). As all dynamic processes take time, time is sometimes counted as a vital resource, along with energy and various nutrients (Herbers 1981; Bunnel and Harestad 1990). Technically this is not necessary, because describing metabolic processes as rates automatically makes time part of the dynamics of other variables. However, the rates of resource acquisition and expenditure may differ; they may be regulated, and thus they can be subject to natural selection. In this context it is convenient to speak of time as a resource to be optimally managed. The life strategies of various species may be reflected in different patterns of basic life processes dynamics; in other words, species strategies differ in their allocation of time. We observe a diversity of activity patterns, changing in space and time, within hours or seasons.

Small mammals are well suited for comparative studies. Their bioenergetics have been thoroughly investigated (see Grodziński and Wunder 1975; Nagy 1987; McNab 1988, 1992; Karasov 1992; Nagy 1994 for reviews). They have a large variety of life strategies even within narrow taxonomic groups (Koteja and Weiner 1993), and many species belonging to different taxonomic groups have convergent life strategies. Such diversity facilitates interspecific comparisons (McNab 1987; Harvey et al. 1991). On the other hand, the vast body of information on the biology of small mammals is unbalanced. While some elements of their life histories are extensively studied, others are poorly known. The majority of behavioural data on small mammals originates from the laboratory and concerns animals living in cages and eating pellets, which makes them useless for ecological-evolutionary discussion. The time allocation of various activities in natural conditions (time budgets) and the efficiency of handling and processing food items in small mammals have rarely been studied, although such standard topics have been the subject of sophisticated analyses for birds and large mammals.

Ecological Studies, Vol. 141
Halle/Stenseth (eds.) Activity Patterns in Small Mammals

Judging from the body size distribution of species, 'small mammals' (terrestrial rodents, insectivores, bats) are the most successful among the Theria. The same seems true for the extant Marsupialia; indeed, their Jurassic ancestors, the Multituberculata, were also 'small mammals'. Why has the small mammal strategy been so successful, given the difficult thermodynamics of small-sized homeotherms – their unfavourable surface-to-volume ratio, low buffering capacity both short-term (low heat storage capacity in relation to the rate of heat dissipation) and long-term (small fat storage capacity in relation to energy budget) – and given the fact that small herbivores also have limited space for a voluminous gut, which is a prerequisite for efficient symbiotic digestion of cellulose? Small animals are easy victims of predators. The explanation is that natural selection operates on the whole life history and not on separate characteristics. The advantage of small size is rapid growth to maturity, allowing high reproductive output to balance mortality (Kozłowski and Weiner 1997); coping with unfavourable thermodynamics is the price to be paid for an otherwise successful life style. One should expect, however, that natural selection has optimised the energy budget within the constraints imposed by the whole strategy of a small mammal.

The aim of this chapter is to discuss energy budget partitioning, focusing on the contribution of activity to the total energy expenditure and the relation between energetics and daily activity patterns in small mammals.

3.2 The Energy Budget and Its Limits

Attempts to estimate energy expenditure in free-living mammals have a long history, particularly in regard to energy flow in ecosystems. In these studies, energy budgets have been assembled from elementary components measurable in the laboratory (Grodziński and Wunder 1975; Karasov 1992). Such a procedure requires acceptance of some weakly supported assumptions, but it gives an insight into the structure of the energy budget. The only way to test such estimates is to measure energy expenditure in the field directly. Here the method of choice has been the doubly labelled water (DLW) technique. This method allows estimation of the cumulative dissipation of carbon dioxide over periods of from one to several days on the basis of stable isotope (^{18}O, ^{2}H) dilution in blood. The field operations – taking blood samples from an animal at the beginning and end of the experimental period – are relatively easy, but processing the samples and isotopic analyses are quite laborious and expensive. However, with improvements in recent years the number of studies concerning animal field energetics has steadily increased, and now they include at least 28 species of small mammals (some of them studied several times independently; Nagy 1987, 1994; Degen 1994).

The great advantage of the DLW method is that it measures (to a known accuracy) the total energy expenditure of an animal living freely in natural conditions. The inherent disadvantage is that not much is known about the composition of each individual energy budget. Only a combination of field activity records and DLW measurements of field metabolic rates (FMR) can help to explain the observed patterns. Although some attempts have been successful (e.g. Karasov 1981; Kenagy et al. 1989), studies to demonstrate and explain the whole variation of energy budgets in the field are still to be done. The available FMR data include mammals with various life styles, food preferences and taxonomic positions (Fig. 3.1). Interspecific as well as intraspecific components contribute to the total variation. The number of entries is still too small, and their selection too incidental, to enable an analysis of causal effects, although such attempts have been made (Nagy 1987, 1994). Although a highly significant allometric regression can be fitted to the data ($R^2 = 0.79$, Fig. 3.1), its informative content and predictive power are quite limited (Kozłowski and Weiner 1997); it merely suggests that larger mammals tend to have higher energy expenditures, but the residual variance is still enormous (FMRs of mammals of the same size may differ severalfold; Fig. 3.1). Nevertheless, the interspecific distribution of the known field metabolic rates of small mammals carries important information, that is, what part of the plane determined by the axes of energy and body mass some individuals (or groups of individuals of some species) may occupy.

The FMRs measured in small mammals are confined to a space between two limits. From below, the field energy budgets are constrained by a line representing 1.6 × the predicted basal metabolic rate (BMR; from McNab's equation for rodents, McNab 1988). This means that an activity component (mechanical, such as locomotion, or chemical, such as thermoregulation) is added to their basal metabolic rate. The distribution of FMRs is constrained from above by a line representing 8 × BMR (one exceptionally high value may be regarded as an outlier). This latter line, arbitrarily fitted to the FMR data, converges with the empirical function for maximum energy assimilation from food (Amax) in mammals (Weiner 1992).

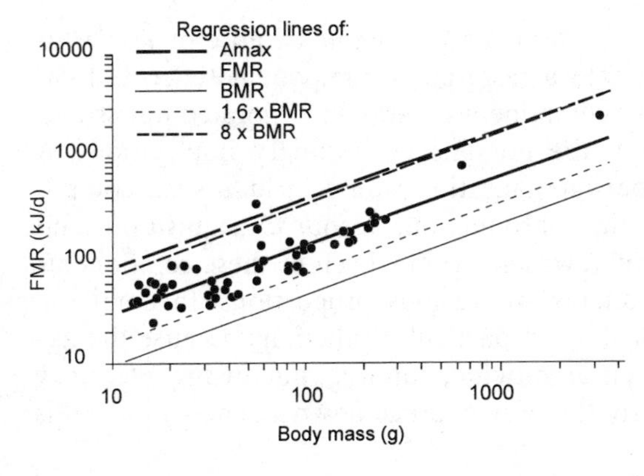

Fig. 3.1. Distribution of field metabolic rates (*FMR*) of rodents as measured with doubly labelled water. Data from Nagy (1987, 1994) and Salsbury and Armitage (1994). Regression for FMR (this study): FMR (kJ/day) = $6.46\ W^{0.666}$ (W = body mass, g), $R^2 = 0.79$. Regression for basal metabolic rates (*BMR*) from McNab (1988), for maximum energy assimilation rates (*Amax*) from Weiner (1992)

The nature of the upper limit of sustained energy budgets in mammals has been debated (Weiner 1987, 1989, 1992; Peterson et al. 1990; Hammond et al. 1994; Konarzewski and Diamond 1994; McDevitt and Speakman 1994a, b; Koteja 1995, 1996; Hammond and Diamond 1997). There is no doubt that the sustained rate of energy expenditure does not exceed a certain level, usually between 4 and 9 × the BMR. Two competing hypotheses to explain these findings invoke either a central limiting mechanism such as the rate of food digestion and absorption, or a combination of peripheral mechanisms, that is, the limited efficiency of energy use in specific systems. The sensitivity of maximum sustained metabolic rates to changes in food quality, and acclimation to chronically elevated energy demands involving an increase of the surface of the alimentary tract, seem to speak for the first option (Weiner 1992).

If the first hypothesis is true, the maximum rate of sustained metabolism should not vary with various kinds of energy loads such as cold exposure, peak lactation or mechanical work. In fact, however, the maximum sustained metabolic rates measured in several species of small mammals under cold exposure or at peak lactation differ (Koteja et al. 1993; Koteja 1996). Moreover, Hammond et al. (1994) demonstrated that the effects of cold and lactation can be additive. It is an open question, however, as to whether this evidence does support the hypothesis of 'peripheral' limitation of energy budgets. Many data show that specific systems, including thermoregulation, can operate at much higher rates, which in turn supports the central limitation hypothesis. A possible explanation may be that animals *are* centrally limited but that current performance may depend more upon optimisation of the whole life strategy (i.e., to maximise whole-life reproductive output, not only the energetic efficiency of one reproductive episode). This subject certainly requires further study. One can agree that all the various life styles of small mammals have their metabolic costs confined within limits which can be quantitatively determined.

3.3 The Structure of Energy Budgets

Any division of an energy budget into compartments is more or less arbitrary. The clearest division is into a respiratory component (energy dissipated as heat) and a productive component (energy incorporated in tissues). This, however, tells little about the possible evolutionary implications. A biologically meaningful compartmentalisation would include such components as maintenance, growth, reproduction, resource acquisition and predator avoidance (Calow and Townsend 1981). Each of these components contains respiration and production in various proportions. In empirical studies it may be difficult to tell apart particular functions because the apparent actions of animals, such as moving, running, burrowing, etc., may serve various vital purposes. Another way to break down an energy budget is to distinguish 'chemical work' from 'mechanical work'. The first may include thermoregulation and tissue production for growth, repair and reproduction, the latter all the animal's doings that manifest themselves as 'activity', as opposed to a resting state. From the point of view of natural selection, activity serves four major functions: survival assurance (running away from predators, territory defence, building shelters such as burrows), reproduction (courtship, mating), movement (dispersal, migration) and acquisition of food. In small mammals, sexually active only seasonally and with a negligible contribution from migrations, the two major components of activity are food acquisition and defence. Thus the questions arise: (1) What is the contribution of 'activity' to the energy budget? (2) What is the relative contribution of various activities? (3) How do the energy relations influence activity patterns in small mammals?

In homeotherms the energy budget allocation depends greatly on body mass. Grodziński and Weiner (1984) compared the partitioning of annual energy budgets of four herbivores differing in size by one order of magnitude (Fig. 3.2). In small mammals, represented in this comparison by the

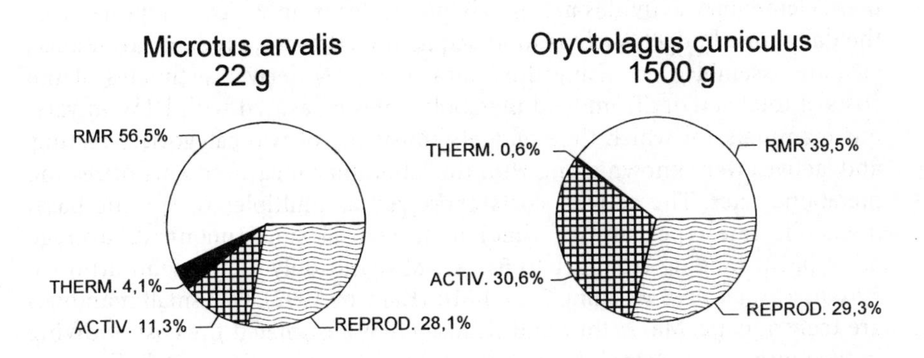

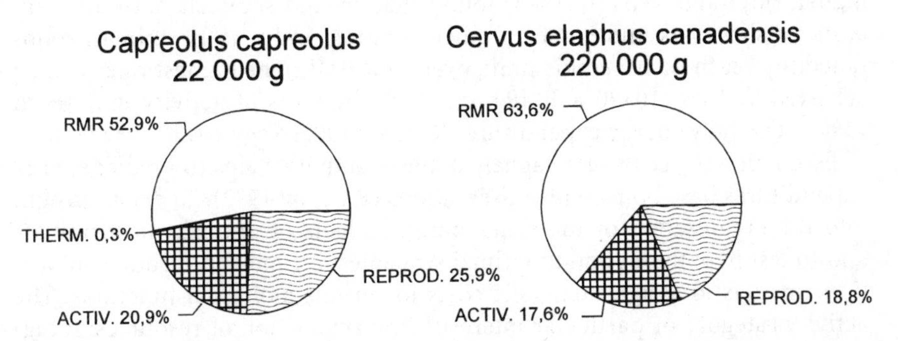

Fig. 3.2. Annual energy budget partitioning in four species of mammalian herbivores differing in size by orders of magnitude (after Grodziński and Weiner 1984). *RMR* resting metabolic rate, *THERM* net energy costs of thermoregulation, *REPROD* costs of reproduction (average female), *ACTIV* costs of locomotor activity

22 g vole *Microtus arvalis*, the share of activity seems relatively small (about 11%), compared with 18–31% in larger mammals (Fig. 3.2). Other authors confirm that mechanical work makes only a small contribution to energy expenditure in mammals. Garland (1983) combined the empirical allometric formulas predicting the costs of traveling a unit distance (see Taylor et al. 1982) with daily movement distances, arriving at the conclusion that in small mammals the cost of locomotion may be as low as 1% of daily energy expenditure, while in large mammals it may increase to 5–15%. Altmann (1987) reanalysed the available data on the cost of locomotion in terrestrial mammals and concluded that the daily costs of moving may be as little as 2% of their total energy expenditure, regardless of body size. The most accurate are field studies in which the total energy budget as well as total daily activity and travelled distances are measured in the same individuals. Karasov (1992) lists a few examples of small mammals, which indicate that their daily costs of locomotion are 4 to 15% of the daily energy expenditure.

Small mammals spend most of their time doing things that cannot be easily attributed to one of the elementary types of activity such as running or flying, for which we have good predictive formulas. On the other hand, these elementary activities are merely molecules from which components of the daily time budget, such as food acquisition, defense, predator avoidance, etc., are assembled. Realising this, Karasov (1992) derived estimates of the costs of total activity from field metabolic rates measured with DLW in various mammals, for which time budgets consisting of two categories, 'resting' and 'active', were known along with the laboratory measurements of resting metabolic rates. The activity costs expressed as multiples of specific basal metabolic rates (per unit of time) in terrestrial small mammals average 4.1 × BMR, ranging from 3.1 to 5.5 × BMR, and only in the semi-arboreal *Peromyscus leucopus* reaching 7.4 × BMR (Karasov 1992). Few small mammals are truly aquatic. MacArthur and Krause (1989) estimated the cost of diving in the muskrat (*Ondatra zibethicus*) at 1.9 to 2.8 × its resting metabolism.

Acute rates of metabolism for various motor activities can be much higher. Hoyt and Kenagy (1988) found that ground squirrels have different costs for slow walk and bursts of quick running, and that the animals spontaneously reach their aerobic limit, over 10 × BMR. The highest rates can be achieved by bats (10–20 × BMR), in which the costs of activity may reach 75% of the daily energy expenditure (Karasov 1992; Nagy 1994).

Estimating the costs of a vaguely defined 'activity' helps to predict energy expenditures based on simple observations (Karasov 1992). To get an insight into the organisation of the time/energy budgets of animals, to formulate and to test hypotheses about optimal strategies of time and resource allocation, one has to assign metabolic costs to various biological functions. The activity category of particular interest is the acquisition of resources: foraging, foraging rates and the costs involved.

3.4 Food Acquisition: Foraging Rates, Foraging Costs

3.4.1 Indirect Estimates Based on Allometry

The problem of the cost and benefit ratio of foraging traditionally attracts the interest of behavioural ecologists. Many empirical studies have been undertaken, too numerous to be cited, of birds, bats and large ungulates. In contrast, terrestrial small mammals have rarely been studied in this respect. The scarcity of relevant information may justify some indirect or approximate estimations.

Based on data on mouth morphology and feeding rates in various mammalian herbivores, Shipley et al. (1994) scaled rates of feeding to body mass. Using two sets of data, collected (1) from the literature and (2) from their own experiments, they produced allometric equations predicting maximum ingestion rates, (1) $I_{max} = 0.45\ M^{0.71}$ and (2) $I_{max} = 0.63\ M^{0.71}$, where I_{max} is the ingestion rate in g dry mass/min and M is body mass in kg. Solved for a 25 g mammal, and assuming the caloric content of dry plant material to be 20 kJ/g, the equations yield 0.65 and 0.91 kJ/min, respectively. Predictions from an allometric equation spanning nine orders of magnitude (from a lemming to an elephant) must be taken with great caution, particularly if the prediction concerns the extremes of this range.

3.4.2 Foraging Rates Estimated from Activity Budgets

A few studies on small mammals directly address the question of the energy ingestion rate and incremental energy expenditure during foraging. Smith (1968; cit. after Ferron et al. 1986) estimated that red squirrels (*Tamiasciurus hudsonicus*) ingest food at rates of 7.5 kJ/min (fungi), 5.8 kJ/min (seeds from Douglas fir cones), 4.7 kJ/min (berries) and only 0.7–1.3 kJ/min (lodgepole cones).

Morgan and Price (1992) measured the rate and the energetic cost of seed collection in semi natural conditions by various species of heteromyid rodents employing the scratch-digging mode of foraging. On average, energy expenditure for scratch-digging was approximately twice the resting metabolism, independent of body mass. There was a distinct allometric relation between body mass and the rate of pouching seeds. The millet seeds used in the experiments weighed 6 mg each. Assuming the caloric density of the seeds at 20 kJ/g, it may be calculated that a 60 g *Dipodomys* spp. collects 0.84 kJ/min, and a 7.5 g *Perognathus longimembris* 0.44 kJ/min. The energetic cost (energy expenditure/energy gain) of such a foraging mode increases with body mass and ranges from 0.04 J/J in *Perognathus* to 0.08 J/J in *Dipodomys*.

For a number of species of small mammals, the rates of energy acquisition during foraging can be estimated from time-energy budgets. Assuming that during the observation period an animal remains in energy balance, one can estimate the rate of energy acquisition by dividing the total energy expenditure (e.g. FMR measured with DLW) by the total time spent foraging. Belovsky (1984) provided data on the duration of foraging activity along with the amounts of food consumed by voles (*Microtus pennsylvanicus*), which made it possible to calculate an energy acquisition rate of 0.27 kJ/min. In a similar way the energy acquisition rates were estimated for granivorous ground squirrels (*Spermophilus saturatus*), insectivorous shrews (*Blarina brevicauda*) and an insectivorous, fossorial mole (*Parascalops breweri*), using data from Kenagy et al. (1989), Buckner (1964) and Jensen (1983), respectively, who also provided the data on energy expenditures.

More estimates could have been obtained using FMR values and time budget data from different sources: for the seed-eating heteromyid rodents *Dipodomys merriami*, *D. deserti* and *P. longimembris* (daily foraging times from Thompson 1985; FMR predicted after Nagy 1994), and for the browsing yellow-bellied marmot (*Marmota flaviventris*; time budget from Melcher et al. 1990; FMR from Salsbury and Armitage 1994).

Truly fossorial mammals, which employ burrowing to feed on the underground parts of plants and/or soil invertebrates, have extremely high costs of foraging (Vleck 1979, 1981; DuToit et al. 1985). Their rate of energy gain during foraging can be approximately estimated from published data. According to DuToit et al. (1985) *Georychus capensis* (189g) burrows at the rate of 4 m/h, penetrating a volume of 0.0078 m^3 of earth, which may contain corms and tubers worth of 65.4 kJ of digestible energy. Thus, the average rate of gross energy gain may reach 1.1 kJ/min. This calculation is based on very rough estimates of field burrowing activity and the density of food items in the soil, but it may give an idea of the order of magnitude of the foraging rate in a fossorial rodent.

Andersen and MacMahon (1981) estimated that *Thomomys talpoides*, a geomyid weighing about 100 g, may gain food at a rate ranging from 0.18 kJ/min (spruce forest) to 0.58 and 0.76 (fir and aspen) to 1.39 kJ/min (meadow). For an adult male it takes two to 30 (unrealistic) working hours to cover maintenance costs including the costs of digging (96 J/min).

Such estimated energy acquisition rates span two orders of magnitude, and refer to animals differing in size by three orders of magnitude (Fig. 3.3). Fitting an allometric regression to the data reveals statistically significant differences between the residuals' averages for particular strategies. Relatively high rates of energy acquisition were observed in ground squirrels feeding on above-ground biomass and hypogeous fungi (browsers) and squirrels feeding on fungi, berries and cones; next to them were invertebrate-collecting shrews and scratch-digging heteromyid seed eaters. The lowest rates of energy acquisition characterise grass-grazing voles and a fossorial insectivore (Fig. 3.3). Such energy acquisition rate estimates must be judged with caution; they are subject to underestimation due to the different meanings of 'foraging activity', which may include only actual eating, or

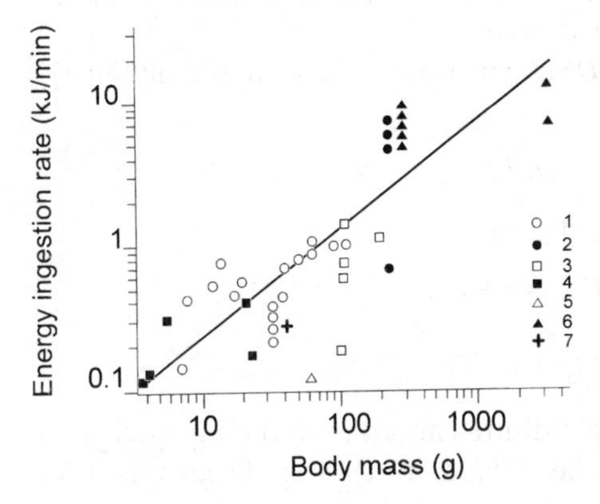

Fig. 3.3. The rates of energy acquisition at various foraging modes in small mammals. Symbols: *1* seed eaters, terrestrial; *2* seed eaters, arboreal; *3* fossorial herbivores; *4* insectivores, terrestrial; *5* insectivores, fossorial; *6* browsers; *7* grazers. Regression: c (kJ/min) = 0.040 $W^{0.76}$ (W = body mass, g), $R^2 = 0.67$

eating plus handling food, or searching, handling, and eating. Particular groups are represented by very uneven numbers of species, and the nature of the information also varies; only a few sources are purposefully made, explicit measurements.

3.5 Constraints on Foraging Activity

3.5.1 The Model

A small mammal has only a limited amount of resources and time to manage. Within these limits there are trade-offs in the allocation of time and resources to different activities. This has been explored by Belovsky (1984), who used linear modelling to compare the strategies of time allocation in three species of mammalian herbivores: the vole, the kudu and the moose. His goal was to check whether the actual composition of diets corresponds with the strategies of time minimisers or nutrient maximisers, under the premise that the animals behave optimally.

Using the information described above one can attempt to define the situation of a small mammal which has to optimise its own time budget, and particularly to define the time window between the minimum activity time necessary to collect enough food and the maximum possible time. Let us denote the fraction of the day spent in foraging activity α, the potential daily rate of energy acquisition in foraging c, the potential daily rate of energy expenditure in foraging f, and the average metabolic rate at rest and in all other non-foraging activities r. The actual daily sums of energy consumed, energy spent in foraging and in other activities, taking into account the time spent in each activity, are denoted by C, F, and R, respectively. The maximum daily energy assimilation from food (= upper limit for sustained energy expenditure; see above, Sect. 3.2) will be called *Amax*.

Total daily energy expenditure (*DEE*) must not exceed total daily energy acquisition ($C = \alpha c$):

$$\alpha c \geq DEE.$$

As *DEE* is the sum of $F = \alpha f$ and $R = (1 - \alpha)\, r$,

$$\alpha c \geq (1 - \alpha)\, r + \alpha f;$$

thus

$$\alpha \geq r \,/\, (c + r - f).$$

On the other hand, daily energy expenditure cannot exceed daily energy assimilation from food: $DEE \leq Amax$, i.e., $DEE \leq (1-\alpha)r + \alpha f$. From this it follows that

$$\alpha \leq (Amax - r) \,/\, (f - r).$$

This situation is presented in Fig. 3.4. The intersection of the line for *C* with that of total energy expenditure (*F*+*R*) determines the shortest period of foraging activity (α_{min}) allowing the energy expenditure to be covered. The intersection of (*F*+*R*) with *Amax* determines the maximum fraction of the day spent in active foraging which does not cause a negative energy balance (α_{max}). At the point of intersection of *C* and *Amax* (α_{opt}) the difference between the amounts of energy ingested and dissipated is greatest, that is, the net energy gain reaches the maximum. The strategy of 'time minimisers' consists in remaining close to α_{min}, whereas the 'energy maximisers' should opt for α_{opt}. It is clear that it does not make any sense to be a 'time maximiser', because each minute spent in foraging beyond α_{opt} decreases the daily total net energy gain.

This simple model allows at least qualitative predictions about the effects of changes in particular elements of the system. Figure 3.5a demonstrates the effect of a reduced rate of energy acquisition from food (*c*). It lengthens

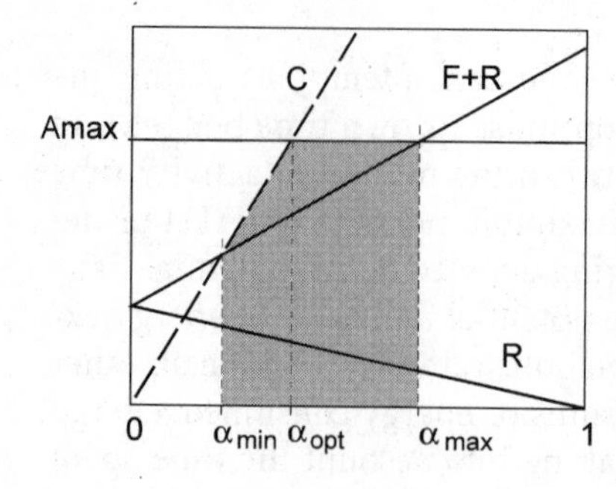

Fig. 3.4. The model of time/energy allocation in mammals. Vertical axis: energy gain or loss; horizontal axis: time spent at foraging (proportion of the day). *Amax* maximum daily energy assimilation, *C* daily sum of energy ingested, *R* daily energy expenditure for rest and non-foraging activities, *F* daily energy expenditure for foraging activity, α_{min} minimum daily activity time at which energy gain from foraging may cover all energy expenditures, α_{max} maximum daily activity time at which energy gain from food covers all energy expenditures, α_{opt} daily activity time at which the net energy gain reaches maximum. *Shadowed area*: space for positive energy budgets

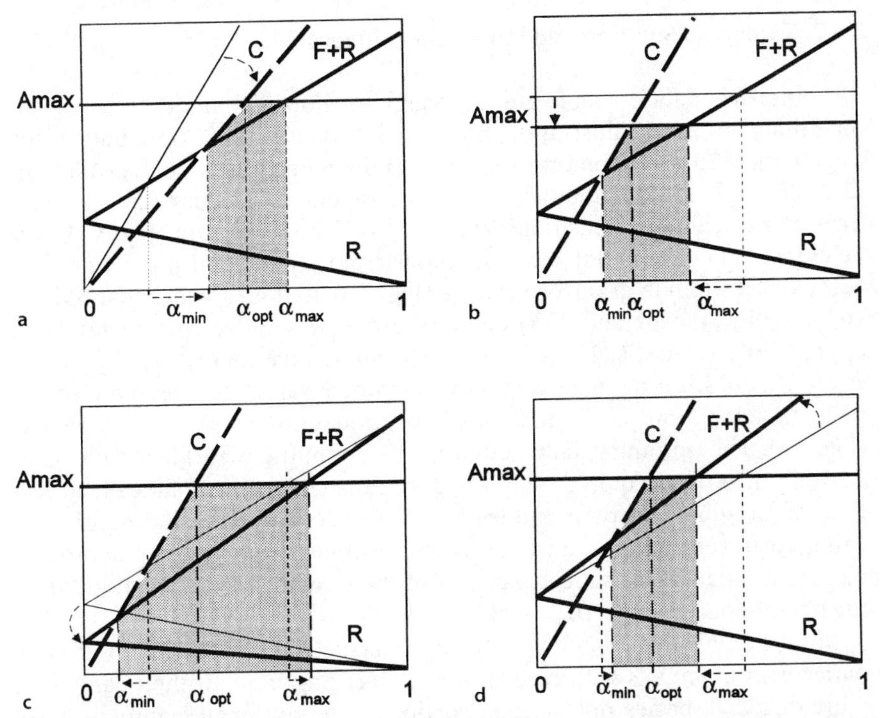

Fig. 3.5. Effects of changes in the rates of energy gain or loss upon minimum, maximum and optimum foraging activity time (symbols as in Fig. 3.4). **a** Decreased rate of energy ingestion (*C*) increases the minimum activity time α_{min}. **b** Decreased rate of energy assimilation (*Amax*) reduces the maximum activity time α_{max}. **c** Decreased rate of energy expenditure for rest and non-foraging activities (*R*) enlarges the time window for activity and improves the net energy gain. **d** Increased foraging costs (*F*) restrict the time window for activity and reduce the net energy gain

both α_{min} and α_{opt}, whereas α_{max} remains unaltered. Reduced *Amax* (e.g. because of poorer digestibility of food) decreases α_{max} and α_{opt} (Fig. 3.5b). In both cases the time window for activity becomes narrower.

If the average rate of non-foraging energy expenditure decreases (e.g. due to reduced costs of thermoregulation, or because of daily torpor), but without changes in the rate of energy expenditure during activity, α_{min} and α_{max} diverges, broadening the space for various activity strategies, while α_{opt} remains unaltered (Fig. 3.5c). With increased costs of foraging (Fig. 3.5d), α_{max} decreases substantially. The effect of increased costs of activity on α_{min} is small, and α_{opt} remains unaltered. Thus, one can expect that with a decreased supply of food (and decreased foraging rate) it is more profitable to work harder than to work longer; this is also true for decreased *Amax*.

3.5.2 Numerical Solution and Empirical Evidence

The empirical data available are too scarce to allow this model to be solved quantitatively. As an illustration, however, let us calculate a time budget for hypothetical 25 g small mammals, a vole and a mouse. Assume that *Amax* in the vole reaches 120 kJ/day, and in the mouse only 100 kJ/day (cf. Sect. 3.2). The rates of energy acquisition predicted from the regression (Fig. 3.3) may be close to 450 J/min, but they can range from about 200 J/min in grazing voles to 800 J/min in granivorous mice (Fig. 3.3). Assume further that BMR is 30 kJ/day in the vole and 20 kJ/day in the mouse, with activity metabolism approaching 4 × BMR (25–100 J/min) and non-active energy expenditure at 2 × BMR. Substituting these values for the model variables gives a minimum daily foraging time of 6.3 h in the vole, and only 52 min in the mouse (Fig. 3.6). The optimum daily activity time amounts to 10.4 h in voles and 2.1 h in mice. The required daily foraging time changes dramatically at low rates of foraging. This pattern does not change very much with different foraging costs (Fig. 3.6). In this numerical example, *Amax* does not directly constrain the foraging time, regardless of the foraging strategy employed by the hypothetical rodent ($\alpha_{max} \geq 24$ h).

Bunnel and Harestad (1990) made an interspecific comparison of activity patterns in mammals with regard to total foraging time. In their sample of more than 200 species, only a small portion concerned small mammals. They computed that on average mammals spend 34% of their time in some activity; foraging takes 22% of the day. Allometric analysis revealed that these proportions are practically independent of body mass. Should this also apply to small mammals, α amounts to 317 min/day, close to the minimum activity

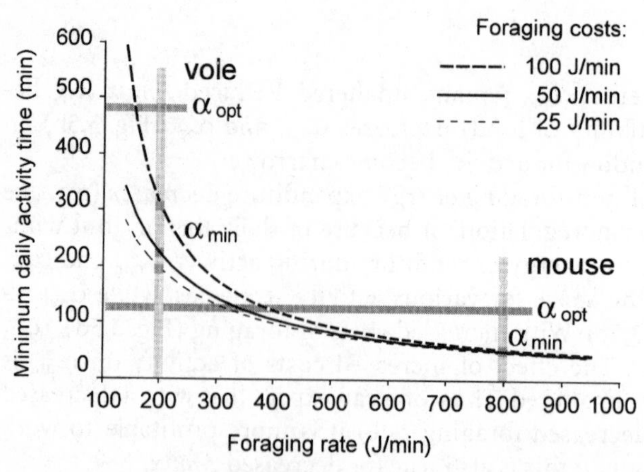

Fig. 3.6. Minimum daily foraging time versus foraging rate at three different foraging costs in a hypothetical 25 g mammal. Minimum (α_{min}) and optimum (α_{opt}) daily activity times for two foraging strategies are shown: a grazing vole (foraging rate of 200 J/min) and a granivorous mouse (foraging rate 800 J/min)

time predicted for our hypothetical 25 g vole. With the foraging rate kept at 450 J/min, the same foraging time would be required if the foraging costs were 538 kJ/day or 17 × BMR, a rate exceeding the aerobic limit and leading to daily energy expenditure of almost 150 kJ, that is above the upper limit of the sustained energy assimilation rate.

These approximate calculations show that a 'normal' small mammal is not very limited by energetics in shaping its time-activity budget. The problem may arise only in mammals with particularly low rates of energy acquisition, such as voles, or with extremely high energy costs of foraging. This may well pertain to bats. Concerning terrestrial mammals, one study (Andersen and MacMahon 1981) shows that gophers (*Thomomys talpoides*) may easily encounter the metabolic constraint.

This model also predicts that a small mammal employing torpor may greatly reduce the minimum foraging time or alternatively may exploit food of a lower quality (e.g. harder to find, to handle, or to digest). This last effect is discussed in detail by Ruf and Heldmaier in Chap. 12.

3.6 Activity Patterns Under High Energy Loads

Distinct circadian patterns of activity are well known in many species of rodents. Nocturnal rodents such as *Phodopus sungorus* or *Peromyscus maniculatus* are active during the night and expend more energy at this time (Fig. 3.7a). This pattern becomes less distinct or disappears entirely when increased energy demand (e.g. due to low ambient temperatures) forces animals to prolong their foraging time (Fig. 3.7a). On the other hand, the ultradian rhythms typical of microtine voles (Gerkema and Daan 1985; Halle 1995) persist with identical timing even after a substantial increase in total energy expenditure (Fig. 3.7b).

Although the proximate factors governing both types of daily activity rhythms may be quite different, one can speculate that the common ultimate factor governing both patterns is related to the organisation of time-energy budgets, as depicted in Fig. 3.4. Selective feeders with constraint C much steeper than constraint $A+R$ have α_{opt} close to α_{min}, and a relatively large difference between $Amax$ and $A+R$ at α_{opt}. This enables these animals to increase the time lag between energy acquisition and energy expenditure; this time lag may be even greater than the time required to digest and assimilate the food consumed. The surplus energy may be stored either as fat or in the form of hoarded food; both ways involve additional energy costs but allow for larger energy stores than just food filling up the gut.

At the low rates of energy acquisition (and high energy expenditure) typical of voles, the difference between $Amax$ and $A+R$ is small (Fig. 3.5a, d); this may preclude gathering any other kind of energy storage than the food currently ingested. These speculations are based on rather weak evidence and

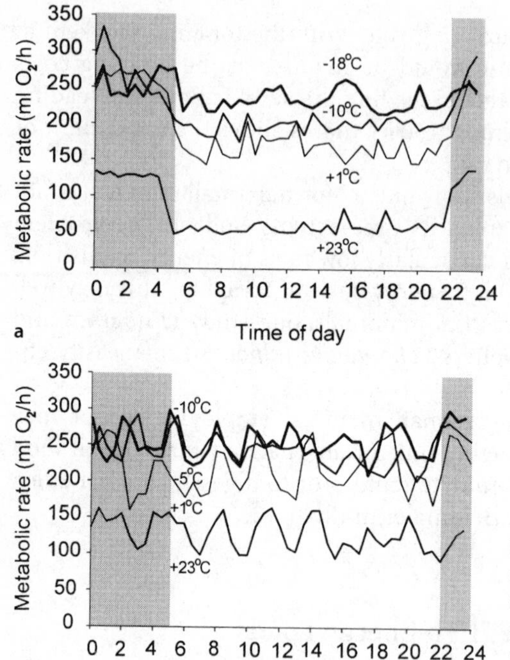

Fig. 3.7. Daily changes in oxygen consumption at various ambient temperatures: **a** in the cricetid *Peromyscus maniculatus*; **b** in the vole *Microtus pennsylvanicus* (from P. Koteja, unpubl.). *Shadowed area*: darkness

can only be regarded as a working hypothesis. They lead, however, to the conclusion that changes in activity patterns cannot fully compensate for adverse changes in environmental conditions. Small mammals with feeding strategies based on low quality food have more rigid activity patterns than do those exploiting more nutritious diets.

3.7 Concluding Remarks

Study of the evolutionary optimisation of activity patterns requires joint application of various approaches. The timing of activity in small mammals obviously depends on the digestion dynamics in the gut (Sibly 1981; Hume 1989), an aspect not discussed in this paper. Similarly, a tremendous effect may be exerted by external forces such as predation pressure (Halle 1993) or climatic conditions (Chappell and Bartholomew 1981). None of these factors acts independently of the others. They are not arranged in a unidirectional chain of causes and effects but rather form a multidimensional space within which the process of evolutionary optimisation of life history proceeds. The variety of mammalian life histories is a good empirical basis for suggesting and testing hypotheses on the evolution of activity patterns. Bioenergetics may provide grounds for estimating the costs and benefits connected with various strategies. To this end, the diversity of energy budgets should be studied with regard to two sources of variation: interspecific, to identify various adaptive strategies, and intraspecific, to determine functional relations and to explain the evolution of different strategies.

Acknowledgements. The critical comments of S. Halle are gratefully acknowledged. Michael Jacobs helped to edit the manuscript.

References

Altmann SA (1987) The impact of locomotor energetics on mammalian foraging. J Zool Lond 211:215–225

Andersen DC, MacMahon JA (1981) Population dynamics and bioenergetics of a fossorial herbivore, *Thomomys talpoides* (Rodentia: Geomyidae), in a spruce-fir sere. Ecol Monogr 5: 179–202

Belovsky GE (1984) Herbivore optimal foraging: a comparative test of three models. Am Nat 124:97–115

Buckner CH (1964) Metabolism, food capacity, and feeding behavior in four species of shrews. Can J Zool 42:259–279

Bunnel FL, Harestad AS (1990) Activity budgets and body weight in mammals: how sloppy can mammals be? Curr Mamm 2:245–305

Calow P, Townsend CR (1981) Energetics, ecology and evolution. In: Townsend CR, Calow P (eds) Physiological ecology: an evolutionary approach to resource use. Blackwell, Oxford, pp 3–19

Chappell MA, Bartholomew GA (1981) Activity and thermoregulation in the antelope ground squirrel *Ammospermophilus leucurus* in winter and summer. Physiol Zool 54:215–223

Degen AA (1994) Field metabolic rates of *Acomys russatus* and *Acomys cahirinus*, and a comparison with other rodents. Isr J Zool 40:127–134

Du Toit JT, Jarvis JUM, Louw GN (1985) Nutrition and burrowing energetics of the Cape mole rat *Georychus capensis*. Oecologia 66:81–87

Ferron J, Quellet JP, Lemay Y (1986) Spring and summer time budgets and feeding behavior of the red squirrel (*Tamiasciurus hudsonicus*). Can J Zool 64:385–391

Garland T Jr (1983) Scaling the ecological cost of transport to body mass in terrestrial mammals. Am Nat 121:571–587

Gerkema MP, Daan S (1985) Ultradian rhythms in behavior: the case of the common vole (*Microtus arvalis*). Exp Brain Res [Suppl] 12:11–31

Grodziński W, Weiner J (1984) Energetics of small and large mammals. Acta Zool Fenn 172: 7–10

Grodziński W, Wunder BA (1975) Ecological energetics of small mammals. In: Golley FB, Petrusewicz K, Ryszkowski L (eds) Small mammals: their productivity and population dynamics, Int Biol Program 5. Cambridge University Press, Cambridge, pp 173–204

Halle S (1993) Diel pattern of predation risk in microtine rodents. Oikos 68:510–518

Halle S (1995) Diel pattern of locomotor activity in populations of root voles, *Microtus oeconomus*. J Biol Rhythms 10:211–224

Hammond KA, Diamond J (1997) Maximal sustained energy budgets in humans and animals. Nature 386:457–462

Hammond KA, Konarzewski M, Torres RM, Diamond J (1994) Metabolic ceilings under a combination of peak energy demands. Physiol Zool 67:1479–1506

Harvey PH, Pagel MD, Rees JA (1991) Mammalian metabolism and life histories. Am Nat 137:556–566

Herbers JM (1981) Time resources and laziness in animals. Oecologia 49:252–262
Hoyt DF, Kenagy GJ (1988) Energy costs of walking and running gaits and their aerobic limits in golden-mantled ground squirrels. Physiol Zool 61:34–40
Hume ID (1989) Optimal digestive strategies in mammalian herbivores. Physiol Zool 62: 1145–1163
Jensen IM (1983) Metabolic rates of the hairy-tailed mole *Parascalops breweri* (Bachman 1842). J Mamm 64:453–462
Karasov WH (1981) Daily energy expenditure and the cost of activity in a free-living mammal. Oecologia 51:253–259
Karasov WH (1992) Daily energy expenditure and the cost of activity in mammals. Am Zool 32:238–248
Kenagy GJ, Sharbaugh SM, Nagy KA (1989) Annual cycle of energy and time expenditure in a golden-mantled ground squirrel population. Oecologia 78:269–282
Konarzewski M, Diamond J (1994) Peak sustained metabolic rate and its individual variation in cold-stressed mice. Physiol Zool 67:1186–1212
Koteja P (1995) Maximum cold-induced energy assimilation in a rodent, *Apodemus flavicollis*. Comp Biochem Physiol 112A:479–485
Koteja P (1996) Limits to the energy budget in a rodent, *Peromyscus maniculatus*: the central limitation hypothesis. Physiol Zool 69:981–993
Koteja P, Weiner J (1993) Mice, voles and hamsters: metabolic rates and adaptive strategies in muroid rodents. Oikos 66:505–514
Koteja P, Król E, Stalinski J, Weiner J (1993) Energy budget limitation and partitioning in rodents of altricial and precocial modes of reproduction. Mesogee 53:7–12
Kozłowski J, Weiner J (1997) Interspecific allometries are by-products of body size optimization. Am Nat 149:352–380
MacArthur RA, Krause RE (1989) Energy requirements of freely diving muskrats (*Ondatra zibethicus*). Can J Zool 67:2194–2200
McDevitt RM, Speakman JR (1994a) Central limits to sustainable metabolic rate have no role in cold acclimation of the short-tailed field vole (*Microtus agrestis*). Physiol Zool 67:1117–1139
McDevitt RM, Speakman JR (1994b) Limits to sustainable metabolic rate during transient exposure to low temperatures in short-tailed field voles (*Microtus agrestis*). Physiol Zool 67: 1103–1116
McNab BK (1987) Basal rate and phylogeny. Funct Ecol 1:159–167
McNab BK (1988) Complications inherent in scaling the basal rate of metabolism in mammals. Q Rev Biol 63:25–54
McNab BK (1992) The comparative energetics of rigid endothermy: the Arvicolidae. J Zool Lond 227:585–606
Melcher JC, Armitage KB, Porter WP (1990) Thermal influences on the activity and energetics of yellow-bellied marmots (*Marmota flaviventris*). Physiol Zool 63:803–820
Morgan KR, Price MV (1992) Foraging in heteromyid rodents: the energy cost of scratch-digging. Ecology 73:2260–2272
Nagy KA (1987) Field metabolic rate and food requirement scaling in mammals and birds. Ecol Monogr 57:111–128
Nagy KA (1994) Field bioenergetics of mammals: what determines field metabolic rates? Aust J Zool 42:43–53
Peterson CC, Nagy KA, Diamond J (1990) Sustained metabolic scope. Proc Natl Acad Sci USA 87:2324–2328
Salsbury CM, Armitage KB (1994) Resting and field metabolic rates of adult male yellow-bellied marmots, *Marmota flaviventris*. Comp Biochem Physiol 108A:579–588
Shipley LA, Gross JE, Spalinger DE, Hobbs NT, Wunder BA (1994) The scaling of intake rate in mammalian herbivores. Am Nat 143:1055–1082
Sibly RM (1981) Strategies of digestion and defecation. In: Townsend CR, Calow P (eds) Physiological ecology: an evolutionary approach to resource use. Blackwell, Oxford, pp 3–19

Taylor CR, Heglund NC, Maloiy GMO (1982) Energetics and mechanics of terrestrial locomotion. 1. Metabolic energy consumption as a function of speed and body size in birds and mammals. J Exp Biol 97:1-21
Thompson SD (1985) Bipedal hopping and seed-dispersion selection by heteromyiod rodents: the role of locomotion energetics. Ecology 66:220-229
Vleck D (1979) The energy cost of burrowing by the pocket gopher *Thomomys bottae*. Physiol Zool 52:122-136
Vleck D (1981) Burrow structure and foraging costs in the fossorial rodent, *Thomomys bottae*. Oecologia 49:391-396
Weiner J (1987) Maximum energy assimilation rates in the Djungarian hamster (*Phodopus sungorus*). Oecologia 72:297-302
Weiner J (1989) Metabolic constraints to mammalian energy budgets. Acta Theriol 34:3-35
Weiner J (1992) Physiological limits to sustainable energy budgets in birds and mammals: ecological implications. TREE 7:384-388

4 Ecological Relevance of Daily Activity Patterns

Stefan Halle

4.1 Increasing Fitness by Activity Timing

The environment of any animal species is a complex set of both abiotic and biotic qualities. Dominant abiotic parameters are light conditions, ambient temperature, relative humidity, precipitation, and wind speed. Biotic components may be classified by the trophic levels at which they occur. On the same trophic level we find conspecifics as mates, as members of a social group, and as competitors. On the same trophic level there are also competitors from other species. Biotic components from different trophic levels are represented by prey, predators, and parasites. The combination of all these factors determines how the world looks for an individual at a specific moment in time.

It is indeed obvious that environmental factors are not constant, but vary considerably over time. Some of these variations are erratic and, hence, unpredictable. There are, however, also distinctive temporal patterns. The most substantial changes in the environment are related to the course of seasons, both with respect to the extent of variation and the number of parameters that change concurrently. As a consequence, long-lived species have developed rather sophisticated adaptations, physiologically as well as behaviourally, to cope with the challenges of a dramatically changing world.

On a much shorter time scale, however, the environment also changes predictably in the course of a 24-h day, and this short-term variability is the concern of the present monograph. The abiotic parameters are under the regime of the everlasting rotation of our planet, resulting in a systematic up and down of parameter values. In fact, the whole drama of evolution of life was acted on a constantly oscillating stage. It is not surprising, therefore, that we find temporal patterns related to the daily periodicity of the environment on almost every level of life, from cell to communities. Because of this, the biotic factors of an individual's world also reveal periodic features, since conspecifics, competitors, parasites, prey, and predators all have their own daily time schedule.

Ecological Studies, Vol. 141
Halle/Stenseth (eds.) Activity Patterns in Small Mammals

The question of how well, or badly, an individual handles the environmental fluctuations is obviously an ecological one, since survival probability and fitness are directly involved. There are "good and bad times for a given activity" and "the critical consideration for the animal in its natural environment ... is that the appropriate activity be performed at the right time in the right place; some variability in timing may be permissible, but it is essential that the animal at least avoid the wrong times for his activities" (Enright 1970). As a matter of fact the task of structuring activity is two-sided since it consists of the total allocation of time to activity, and of the temporal order of behavioural sequences (Daan 1981).

The easiest way to tackle this problem is to follow a simple rule like 'if a specific time was suitable for doing a specific thing on one day, choose the same time to do the same thing next day.' Following such a rule would result in learning suitable time windows by experience, provided that the animal is able to assess time and also has some kind of time memory. For both abilities a clock mechanism is essential. Its development and perfection is driven by natural selection since individuals who always choose the appropriate time are clearly better off than nitwits who always pick the wrong time. How strong the selective pressure is may be demonstrated by the amazing paradox that the precision of biological clocks is much higher than that of the environment, although the first have developed as an adaptation to the latter (Daan and Aschoff 1982).

Time memory and time learning are important first steps towards activity patterns. However, environmental characteristics also have a profound effect (Daan 1981). In a highly predictable environment, i.e., with a strong day-to-day correlation of environmental parameters, a genetically fixed temporal program will probably be the evolutionarily stable strategy (ESS) because it guarantees that activities will always be performed at the right time. However, there are also costs related to strictly fixed patterns because of the lost opportunity of exploring other time windows. Therefore, fixed programs are particularly advantageous when errors are fatal and prevent animals from learning. On the other side of the spectrum are highly unpredictable environments where fixed programs are useless. In this situation direct responses to the actual conditions is probably the better strategy.

Environmental predictability, or unpredictability, is not an absolute term but will depend on the behaviour concerned. For some kinds of activity a fixed program may be a good choice, while other aspects of the behaviour should be rather flexible. Obviously, this distinction will vary among species, resulting in a remarkable variety of activity pattern characteristics we can observe even in closely related species. However, species-specific programs are individually modulated and most often include at least some degree of flexibility. Daily activity patterns might, therefore, be seen as adaptive sequences of daily routines that meet the time structure of the environment, shaped by evolution, but additionally fine-tuned by flexible responses to the actual state of the environment. So the behaviour we observe in the field is "the result of a complex, interactive mixture of endogenous and exogenous factors" (Enright 1970).

4.2 Autecological Advantages of Appropriate Timing

Activity patterns can be considered as having two major aspects: the first is the actual timing of behaviour, i.e., when to do what, and the second is the temporally variable tuning of physiological parameters in relation to the surroundings. Most older studies on activity rhythms focus on the second aspect of how the internal milieu is synchronised with the environment (see reviews by Daan and Aschoff 1982, and DeCoursey 1990). Physiological rhythms give the internal things-to-do a reasonable order, a feature coined 'temporal compartmentalization' by Daan and Aschoff (1982). There is no evidence that internal temporal order is of value as such, rather it seems meaningful only against the background of the changing external world. Hence, physiological rhythms are basically a concern of ecology. Physiology, behaviour, and the environment are parts of a complex system in which interactions are optimal when all components dance in step. The optimisation process to achieve integrity must have been driven by natural selection.

DeCoursey (1990) gives a comprehensive overview of six relevant physiological rhythms, distinguished by their period length. In ascending order these are ultradian, tidal, daily, semi-lunar, lunar, and annual rhythms. Which of these, or which combinations, apply to a certain species depends on its ecological niche and general living conditions. For circadian rhythms, which are the focus of our contemplation, many examples can be found to demonstrate their ecological relevance: a regular sleep/wake rhythm gives order to activity and rest. Besides having complex and still not fully understood functions, sleep also reduces energy expenditure and mortality risk during times of the day when the net effect of activity would otherwise have been negative (Meddis 1975). To start activity, alarm clocks wake the animal up at the right time. Reliable and self-sustained clocks are of particular importance when no information about the state of the environment can be obtained directly, e.g. with bats resting in caves or small rodents dwelling in subterranean burrows. In hares, the mother and her offspring meet only once a day for a few minutes for suckling, and these meetings happen at a strictly fixed time of day. However, apart from such amazing examples of behavioural timing, daily and seasonal environmental changes have even more profound effects on the life, and even design, of mammal species.

4.2.1 Shiftwork in the Habitat

The most obvious and most predictable change in the environment is the constant alternation between day and night. The world does not only look very different during the two phases, it actually is so, with respect to illumination, temperature, and humidity. Moreover, the spectrum of other species that can be met, be it predators, prey or competitors, is very much dependent on the time of day. It is not surprising, therefore, that many species have specialised to only one of the two phases (Park 1940; Ashby 1972; Daan 1981). So the physical space of any given habitat is populated by different residents at day and by night, resembling a factory in which the same facilities are used by the day and the night shifts who do not, or just for a short moment, meet.

However, there is one important difference between shift workers and free-living animal species: animals are, in the long run, shaped by natural selection (it is, however, tempting to speculate how humans would change their attitudes when constantly restricted to either day or night). Individuals are favoured when they are better adapted to their environment, and since environmental conditions are so different at day and night, evolution will foster different traits of adaptation at the two phases. In particular, this applies to the sensory outfit, which has to be optimal in order to maximise predator avoidance and/or hunting success.

Knowing nothing else about a mammalian species, it will often be enough to have a close look at its ears, eyes, and concealment coloration to infer a nocturnal or diurnal way of life (Park 1940; Ashby 1972; Daan 1981). The eyes of exclusively nocturnal species are large, have mainly rods, a large lens in relation to the focal length, and a large cornea. Their ability to see details is, therefore, rather limited, but a relatively bright image can be perceived at low levels of illumination. Night vision might further be enhanced by a tapetum lucidum in highly specialised nocturnal species, especially night hunters. Nocturnal species are further characterised by large inner ears, large and highly movable auricles, and often dull coloration. In contrast, the design of diurnal species is in many respects just the opposite, with high visual acuity in daylight, but poor vision at night. Auricles may also be large in diurnal species, but this is often functionally related to thermoregulation. Coloration of daytime species sometimes appears astonishingly spectacular at first view, but in fact frequently turns out to achieve optimal camouflaging by smartly mimicking the play of light and shadow.

The distinction between day and night shifts is also prevalent with behaviour. For instance, intraspecific communication in diurnal species is often based on visual cues, while nocturnal species primarily use smell and noise. The mode of movement and even the choice of habitat is affected by the activity phase, since diurnal species are much more dependent on visual cover. Where animals are found may be different by day and by night, because activity occurs where food is abundant, while resting is confined to safe and sheltered places (Daan 1981). If the two requirements are not located in close vicinity to one another, commuting is a necessary part of the daily routine. Even the clock itself, which triggers activity behaviour, works differently in diurnal and nocturnal species, demonstrated by classical chronobiological experiments with manipulated photoperiods (Daan and Aschoff 1975; DeCoursey 1989): in diurnal species the times of onset and end of activity are directly related to the LD ratio of the entraining light schedule, whereas nocturnal species are highly sensitive to single light pulses and commonly show light-sampling behaviour at activity onset.

All this evidence may give the impression that the blueprints for day and night mammals are essentially different, being a consequence of evolutionary adaptation after a 'choice' for one half of the day has been made. However, we have to be aware that there are also important exceptions which moderate the concept of rigorous separation. Firstly, there are the few species that are routinely active both day and night. This is the case in shrews and voles, for instance, and is related to metabolic peculiarities (see Chaps. 11 and 13). With respect to sensory outfit and behaviour such species have to be jacks of all trades, but inevitably are not perfectly adapted to either phase. Secondly, there are changes in the activity phase related to age. It is not uncommon, for example, that specific age stages are preferably active at times of the day avoided by other age classes (see below).

Finally, it may be suspected that even within one individual of one specific age some temporal specialisation for certain activities, like exploration in particular, may occur. It is possible, and there are some provisional hints from field observations that, let's say, even a strictly nocturnal species performs some special activities also during daytime (Halle 1988b). Another possibility for temporal differentiation is space use, meaning that individuals use different parts of their activity range at different times of the 24-h day. However, to quantify inter- and intra-individual behavioural differentiation needs detailed data from continuous surveys of individual activity patterns in the field. Such data have been largely unavailable until now, mainly for methodological reasons (see Appendix).

4.2.2 A Matter of Season

Day and night are different aspects of the same world, but they are by no means constant. Rather, the length of the two phases vary considerably with season, enlarging and reducing the time windows for activity in an oscillating manner. This is especially true in high northern latitudes above the polar circle, where the attainable activity phase for nocturnal and diurnal species, respectively, shrinks to zero alternately. It is quite obvious that a species cannot insist on its agreed quota of time for activity, because this would mean that it could not perform any activity for up to some months. So once again, we have to realise that the strict separation between a diurnal and nocturnal world, although correct in general terms, is softened by other features of the environment.

The superimposition of the daily and annual rhythms in the environment brings about a highly dynamic system, and this is what animals really have adapted to. Thus, seasonal changes in the allocation of time to activity are also part of the temporal optimisation process. Free-running circadian rhythms under constant experimental conditions and phase response curves (PRC), which are both distinct focuses of classical chronobiological work, give insight into a most sophisticated mechanism that allows animals to re-adjust their circadian clocks to the ever changing photoperiod. Although the clock depends on metabolism as far as we know (see Chaps. 2 and 3), it is not affected by temperature. This physiologically demanding task is a obligatory precondition for correct entrainment with the environment in all seasons (cf. Bartness and Goldmann 1989).

When circadian clocks emerged as a new idea not too many decades ago, people were rather reserved in accepting it. It was argued that there are so many signals provided by the environment that animals, and plants, are more likely to respond to these cues directly rather than evolving a most complicated and superfluous time assessment mechanism. However, seasonal changes of the photoperiod in particular gave strong evidence that clocks are indeed a necessity. The compulsion to cope with the course of seasons may also be the solution for the neat logical problem stressed by Enright (1970) that biological rhythms are persistent for many 24-h cycles, although the environment in which this feature evolved reliably gives two strong signals, sunrise and sunset, each 24-h cycle.

Many aspects of a mammal's physiology and behaviour vary systematically with season: moulting, deposition of fat reserves, food storing, and, in fact first of all, the reproductive state. It has been verified by numerous experiments that the photoperiod is at least one important, if not the principal, factor that triggers these changes. Since the photoperiod is the same in spring and autumn some additional information must tell the animal which time of the year it is in order to be correctly synchronised with the environment. So, what animals really have to assess is whether day length is increasing or decreasing, which means that today's phase length must be stored and compared with phase length tomorrow. Such comparison cannot be readily archived from direct responses to environmental signals but requires time assessment and time memory, two features that are dependent on a continuously ticking clock. Hence hysteresis effects may be expected and in fact are verified by the observation of different activity timing at the same photoperiod before and after the solstices (Daan and Aschoff 1975).

Other complications that would confound simple light-sensitive on/off switches are the numerous perturbations from environmental conditions. There are bright and cloudy days, there are bright and dark nights depending on lunar phase and, again, cloud cover, and there is the profound effect of vegetation cover which also changes in the course of seasons. All these features acting together prevent genetic fixation of a definite threshold of light intensity to distinguish between day and night. Take the obvious example of a woodland species dwelling in a deciduous forest. Day length is longer in summer than in winter, but at the same time the daily changes in light intensity are much more pronounced in winter than in summer due to the lack of foliage shadowing. Hence, there must be different thresholds to begin and terminate activity in summer and winter to achieve the same timing in relation to sunrise and sunset, and this requires a quite complex mechanism (DeCoursey 1989). In fact, the threshold characteristics are an adaptation to the type of habitat and, therefore, directly related to the species' ecology.

Another example where the need for an ever-running clock is self-evident are hibernating species. To enter hibernation is a complex process, involving both physiological and behavioural traits. Achievement of preparedness for hibernation, with respect to fat reserves for instance, is partly controlled by the photoperiod. After the animal finally falls asleep, some signal has to break the lethargy when it is time to wake up again. For this purpose a direct environmental signal is not a good choice, especially when the hibernaculum is a sheltered and encapsulated place. Furthermore, in species with delayed nidation pregnancy in a strict sense starts when the blastocyte makes contact with the uterus which may be months after copulation. In some species with this peculiar modification of the reproductive cycle this happens during hibernation, or just after the spring wake-up. In both cases, the female reproductive machinery has to be kick-started during hibernation, and this has to happen at the right time to guarantee timely birth of the young. So the clock has to keep running during hibernation, just as it would not be wise to stop our alarm clock when falling asleep. Futhermore, the biological clock has to run properly, which is amazing when we again realise that all physiological parameters are extremely slowed down in torpid animals, and that the clocks are basically physiological mechanisms.

4.3 Inter-individual Aspects

The questions and problems mentioned above are all related to the autecology of individuals and have already been studied in classical chronobiology approaches. Our ecological approach looks at time-dependent interactions among individuals, either among conspecifics or among different species. In the remainder of this chapter I will put forward some thoughts about these topics which I consider important. Most herein is meant as a very general contemplation. However, my bias towards small mammals, and microtines in particular, is quite obvious and I will not even try to conceal it.

4.3.1 Community Life

Consider an individual that is synchronised with its environment by some species-specific entraining mechanism that has evolved as an adaptation to its particular ecological conditions. This has the interesting consequence that the individual does not have the world alone. Rather, it is active together with all the other conspecifics around, because as members of the same species they all respond to the same environmental cues and follow basically identical time schedules. As outlined below, synchronous activity may involve both positive and negative effects. In any case, however, the activity of conspecifics will make the process of temporal adaptation more complicated. The selective value of a specific behavioural strategy is not entirely determined by the individual-environment interaction alone, but also depends on what the neighbours are doing. To put it another way: the behaviour of the conspecifics is part of the complex environment the individual has to deal with.

Conspecifics that are active in the same place and at the same time must first be seen as competitors for both resources and space. As a result, agonistic interactions will bring about some system of spatial organisation in many species. In fact, spacing systems are one of the best and most reliable forms of information we have on the social organisation of animal communities, including their mating systems. For this reason it is obvious why the literature on spacing is so extensive, especially in small mammals in which direct observations are always difficult, and often simply impossible. However, without considering temporal organisation one will not understand the spatial organisation thoroughly.

4.3.1.1 Spacing Behaviour

The timing of activity will be crucial in determining whether two neighbouring territory holders physically meet at the borderline, or whether they only keep in touch by olfactorial cues (Fig. 4.1). There are observations that hint at specific time schedules for home range patrolling and scent marking to facilitate maintenance of the arrangement of territories (Daan and Aschoff 1982). It may be economically more reasonable for neighbours to keep the claims clear by meeting from time to time and having the opportunity for limited wrangling instead of permanently sneaking into the neighbours' keep-out area when the territory holder is not present and ending up in a big conflict when finally caught. This idea was first presented by Kenagy (1976) who observed synchronous onset of activity in kangaroo rats although there was no limited time window for food availability and animals live solitarily in their burrows. Similar mechanisms were hypothesised by Lehmann and Sommersberg (1980) for the synchronised short-term activity bouts in populations of common voles.

Further, it makes a difference with respect to the sociobiological inferences if the area where home ranges overlap is used by several individuals at

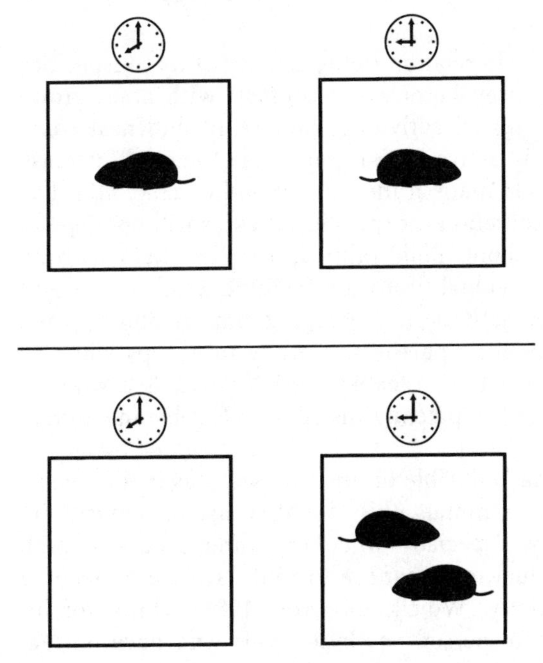

Fig. 4.1. Dependence of social interaction upon temporal behaviour. In the *upper panel* a specific habitat area (schematically indicated by the *square*) is used by one individual at time t1 and by another individual at time t2, thus direct social interaction is avoided by activity timing. The *lower panel* shows synchronous activity in which the habitat lies empty at time t1, but is used by both individuals concurrently at time t2. Synchronous activity encourages social contacts, but also increases intraspecific strife

the same time, or subsequently by only one individual at a time. If concurrent activity occurs, the overlap-area may be interpreted as no man's land with free access for everyone and where spatial partitioning is relaxed. With limited time windows for individuals the overlap-area is more a kind of shared territory with some temporal agreement for the order of use. In the first case all visitors probably enjoy equal rights while in the latter case a hierarchy of area utilisation is more likely.

Synchronous activity has the consequence that space, which is a limited resource in territorial species, is further reduced because an individual only gets what it can keep free from the others. This is particularly surprising in microtine rodents which are characterised by synchronised ultradian rhythms (see Chap. 11). All, or at least most, of the population members are active together for about 1 h, which causes considerable crowding especially at high population densities. The same fiercely contested habitat lies, however, deserted for the next 2 h. An individual being active out of phase with the rest of the population would have all the food and space it wants because the territory holders have disappeared into their burrows. However, this strategy does not seem to be followed widely, indicated by the observation that the short-term rhythm of vole activity is distinctly stable on the population level (Halle 1995). In such an obvious case, synchronous activity, and possibly even crowding, must have some serious selective advantages to balance the spatial confinements.

4.3.1.2 Social Contacts

By timing its own activity phase in relation to the activity of its conspecifics an individual can consciously provoke or avoid contacts with other group members. For instance, when sexual activity is preferably shown at some limited period of the day, being active at that particular time will increase the chance of meeting a possible mate in the same mood (Daan 1981). The evolution of such temporal reservation for specific activities will not depend on group selection arguments as one might think at first view. Rather, group members can increase their individual fitness by avoiding wasting time and losing opportunities for other activities by escaping time windows when they frequently encounter unwilling partners. Activity in groups will also enhance the transition of information, a feature recently discussed with respect to foraging when the food is patchily distributed (Krebs and Inman 1994).

On the other hand, if timing is flexible on the fine scale, as is the case at least to some extent in most mammals, this is a very useful opportunity when living in a community. Especially when crowding occurs at high population density, an individual may manage to find its place by using a slightly different time for activity (Wójcik and Wołk 1985). Think, for instance, about a subadult or a disperser, i.e., individuals who have not acquired a home range on their own yet. In many species, including our own, individuals without a residence are social underdogs and have to look for unoccupied niches to make a living. For them there is the possibility of finding a temporal niche by avoiding the times when the territory holders can actively chase them around. Calhoun (1975), for example, found in laboratory colonies that subordinate house mice and rats elude dominant individuals by choosing a different time for activity. A consequence of this strategy may be that the most preferable time of the day is not available for them for activity, and that they are thus forced to take a higher risk, with respect to predation for instance.

4.3.2 Tied by Conflict

The share of predation risk is another field where the selective value of individual behaviour is certainly dependent on what the other members of the group do. The temporal aspect of predator-prey interaction is that ecological aspect of activity timing which has gained most interest in the past, especially in microtine rodents as they are prey par excellence (e.g. Rijnsdorp et al. 1981; Raptor Group 1982; Halle 1988a, 1993; Gerkema and Verhulst 1990). In fact, predator-prey interaction is probably the most obvious example for the relevance of ecological approaches and will, therefore, be treated here in some more detail.

4.3.2.1 Prey

Imagine a prey individual resting in a relatively safe refuge where the risk of predation is relaxed, as is the case with a small rodent in its subterranean burrow, for instance (Fig. 4.2). There is indeed a constant risk due to specialist predators that manage to intrude into the burrows, or that simply dig away the shelter and open the burrow violently. This source of risk might be reduced by the manner of burrow construction and/or the choice of burrow site. It cannot, however, be affected by activity timing and is hence not a concern of the present contemplation. So, as a general rule, we can state that mortality risk in the shelter will normally be lower than outside, at least because the spectrum of relevant predators is curtailed.

Nevertheless, from time to time the individual has to leave its refuge in order to perform activities that are not possible within the burrow, including foraging, home range patrolling, scent-marking at the borderlines, search for a mate, exploration, and so on. When active on the surface and in the open in particular, the situation for prey changes dramatically because the individual is now exposed to all the predators in that area. Moving around further increases the chance of colliding with a predator or attracting its attention, respectively. So, again as a general rule, times of activity will normally be times of distinctly increased mortality risk.

The timing of prey activity has significant potential to affect the extent of predation risk (see Lima and Dill 1990 and references given therein). In fact, activity timing is under the regime of an optimisation process in which predation risk should be minimised and opportunities for foraging and all other

Fig. 4.2. Dependence of mortality risk upon temporal behaviour of prey. The *upper panel* shows a time interval in which prey is shielded from predation risk by resting in a sheltered burrow. However, prey is forced to leave its relatively save refuge from time to time for foraging (*lower panel*) and is thus exposed to predation and, possibly, harsh weather conditions. Foraging also involves higher energy expenditures due to locomotion and thermoregulation

sorts of activity should be maximised (cf. Vásquez 1994). This is verified, for instance, by the often confirmed observation that nocturnal species, particularly those living in open habitats like deserts, change both activity timing and general foraging behaviour on moonlit nights (Clarke 1983; Travers et al. 1988; Kolb 1992; Kotler et al. 1993; Hughes et al. 1994) although such risk avoidance is associated with real costs (Vásquez 1994). Nocturnal species may even compensate for the lost opportunity to perform activity in darkness by increasing activity at twilight (Daly et al. 1992, Chap. 8).

By choosing a specific time of the day for activity, the spectrum of predators is also chosen, because only predators that are active at the same time are a source of danger. During the daytime, raptors and snakes are the natural enemies; at night the prey has to deal with carnivore mammals and owls. Particularly critical times are dawn and dusk. During dawn, for example, there is a good chance of encountering both late nocturnal and early diurnal predators as well as those specialised for hunting at twilight.

For the task of timing activity with respect to predation risk it is important to realise that it is hardly ever, or very seldom, one particular predator species, or one particular functional group of predators that the prey has to take into account. Unfortunately, this is what most case studies on activity patterns in small mammal prey and their predators do (e.g. Mikkola 1970; Erdakov 1981; Rijnsdorp et al. 1981; Raptor Group 1982; Dickman et al. 1991). In reality, however, prey will be exposed to a diverse range of predator species, each with its own characteristic time schedule for hunting. Thus, the diel pattern of predation risk must be analysed in terms of an integrated approach which considers the pooled activity pattern of the entire predator community in the respective habitat (Halle 1993).

Discrimination of predation risk among different time windows is an extremely complex and demanding enterprise when thoroughly scrutinised. The activity patterns of predators, their degree of specialisation on a particular prey, and their actual density in a given area must all be taken into account. A largely simplified example may demonstrate the point: given that there is only one diurnal predator species present, prey could simply shift activity to night-time to escape predation risk. However, as soon as there is at least one diurnal and one nocturnal predator, the prey has to appraise the risk sources. For instance, if the nocturnal predator specialises on a particular prey while the diurnal predator only searches for food in general, night would be the more risky period. Nevertheless, if the diurnal and generalist predator greatly outnumbers the nocturnal specialist, then daytime may become more risky. The trade-off becomes increasingly puzzling when there are several diurnal, nocturnal, and crepuscular predator species, as is the case for most prey and in most types of habitat (for a detailed analysis of diel predation risk patterns for microtines see Halle 1993).

Given that gauging risk actually indicates times of the day that are relatively safe compared with others, an individual prey could easily increase its fitness by concentrating activity within this time window and avoiding the more dangerous hours. However, things are more complicated because it is not only the density and activity phases of all predators that determine the times of high and low predation risk. Rather, the activity timing of conspecifics has a profound additional effect (see Daan and Slopsema 1978; Raptor Group 1982). When many prey specimens are active together in synchrony, each single individual is relatively safe because there is a fair chance that someone else will be taken instead (the dilution effect). Beyond that there is the opportunity of enjoying increased overall vigilance and exchange of warning signals, especially when members of a group are not only active at the same time but also keep in touch (Gerkema and Verhulst 1990). Finally, the predator may be confused by a large number of swarming prey items which possibly may lower the attack rate as well as overall hunting success.

On the other hand, there are also features that may result in an increase of individual predation risk with synchronous activity. Predators may, for example, concentrate on time windows of high profitability, i.e., when most prey is active (see below). Synchronous activity of many prey individuals may further attract the predator's attention for a specific plot as a hot spot for hunting. So there are apparently two alternative strategies for a prey individual: to be active together with all the others during the rush hours and to bet on the dilution effect, or to sneak out of the dangerous periods by choosing just the times when the majority of prey rests in their refuges and predators are somewhere else. Which strategy is more profitable in the long run will depend on many factors, including prey and predator behaviour as well as frequency-dependent selection.

It should be mentioned that all of the above does not only apply to conspecifics, but also to alternative prey, which makes things even more complicated. Also, the type of habitat might have an effect, especially with respect to the distribution of vegetation cover and shelter. There is evidence from experimental field studies that the activity pattern of prey may change as a response to landscape structure: root voles were more active during night-time and activity on the population level was less synchronous in a fragmented habitat where individuals were forced to cross an open matrix more frequently (Halle, unpubl. data).

4.3.2.2 Predators

For prey it seems easy to understand that having access to a clock is advantageous in order to minimise predation risk. In predators this may not be as obvious at first sight, especially since it is often said that losing a life is a much stronger selective force for prey than missing a dinner for predators. On the other hand, we have to realise that searching for prey is a rather expensive job, both with respect to energy costs (think about hovering kestrels, see Daan and Aschoff 1982) and missed opportunities, and that these costs are minimised by evolutionary processes. Take the theory of optimal foraging, for instance, which relays on the basic assumption that success rate,

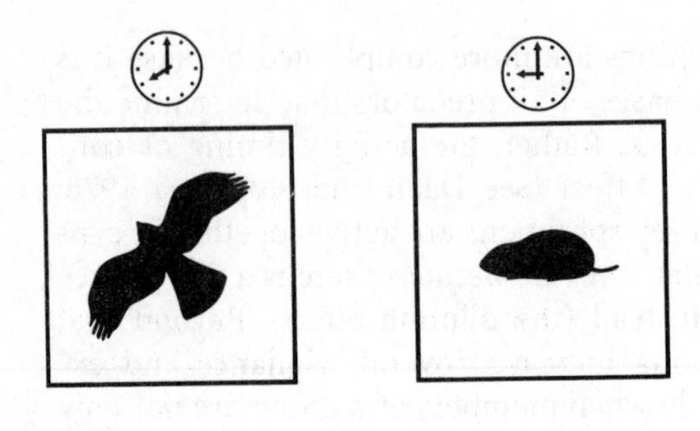

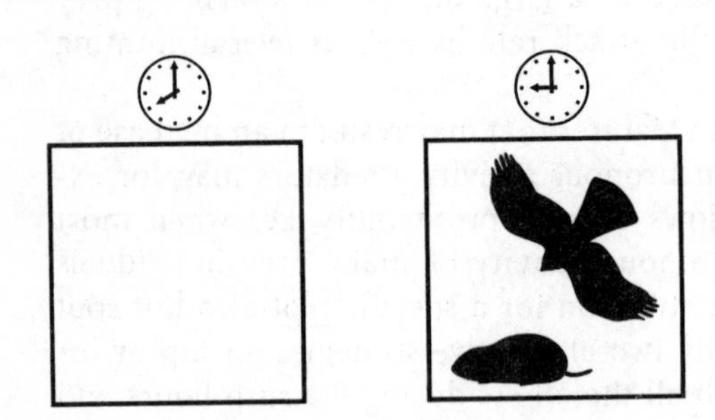

Fig. 4.3. Dependence of predator-prey interaction upon temporal behaviour. In the *upper panel* a specific habitat area (schematically indicated by the *square*) is used by a predator at time t1 and by prey at time t2. Although both species are present in the habitat, direct interaction cannot occur due to divergent activity timing. Interaction can only happen when predator and prey meet at the same place and at the same time, as shown in the *lower panel*

measured as prey items per time, is optimised by natural selection (Krebs and Inman 1994). Gauging encounter rates and the trade-off with travel time all depends on assessing time, and hence requires a clock.

However, predators cannot only search for prey at the wrong place. They can also search an actual profitable plot, but at the wrong time (Fig. 4.3). For a predator the activity pattern of its prey is a temporal pattern of food availability. To keep foraging costs low the predator would be well advised to save its strength for the time when it is worth the effort. Calculations show that energy-saving by convenient timing may appear low, but has a remarkable effect due to day-to-day permanence (Daan and Aschoff 1982). There are field observations indicating that at least some predators are in fact able to learn temporal patterns (Erdakov 1981; Rijnsdorp et al. 1981; Raptor Group 1982). This skill dramatically changes the estimates of individual predation risk for prey with synchronous and asynchronous activity, respectively (see also below).

Although predators are often closely related to one specific prey species as their staple food, the predator also has to deal with a variety of possible prey items. The pooled activity patterns of all alternative prey results in the relevant pattern of food availability. An interesting problem occurs when a predator switches prey due to low densities of its preferred prey in a type 3 functional response. The two prey items will hardly have identical activity patterns, so switching prey will include readjustment of the temporal hunting strategy. A field study looking for such kinds of behavioural reorganisation due to changing prey is not known to me, but would require detailed data on both predator and prey activity as well as on prey abundance.

Another aspect, also largely unexplored, is activity synchrony of prey in different plots or patches. Is rodent activity, for instance, low all over at one specific time, or can a predator get a meal just by moving to another plot? It may be suspected, and again there is some evidence from experimental studies (Halle, unpubl. data), that sunrise and sunset are probably such strong entraining signals that prey activity on plots within the range of an individual predator will be largely synchronous. However, this aspect is important for understanding the conditions for a hunting predator and should be studied in detail.

4.3.2.3 Temporal Coevolution?

An obvious precondition for an interaction between predator and prey is that both parties meet at the same place *and* at the same time; if not, there will be no predator-prey interaction. Prey tries to avoid predators by adroit timing of its activity, avoiding the bad times when the probability of encountering the enemy is high. Predators, on the other hand, try to maximise hunting efficiency by doing just the same, i.e., remembering the time when they hunt successfully. The delicate feature of this situation is that the same event is a disaster for one species and a bargain offer for the other. Both parties have at least some behavioural flexibility for adaptive fine-tuning of the activity phase. Moreover, for both, there is selective pressure to find a way out of the dilemma, since survival probability is affected. As a consequence, any convenient solution for the temporal optimisation problem would easily establish as ESS. With that we have a classical set-up for an arms race in temporal behavioural strategies, just as with many other aspects of the complex relationship between predators and prey. This has been formulated in an even more provoking manner, which says that coevolution may occur between predator and prey with respect to activity timing because species are tied together within their relationship in the long run (Daan 1981; DeCoursey 1989).

Let us again consider microtine rodents with their peculiar short-term rhythm as an illustrative example (Fig. 4.4). Imagine two relatively short time intervals (let us say 1 h) with comparable vole activity during both intervals (step 1 at top of Fig. 4.4). For predators, time is homogeneous with respect to capture probability, so they will use all the time evenly for hunting. Prey then decide that something should be done with the annoying predation risk: some especially clever individuals start to concentrate activity in the first of the two time intervals to hide in the crowd. When temporal concentration first starts, which may also happen due to some other factor like social interactions or simply by chance, it easily becomes a runaway process because synchronous individuals are relatively safe in numbers while those that are out of phase bear all the risk alone during the second time interval.

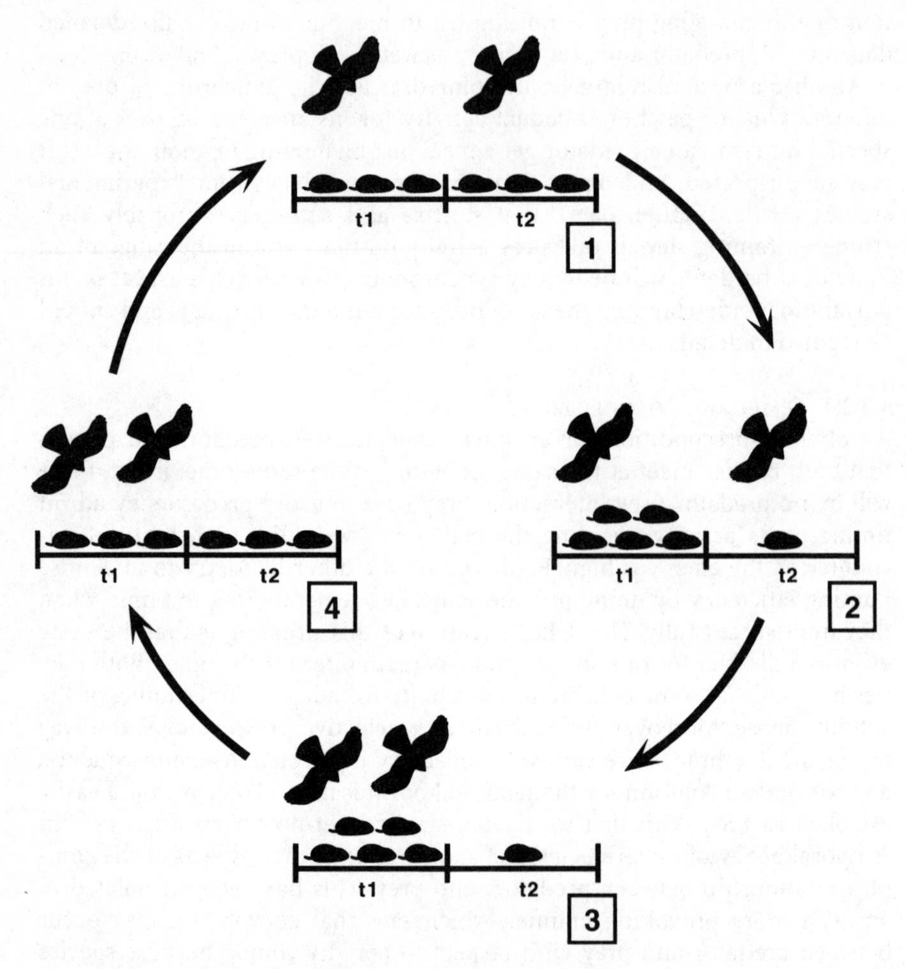

Fig. 4.4. Arms race in activity timing between predator and prey. At step *1* prey activity is the same in both of two consecutive time intervals *t1* and *t2*, and hunting activity of predators is evenly distributed. In step *2* prey activity is concentrated in interval *t1*. This leads to lower individual predation risk due to safety-in-numbers during *t1*, and markedly increased risk for asynchronous individuals during *t2*. As a response to variation in food availability predators may concentrate hunting effort to times of high prey activity (*t1* in step *3*), causing an immediate reversal in predation risk for prey: asynchronous individuals now suffer less losses because they are active at times of low predator activity. Decreased mortality helps the strategy of asynchronous activity to spread in the prey population, so after some time activity will be the same again at both time intervals (step *4*). Temporal variation in food availability is no longer given, hence predators will, with some delay, utilise all disposable time for hunting, leading to the starting point of the circle (step *1*). None of either predator or prey temporal strategies will stabilise as ESS to break the capacity of oscillation unless an additional advantage of one or the other time schedule is involved

The more activity concentration progresses, the more relative predation risk increases for asynchronous individuals (step 2 in Fig. 4.4).

However, for the predator time is at this stage no longer homogeneous with respect to capture probability. Rather, there is a time where prey is abundant, and another time when prey is scarce. Following the principal instruction of saving energy the predator should try to trace the prey temporally. As soon as hunting effort becomes concentrated in the first time interval as a response to the temporal pattern of food availability (step 3 in Fig. 4.4), the balance of predation risk flips. The few asynchronous individuals are suddenly safer than the synchronous majority because they are active at times of relatively low predation risk (cf. Raptor Group 1982). Due to decreased mortality rates, the behavioural strategy of asynchronous activity will spread in the prey population, so after some time activity will be the same again at both time intervals (step 4 in Fig. 4.4). At this stage there is no longer any reason for predators to confine foraging to a constrained time window, so with some delay all disposable time will be utilised evenly again, leading to the starting point of the circle (step 1 in Fig. 4.4).

As we have seen, the system has a strong capability of oscillating and will not stabilise at one of the steps unless an additional factor favours a specific strategy for either predator or prey. In microtines, for instance, synchronous ultradian rhythms are stable as far as we know, so an additional advantage, apart from predation risk minimisation, has to be assumed. It is by no means verified that mechanisms like the one outlined above are really at work in the field, and if so how common they are. However, the possibility of such kind of interaction can hardly be disclaimed. Presumably there will be no general answer, rather temporal coevolution may be a crucial factor in some predator-prey relations, but not in others. Strong candidates are pairs of prey species under high predation pressure, as in fact many small mammals are, and highly specialised predators.

4.3.3 The Temporal Niche

Competition for resources not only occurs within, but of course also among species. Following the niche concept in the sense of Hutchinson (1957) long-term coexistence of two species is only stable if they differ in at least one important ecological characteristic, i.e., in one niche dimension. As stressed above, specialisation for day and night, respectively, as the main activity phase shapes the design of species markedly. One may presume, therefore, that the same ecological position may be filled with a diurnal and a nocturnal species with diminished competition due to activity timing (Fig. 4.5). If so, daily routines for activity would be one of the factors shaping the community structure, leading to something that could be termed 'temporal niche'.

The idea that time for activity is one of the relevant niche dimensions, along with habitat and food type, dates back to Park (1940). It was taken up,

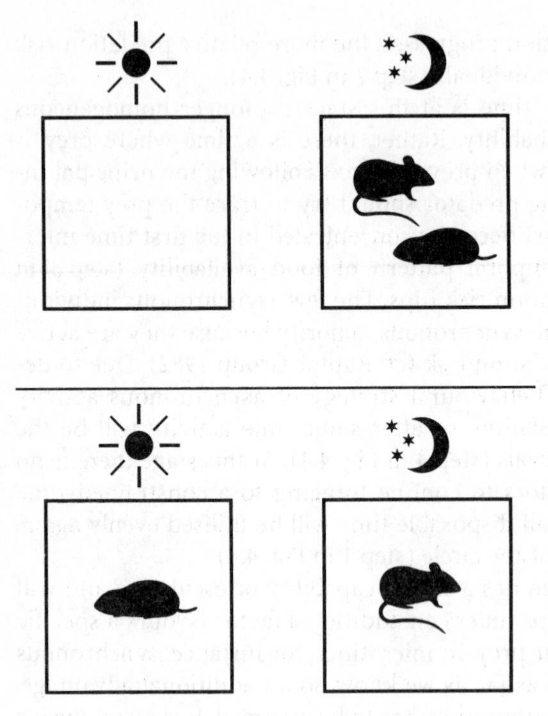

Fig. 4.5. Dependence of interspecific competition upon temporal behaviour. In the *upper panel* a specific habitat area (schematically indicated by the *square*) is used by two species with considerable niche overlap at the same time of the day, resulting in interference competition. In the *lower panel* niche partitioning is archived by different activity phases, one species being active during the day, the other by night. The use of different temporal niches induces a shift from interference to exploitation competition and may allow stable coexistence of the two species

for instance, by Miller (1955) and Pearson (1962), but was explicitly treated in detail in the classic paper by Schoener (1974). By reviewing 81 studies on competition among species from various taxa Schoener concluded that activity phase is the least important of the three dimensions: in 90% of the surveyed cases species were separated by habitat, in 78% by food, and in 41% by time. In only 5% of the surveyed groups was time the most important niche dimension while 55% were primarily separated by habitat. The suggested explanation for this pattern was that in choosing the foraging habitat, a species is weighing the energy gain in one type of habitat against the gain in another. In contrast, when deciding between activity and non-activity the species is weighing some yield, although lowered by competition, against nothing. In fact, being not active even causes a negative net gain because resting metabolism is not free. So behavioural discretion with respect to activity timing is considerably constrained by very restricted options. Remmert (1976) also doubted that the temporal niche exists at all because of two other reasons. First, an ecological niche may seem identical during day and night at first sight but reveals substantial differences when analysed in detail. Second, temporal flexibility as a short-term response to the environmental conditions seems to be favoured by evolution, as indicated by the fact that activity timing in animals is rather plastic.

It is quite obvious that competition cannot be totally relaxed by activity timing. In most instances, a food resource of one species will be reduced by another, irrespective of the time when the bite is taken away (Daan 1981). However, timing of activity can be used to evade interference, i.e., physical confrontation. Activity timing permits exploitation to occur which is a more gentle form of competition because species merely respond to the depressed level of resources without direct interaction. Furthermore, Schoener (1974) has outlined some special circumstances in which temporal partitioning may nonetheless be a relevant factor of niche separation, for example in terrestrial ectotherms, which are not a subject of this book. Another special case are, however, predators, and in this category we find several species of small mammals (see Chaps. 5, 6, 13, 14). Even more important, Schoener stated: "Only where ability to process food is limited relative to risk of being eaten during feeding should temporal specialization be marked." This definitely applies to a whole bunch of small mammal species, hence for the group of animals focused on in this book, the temporal niche concept appears to be rather substantial.

Experimental verification of competition at work in currently coexisting species is a rather hopeless undertaking. Due to the 'ghost of competition past' (Connell 1980), one could even frame the somewhat impudent argument that the best evidence for the importance of competition is when you can never observe it. Given that the avoidance of niche overlap is really a significant factor determining community structure, evolution would shape species such that competition is avoided. Only those species which in one way or another achieve niche separation coexist, while strong competitors will normally not live in syntrophy. However, this serious paradox not only applies to the temporal niche but to all aspects of interspecific strife.

Nevertheless, there is growing evidence for temporal niche separation in small mammals. Some are from studies in which the activity patterns of competing species were compared directly (Miller 1955; Brown 1956; Pearson 1962; Shkolnik 1971; Wójcik and Wołk 1985; Bruseo and Barry 1995). Kotler and co-workers have presented a particularly thorough analysis of temporal partitioning in two competing desert gerbils living in a habitat in which food availability varies on a daily basis (e.g. Kotler et al. 1993). Moreover, Mitchell et al. (1990) found that activity times in one of the species were directly affected by the density of the other. As already suggested by Daan (1981), the strongest evidence would come from an experimental approach, like the desert gerbil system: when we accept that time represents an ecological dimension it may be regarded as a limited resource that can be used up. Consequently, species compete for the most attractive temporal niche and the superior competitor probably monopolises the preferred time window. When the despot is taken away from the system the inferior competitor may have the chance to readjust its activity phase and to shift to the formerly occupied temporal niche. In fact, this kind of response, which follows from theoretical considerations, was shown to occur in gerbils by Ziv and co-workers (Ziv et al. 1993 and Chap. 9; cf. Glass and Slade 1980).

However, in performing the reverse experiment, i.e., introducing voles (*Microtus ochrogaster*) into resident populations of competing lemmings (*Synaptomys cooperi*) and vice versa, Danielson and Swihart (1987) found a spatial instead of a temporal response. Both species kept to their original nocturnal time schedule but avoided each other by using areas not occupied by the other species. In particular this was true of introduced lemmings which were considered the most subordinate individuals. Thus, it remains an open question how often and under which circumstances temporal partitioning is the strategy of choice for interference avoidance.

4.4 Concluding Remarks

It is obvious to everyone that the environment changes considerably in the course of a 24-h day, both with respect to biotic and abiotic factors. As there is hardly any environmental differentiation that is not addressed by adaptive traits it would in fact be most amazing if this particular feature was overlooked by evolution. The daily time schedule of any animal, and probably also plant, is, therefore, part of its ecology and one component in determining its fitness.

In my view, the major division of responses to the diel pattern in the environment is between the internal milieu of an individual on the one hand, and individuals' interactions with other individuals on the other. The first category is primarily related to the abiotic changes in the environment and represents the outcome of physiological optimisation processes. However, daily internal periodicity does not fall entirely in the domain of pure physiology because these rhythms may only be understood against the background of changes in the environment. So this field of study is, or should be, part of autecological research and physiological ecology in particular. As long as investigations of circadian rhythms stop at the stage of description, although imaginative, little can be said about the rhythms' functional role. To answer this question we must ask how the internal machinery was shaped by the external world, and how natural selection empowered its development.

While the first category is related to the way an individual deals with its physical surroundings, the second is concerned with the temporal arrangement among individuals, whether of the same or different species. For instance, there can be little doubt that, to some extent, the actual ecological situation of an individual is governed by its conspecifics. Population density affects the availability of all kinds of resources, including food, space, and mates. Obviously conspecifics determine the social environment in which the individual grows up and lives, which in turn is crucial in determining its reproductive fate, longevity, and fitness. In the case of frequency-dependent selection, even the adaptive value of a phenotype depends on the characteristics of the neighbours.

Alternative temporal strategies may allow an individual to find its place and time within the population. In fact there is a classical trade-off between the advantages of the most preferable time window and the disadvantages of maximum intraspecific competition at just that time (think about yourself when visiting a restaurant at dinner time or when trying to book holidays). However, the temporal distribution of other requirements, such as food sources (or weather expectations and school holidays in the case of holiday booking), may constrain the range of temporal options.

Spatial and temporal organisation of a population are two sides of the same coin and belong inseparably together, thus one should always be studied along with the other. Although this demand may sound self-evident, a look into current literature on spacing behaviour confirms that the claim is by no means trivial. The two aspects are scarcely, if ever, integrated in experimental designs. An ecological approach would have to link population dynamics and spacing with activity rhythms, an attitude only taken by a mere handful of students of behaviour so far. As a consequence we do not know much about the presumable importance of activity timing for community organisation in animals.

The presence or absence of other individuals is also essential for predator-prey interactions, so much so that arms races and coevolution are common features of species evolutionary history. To cite DeCoursey (1989), the observation of light-sampling behaviour in den-dwelling animals "suggested that circadian pacemakers have evolved as precision instruments, fine-tuned by selective pressures over a long course of predator-prey interaction." This probably applies not only to the temporal behavioural strategy of prey, but of both parties because the predator also has to solve a precarious problem, related to temporal variation of food availability. The most interesting and fascinating aspects of this game arise from the close coupling of behavioural traits in prey and predators: "It may seem paradoxical that a predator and its prey should exploit essentially the same mechanism as a strategy against each other. The better known crypsis of both hunter and hunted illustrate that such sharing of defence mechanisms is not unique" (Rijnsdorp et al. 1981). Hence, simple statements such as "outside very specialized cases, predation is the ecological factor most influencing small mammal periodicity" (Pearson 1962) or "escape from predation therefore seems to be an important factor which may influence the activity pattern" (Peterson and Batzli 1975) may in general be close to the truth. However, the underlying decision process is anything but easy because assessing predation risk is a most difficult task.

Virtually nothing is known about diel variation in host-parasite interactions, let alone pathogens. That we know nothing, however, does not mean that this topic is of no relevance. Only recently has ecology started to think about parasites and pathogens as one of the main forces driving evolution. It is quite imaginable that temporal variability in the environment results in different probabilities of parasite attack or pathogen infection, and that selective pressure drives the host to avoid these time windows. The set-up would be very much the same as in predator-prey interactions, but would be even more rebellious against scientific scrutiny, especially under field conditions.

The last aspect of temporal interactions among individuals, i.e., interspecific competition and the temporal niche, has earned most discredit in the past, but now seems to be winning back lost ground. From the small mammal systems studied so far it seems that competition at work is more likely to be traceable when fine-tuning of behaviour and the realised Hutchinsonian niche are concerned. Diurnalism versus nocturnalism, in contrast, are more related to the fundamental niche. This basic temporal partitioning happened so long back in niche separation that there is no way left to prove assumptions experimentally in currently existing systems. However, the ghost of competition past is not a special problem with temporal niche separation but occurs in many, if not all, aspects of interspecific competition.

A second inference from the temporal niche concept is that time might be considered a resource by definition, since it can be used up like any other item of limited amounts. If all available time windows are filled by one or another species, every additional species has to share the resource with a competitor. Obeying the exclusion principle, coexistence will only be stable when the two species differ in some other substantial characteristic. This implies that struggle for the most attractive times for activity may occur, i.e., the times of the day with the lowest predation risk, or times which allow for most efficient foraging. As outlined above, there are experimental findings that make such mechanisms likely, especially under the peculiarities of desert ecosystems.

So no doubt the daily activity pattern is of ecological relevance. Studies of this subject include all the ingredients of an important and intellectually challenging topic of behavioural ecology. However, when we are honest we have to subscribe to DeCoursey's (1989) warning: "It is always easier to 'armchair' evolutionary theories than to collect the requisite ecological cost/gain facts for testing these hypotheses", but also in this respect the study of daily activity patterns is representative of behavioural ecology. Obviously, temporal responses to population dynamics, predator-prey interactions, and interspecific competition cannot be examined in the laboratory with activity cages. So, if we are interested in understanding the constraints under which the specific solution of a particular species to the activity timing problem has evolved, and we should be, we have to do studies in the field, or at least in sufficiently large enclosures. Hence, students in this area of research have to rely on advanced field technology when entering unknown scientific terrain. This should be tempting enough to follow the trail.

References

Ashby KR (1972) Patterns of daily activity in mammals. Mammal Rev 1:171–185

Bartness TJ, Goldmann BD (1989) Mammalian pineal melatonin: a clock for all seasons. Experientia 45:939–945

Brown LE (1956) Field experiments on the activity of the small mammals, *Apodemus*, *Clethrionomys* and *Microtus*. Proc Zool Soc Lond 126:549–564

Bruseo JA, Barry RE (1995) Temporal activity of syntopic *Peromyscus* in the Central Appalachians. J Mammal 76:78–82

Calhoun B (1975) Social modification of activity rhythms in rodents. Chronobiologica 2:11–13

Clarke JA (1983) Moonlight's influence on predator/prey interactions between short-eared owls (*Asio flammeus*) and deermice (*Peromyscus maniculatus*). Behav Ecol Sociobiol 13:205–209

Connell JH (1980) Diversity and the coevolution of competitors, or the ghost of competition past. Oikos 35:131–138

Daan S (1981) Adaptive daily strategies in behavior. In: Aschoff J (ed) Biological rhythms. Handbook of behavioural neurobiology, vol 4. Plenum Press, New York, pp 275–298

Daan S, Aschoff J (1975) Circadian rhythms of locomotor activity in captive birds and mammals: their variations with season and latitude. Oecologia 18:269–316

Daan S, Aschoff J (1982) Circadian contributions to survival. In: Aschoff J, Daan S, Groos GA (eds) Vertebrate circadian systems. Springer, Berlin Heidelberg New York, pp 305–321

Daan S, Slopsema S (1978) Short-term rhythms in foraging behaviour of the common vole, *Microtus arvalis*. J Comp Physiol A 127:215–227

Daly M, Behrends PR, Wilson MI, Jacobs LF (1992) Behavioural modulation of predation risk: moonlight avoidance and crepuscular compensation in a nocturnal desert rodent, *Dipodomys merriami*. Anim Behav 44:1–9

Danielson BJ, Swihart RK (1987) Home range dynamics and activity patterns of *Microtus ochrogaster* and *Synaptomys cooperi* in syntopy. J Mammal 68:160–165

DeCoursey PJ (1989) Photoentrainment of circadian rhythms: an ecologist's viewpoint. In: Hiroshiga T, Honma K (eds) Circadian clocks and ecology. Hokkaido University Press, Sapporo, pp 187–206

DeCoursey PJ (1990) Circadian photoentrainment in nocturnal mammals: ecological overtones. Biol Behav 15:213–238

Dickman CR, Predavec M, Lynam AJ (1991) Differential predation of size and sex classes of mice by the barn owl, *Tyto alba*. Oikos 62:67–76

Enright JT (1970) Ecological aspects of endogenous rhythmicity. Annu Rev Ecol Syst 1:221–238

Erdakov LN (1981) Adaptability of activity rhythms in a least weasel (*Mustela*, Mustelidae). Izvest sib otd Akad Nauk SSR, Ser Biol Nauk 3:140–144

Gerkema MP, Verhulst S (1990) Warning against an unseen predator: a functional aspect of synchronous feeding in the common vole, *Microtus arvalis*. Anim Behav 40:1169–1178

Glass GE, Slade NA (1980) The effect of *Sigmodon hispidua* on spatial and temporal activity of *Microtus ochrogaster*: evidence for competition. Ecology 61:358–370

Halle S (1988a) Avian predation upon a mixed community of common voles (*Microtus arvalis*) and wood mice (*Apodemus sylvatcus*). Oecologia 75:451–455

Halle S (1988b) Locomotory activity pattern of wood mice as measured in the field by automatic recording. Acta Theriol 33:305–312

Halle S (1993) Diel pattern of predation risk in microtine rodents. Oikos 68:510–518

Halle S (1995) Diel pattern of locomotor activity in populations of root voles, *Microtus oeconomus*. J Biol Rhythms 10:211–224

Hughes JJ, Ward D, Perrin MR (1994) Predation risk and competition affect habitat selection and activity of Namib Desert gerbils. Ecology 75:1397–1405

Hutchinson GE (1957) Concluding remarks. Cold Spring Harbor Symp Quant Biol 22:415–427

Kenagy GJ (1976) The periodicity of daily activity and its seasonal changes in free-ranging and captive kangaroo rats. Oecologia 24:105–140

Kolb HH (1992) The effect of moonlight on activity in the wild rabbit (*Oryctolagus cuniculus*). J Zool Lond 228:661–665

Kotler BP, Brown JS, Mitchell WA (1993) Environmental factors affecting patch use in two species of gerbilline rodents. J Mammal 74:614–620

Kotler BP, Brown JS, Subach A (1993) Mechanisms of species coexistence of optimal foragers: temporal partitioning by two species of sand dune gerbils. Oikos 67:548–556

Krebs JR, Inman AJ (1994) Learning and foraging: individuals, groups, and populations. In: Real LA (ed) Behavioral mechanisms in evolutionary ecology. University of Chicago Press, Chicago, pp 46–65

Lehmann U, Sommersberg CW (1980) Activity patterns of the common vole, *Microtus arvalis* – automatic recording of behaviour in an enclosure. Oecologia 47:61–75

Lima SL, Dill LM (1990) Behavioral decisions made under the risk of predation: a review and prospectus. Can J Zool 68:619–640

Meddis R (1975) On the function of sleep. Anim Behav 23:676–691

Mikkola H (1970) On the activity and food in the pygmy owl *Glaucidium passerinum* during breeding. Ornis Fennica 47:10–14

Miller RS (1955) Activity rhythms in the wood mouse, *Apodemus sylvaticus* and the bank vole, *Clethrionomys glareolus*. Proc Zool Soc Lond 125:505–519

Mitchell WA, Abramsky Z, Kotler BP, Brown JS, Pinshow BP (1990) The effect of competition on foraging effort: theory and a test with desert rodents. Ecology 71:844–854

Park O (1940) Nocturnalism – the development of a problem. Ecol Monogr 10:485–536

Pearson AM (1962) Activity patterns, energy metabolism, and growth rate of the voles *Clethrionomys rufocanus* and *C. glareolus* in Finland. Ann Zool Soc Vanamo 24:1–58

Peterson RM, Batzli GO (1975) Activity patterns in natural populations of the brown lemming (*Lemmus trimucronatus*). J Mammal 56:718–720

Raptor Group (1982) Timing of vole hunting in aerial predators. Mammal Rev 12:169–181

Remmert H (1976) Gibt es eine tageszeitliche ökologische Nische? Verh Dtsch Zool Ges 1976: 29–45

Rijnsdorp A, Daan S, Dijkstra C (1981) Hunting in the kestrel, *Falco tinnunculus*, and the adaptive significance of daily habits. Oecologia 50:391–406

Schoener TW (1974) Resource partitioning in ecological communities. Science 185:27–39

Shkolnik A (1971) Diurnal activity in a small desert rodent. Int J Biometerol 15:115–120

Travers SE, Kaufman DW, Kaufman GA (1988) Differential use of experimental habitat patches by foraging *Peromyscus maniculatus* on dark and bright nights. J Mammal 69:869–872

Vásquez RA (1994) Assessment of predation risk via illumination level: facultative central place foraging in the cricetid rodent *Phyllotis darwini*. Behav Ecol Sociobiol 34:375–381

Wójcik JM, Wołk K (1985) The daily activity rhythm of two competitive rodents: *Clethrionomys glareolus* and *Apodemus flavicollis*. Acta Theriol 30:241–258

Ziv Y, Abramsky Z, Kotler BP, Subach A (1993) Interference competition and temporal and habitat partitioning in two gerbil species. Oikos 66:237–246

Section III Empirical Findings

Empirical Findings – Introduction

Stefan Halle and Nils Chr. Stenseth

The following ten chapters provide an overview of recent studies dealing with ecological aspects of small mammal activity. The chapters are a collection of reviews, some with a special focus on particular studies to illustrate a specific point of interest. This part of the book is intended to serve as a reference for the empirical data base that is currently available to develop ideas and hypotheses for a behavioural ecology approach to activity behaviour. However, it is also intended to demonstrate the enormous diversity of activity patterns in this animal group, which presumably is related to a corresponding diversity of ecological habits.

It was not possible to consider all small mammal groups for which ecological studies on activity behaviour have been conducted. For our selection, two criteria were decisive. First, we wanted to cover as large as possible a variety of ecological situations. Therefore, species groups of different size and feeding habits were included, especially when living in different habitat types and within different social structures, and when confronted with different environmental challenges. Second, with a few exceptions, the reported surveys are field studies that were primarily designed to interpret activity with respect to ecological considerations, and that were made possible by modern investigation techniques.

The chapters are roughly ordered by the size of the concerned species. The first two chapters are on relatively large predators. Zielinski in Chap. 5 deals with weasels and martens in northern latitudes. His focus is on the specific metabolic problems of this group due to body size and shape, and on the temporal relation between predators and prey. In Chap. 6 Palomares and Delibes provide a literature review on herpestids and viverrids in southern latitudes with a special focus on the question of how different feeding habits and habitat preferences are reflected by the activity pattern. In addition they present results from a comparative radiotelemetry study of two sympatric species to investigate specific adaptations.

The following six chapters are on granivore and graminivore rodent species. The bias towards rodents in this book partly reflects the availability of data from laboratory and field studies, but is also due to the fact that rodents are the most species-rich group of mammals. Wauters in Chap. 7

Ecological Studies, Vol. 141
Halle/Stenseth (eds.) Activity Patterns in Small Mammals

specifically focuses on seasonal changes in the food situation and the consequences for diel activity patterns in Eurasian red squirrels. In Chap. 8 Daly et al. ask how predation risk shapes the activity pattern of nocturnal kangaroo rats in desert habitats. Special aspects for their considerations are intraspecific behavioural variations and the response to moonlight in an almost coverless habitat. Also, Ziv and Smallwood deal with desert species (i.e., gerbils and pocket mice). In Chap. 9 they report on findings from field experiments to verify that a temporal niche in fact exists, and show that even interspecific competition for the most attractive times of the day may occur.

In Chaps. 10 to 12 species of roughly the same body size, but having very different activity patterns, are discussed. Flowerdew in Chap. 10 reviews studies on monophasic nocturnal wood mice with a granivore/insectivore feeding habit. He shows that there are a lot of hints at social and ecological influences, but also stresses that our knowledge is still fragmentary, which is at least partly due to methodological problems. In Chap. 11 Halle provides an overview of what is known today about the special case of synchronous ultradian activity in voles. Special points of interest are the relations between different time scales of rhythmic behaviour, between the individual and the population level, and between voles and other species, i.e., predators and competitors. In Chap. 12 Ruf and Heldmaier present laboratory data about the remarkable strategy of daily torpor in Djungarian hamsters, which probably has evolved as a special metabolic adaptation to harsh environmental conditions.

The last two reviews deal with the smallest of the small mammals. Chapter 13 by Merritt and Vessey is on shrews which reveal an ultradian rhythm similar to voles, but in this case triggered by a short-term hunger cycle. A special focus is on the particular metabolic constraints resulting from small body size and winter activity. Erkert in Chap. 14 provides an overview of the extensive literature on activity behaviour in bats. He shows, using many examples, that environmental factors, of which the activity of insect prey is a very important one, are able to shape the fine-structure of a basically nocturnal activity pattern.

5 Weasels and Martens – Carnivores in Northern Latitudes

William J. Zielinski

5.1 Introduction

The first mammals were probably nocturnal (McNab 1978; Kemp 1982) and many extant orders of small mammals have retained this trait. This behaviour is presumably due to the measure of protection from predation that darkness provides (Crawford 1934; Falls 1968; Daan and Aschoff 1982). Large, herbivorous mammals are less restricted to the nocturnal phase (e.g. Bunnell and Harestad 1990) and have been classified as 'nocturnal-diurnal' (Charles-Dominique 1978). This is either because their large size renders them less vulnerable to predators, because the low energy nature of their folivorous diet demands extended periods of food consumption, or because food search time increases with increasing body size (Harestad and Bunnell 1979; Belovsky and Slade 1986). The carnivorous mammals, however, exhibit a wide diversity and flexibility of activity patterns (Kavanau 1971; Ewer 1973; Curio 1976; Daan 1981; Gittleman 1986). Peak activity times vary among the Carnivora and within individual species studied at different locations or during different seasons. For example, *Mustela erminea* has been described as nocturnal (Figala and Tester 1992), nocturnal in winter but diurnal in summer (Bäumler 1973; Debrot et al. 1985), crepuscular (Müller 1970) and mostly diurnal (Erlinge and Widen 1975; Erlinge 1979).

The activity patterns of mammalian carnivores are influenced by a number of factors, including: diel temperature variation (Schmidt-Nielsen 1983), interference from competitors (Carothers and Jaksic 1984), limitations of the visual system (Walls 1963; Dunstone and Sinclair 1978), risk of predation (King 1975), social behaviour (Ewer 1973; Gittleman 1986), and behavioural thermoregulation (Chappell 1980). However, what makes carnivores unique is the fact that their foods, unlike that of herbivores, have their own circadian cycles of availability and vulnerability (Curio 1976; Zielinski 1986a). The foods of herbivores, although patchy in space, are relatively stable and predictable in time. Perhaps this distinction is the reason why the activity patterns of mammalian carnivores can be described so differently by different authors while the activity patterns of mammalian herbivores tend to be less

Ecological Studies, Vol. 141
Halle/Stenseth (eds.) Activity Patterns in Small Mammals

variable. If small carnivores feed on different proportions of prey types in different locales, each with different circadian phases of maximum vulnerability, we would expect considerable intraspecific variation in modal activity times. How the circadian rhythms of prey vulnerability influence predator activity will be a central issue of this review.

Mustelids are the most ecologically diverse family within the Carnivora (Wozencraft 1989), including dietary specialists and generalists that range from strictly carnivorous to omnivorous. The subfamily Mustelinae (e.g. weasels, mink, polecats, ferrets, martens, sable, fishers) originated in the Miocene (Martin 1989; Anderson 1994) and the earliest forms occupied forest habitats and probably lived in much the same way martens do today. During the Plio-Pleistocene the cooler, drier climate favoured the establishment of grasslands and small mammals, especially voles (*Microtus* spp.), began to radiate into forms that occupied the forest-steppe environment (Webb et al. 1983; Martin 1989). Although the genus *Martes* was clearly defined by this time, intermediates between *Martes* and *Mustela* were becoming established (Anderson 1970). The development of the grassland biome provided opportunities for predators that were small enough to pursue voles and lemmings (*Dicrostonyx* spp., *Lemmus* spp.) in their burrows, tolerate the fluctuation in prey numbers typical at northern latitudes, and sustain themselves through the harsh winters (King 1989).

5.2 Metabolic Consequences of Mustelid Size and Shape

Weasels, especially the Holarctic *Mustela nivalis* and *M. erminea*, are a paradoxical group; they occupy some of the coldest regions of the world but manage to do so with short fur, very little fat storage capacity and a surface area/volume ratio that favours heat loss (Brown and Lasiewski 1972; King 1989). Arctic weasels can have a basal metabolic rate two to three times greater than those in Wisconsin (Scholander et al. 1950). Weasels do not migrate, hibernate, or enter torpor during periods of cold or food shortage. They must continually fuel their bodies or die. Long, thin mustelids have basal metabolic rates that can be two to six times higher than that of nonelongate mammals of equivalent weight (Brown and Lasiewski 1972; Iversen 1972; Chappell 1980; Sandell 1985). Least weasels consume from 30 to 40% of their weight in food a day and expend an estimated 21% of the mean average daily metabolic rate on activity; a greater percent than some shrews (Moors 1977; Gillingham 1984). Their elongate body shape and small size impose constraints that influence the ecology of small mustelines in myriad ways (Fig. 5.1).

Because their activities are focused on confined spaces that are refuges for their prey, weasels must maintain their thin body form, limiting the storage of fat to slight dips in the contours of their body outline (King 1989) and

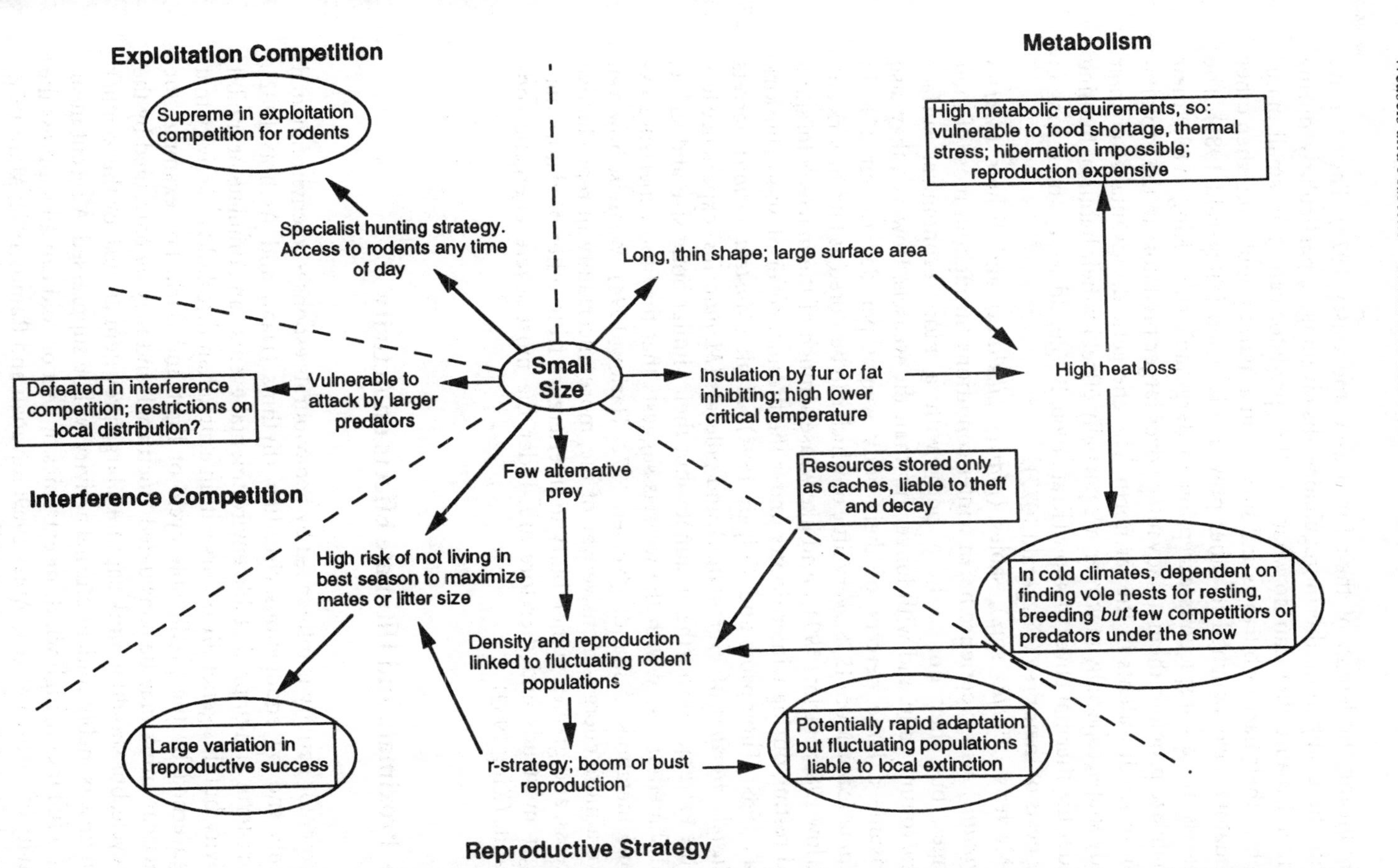

Fig. 5.1. The advantages (*circled*) and disadvantages (*boxed*) of small size to the weasels as a group. Four main subject areas (metabolism, exploitation competition, interference competition, reproductive strategy) are linked together at various levels. (From King 1989)

minimising the length of their fur (Casey and Casey 1979). They survive with these apparent thermoregulatory disadvantages, particularly during winter, because they can do what no other predator can: hunt small mammals in their narrow burrows and survive on a sparse population when other predators have switched to other prey or dispersed (Fitzgerald 1981). The weasels are a conspicuous exception to Bergman's rule (King 1989); their small size provides them an advantage over larger predators in the exploitation of small rodents in the far north. The energetic disadvantages of their body shape, especially in winter, are partially offset by their habits of hunting under the thermal cover of snow (Kraft 1966; Fitzgerald 1981) and resting in the nests of their prey (Fitzgerald 1977).

Martens (*Martes* spp.), sables (*Martes zibellina*) and fishers (*Martes pennanti*) also experience harsh winter conditions throughout most of their ranges, but they do so in a body with a more economical surface area/volume ratio and with longer fur than the weasels. However, they also consume a large percent of their body weight per day (about 25% in *M. americana*, More 1978), very little of which can be stored as fat (Buskirk and Harlow 1989; Harlow 1994). And like weasels, much of the marten's foraging and resting during winter occurs under the thermal cover of snow (Buskirk et al. 1989; Thompson and Colgan 1994). Mink (*Mustela vision*), ferrets (*Mustela nigripes*, *M. eversmanni*), and polecats (*M. putorius*) experience less of the extremes of northern climates but their similar body size and shape, and shared ancestry with the weasels suggests that they too experience extreme metabolic demands (Iversen 1972; Harlow 1994). Weasels, however, have a more energy-intensive way of life, more uncertainty in reproductive success, and greater vulnerability to interference from other predators than larger mustelids and therefore are under the most severe energetic constraints (King 1989).

5.3 Proximate and Ultimate Effects on Activity

Locomotor activity is influenced by behavioural responses to environmental events and by endogenous circadian rhythms (Daan and Aschoff 1982). Short-term and unpredictable environmental events can stimulate or inhibit activity, but this generally occurs within a circadian threshold of activity that is governed by the predictable cycle of light and dark. For example, the American mink may be considered 'nocturnal' and may be entrained by the LD cycle, but should a mink capture a large prey item at dusk or the weather turn unseasonably cold, nocturnal activity will be suppressed. A recent meal, rain, a full moon, high winds or extremes of heat or cold can all suppress the activity of carnivores (e.g. Ables 1969; Kavanau and Ramos 1972; More 1978; Buskirk et al. 1988; Taylor 1993; Beltrán and Delibes 1994).

Ecological factors affect how an animal allocates the energy that is acquired into various activities. Survival and reproduction will presumably be maximised when the amount, duration, and distribution of particular activities are allocated in an optimal fashion (Schoener 1971; Daan and Aschoff 1982). The general activity pattern (i.e., nocturnal, diurnal, crepuscular, ultradian, arhythmic) is a phenotypic expression of the way that environmental constraints, and behavioural responses to them, influence endogenous rhythms. A rhythmic circadian pacemaker may structure the temporal pattern of activity but the ultimate expression is influenced by the physical environment, trophic community interactions, and social factors (Aschoff 1964).

It is important to realise that characterising a species as 'nocturnal' simply means that the highest probability of discovering the animal active occurs during the hours of darkness. Any time of an animals' internal day is characterised by its own distribution of probabilities of the occurrence of both spontaneous behaviours and responses to environmental stimuli (Daan 1981). The most common activity time for small mustelids appears to be during the night, followed by crepuscular activity, with diurnal activity much less common (Table 5.1; Gittleman 1986). However, even the phase of the cycle that has the highest probability of activity will include periods of inactivity because foraging and other locomotor activities frequently occur in bouts (e.g. Erlinge 1979; Powell 1979; Thompson and Colgan 1994). Unfortunately, the data on activity patterns in the Mustelidae are few and generally unsuited for quantitative analysis. Rarely is activity the sole focus of a field study of a small mustelid and the activity patterns of captive animals may not be representative of those in the wild (Kavanau 1969; Kavanau et al. 1973; Davidson 1975; Zielinski 1986b). Although the nature of the data make intra-familial comparisons difficult, in this section I examine the variety of reasons that could favour nocturnality in the Mustelidae and also try to explain some of the deviations from this basic pattern.

5.3.1 Foraging Time, Meal Patterning and Digestive Constraints

The metabolic requirements of weasels demand that they eat frequent, energy-rich meals (King 1989). Musteline carnivores are active, on average, about 25% of the 24-h day (Table 5.2, see page 104), ranging from a low of 7 to a high of about 60%. This appears to be somewhat less than is typical for other mammals in the *Mustela* and *Martes* weight categories (Bunnell and Harestad 1990). Outside the breeding season, searching for food probably accounts for most of this activity, if small mustelids are similar to other mammals which engage in 'foraging' during 70% of the time they are active (Bunnell and Harestad 1990).

Because the individual prey of weasels and martens can often exceed the amount that can be consumed in a single meal, excess is cached for later consumption (Oksanen et al. 1985; King 1989; Henry et al. 1990). Thus, foraging

Table 5.1. Summary of studies describing musteline activity patterns

Author (year)	Season	Activity pattern[a]	Number (M:F)[b]	Location[c]	Method
Mustela nivalis					
Kavanau (1969)	Summer	Nocturnal/arhythmic[d]	2 (1:1)	Michigan, USA	Captivity
Price (1971)	Summer	Nocturnal	4 (2:2)	Michigan, USA	Captivity
King (1975)	Annual	Diurnal	10 (7:3)	England	Field – trapping
Erdakov (1981)	–	Ultradian	6 (3:3)	Northern Russia	Captivity
Zielinski (1986a)	Summer	Arhythmic	3 (3:0)	Illinois, USA	Captivity
M. erminea					
Robitaille (unpubl.)	Annual	Nocturnal	7 (5:2)	Quebec, Canada	Field – observation
Bäumler (1973)	Winter	Nocturnal	–	Germany	Field – snow tracking
	Summer	Diurnal	–	Germany	Field – observation
Erlinge and Widen (1975)	Fall	Diurnal	7 (3:4)	Sweden	Field – telemetry
Erlinge (1979)	Winter	Nocturnal	4 (4:0)	Sweden	Field – telemetry
	Spring/summer	Diurnal	4 (4:0)	Sweden	Field – telemetry
Nams (1981)	Fall/spring/summer	Nocturnal	–	NWT, Canada	Captivity
Debrot et al. (1985)	Spring/summer/fall	Diurnal	–	Switzerland	Field – trapping
Zielinski (1986b)	Summer	Nocturnal	1 (1:0)	Minnesota, USA	Captivity
Zielinski (1988)	Summer	Nocturnal/crepuscular	3 (3:0)	Minnesota, USA	Captivity
Figala and Tester (1992)	Annual	Nocturnal	1 (0:1)	Germany	Captivity
M. frenata					
Kavanau and Ramos (1975)	Summer	Nocturnal/diurnal	5 (3:2)	Nevada/California, USA	Captivity
M. vison					
Gerell (1969)	Annual	Nocturnal	6 (5:1)	Sweden	Field – telemetry
Melquist et al. (1981)	Fall/winter	Nocturnal/diurnal	6 (–)	Idaho, USA	Field – telemetry
Zielinski (1986b)	Summer	Nocturnal	6 (0:6)	Illinois, USA	Captivity
Zielinski (1988)	Summer	Nocturnal/crepuscular	3 (0:3)	Illinois, USA	Captivity
M. putorius					
Danilov and Rusakov (1969)	–	Nocturnal/diurnal	–	Russia	Field
Bäumler (1973)	Winter	Nocturnal	–	Germany	Field – snow tracking
Blandford (1987)	–	Nocturnal/crepuscular	–	Wales	Field – telemetry
Lode (1995)	Annual	Nocturnal	9 (5:4)	France	Field – telemetry

Table 5.1 *(continued)*

M. putorius furo (domestic ferret)					
Donovan (1987)	–	Diurnal[e]	6 (0:6)	–	Captivity
Stockman et al. (1985)	–	Arythmic	12 (6:6)	–	Captivity
M. nigripes					
Henderson et al. (1974)	Annual	Nocturnal	–	South Dakota, USA	Field – observation
Martes americana					
More (1978)	Winter	Diurnal	14 (8:6)	NWT, Canada	Field – telemetry
	Summer	Crepuscular	14 (8:6)	NWT, Canada	Field – telemetry
Hauptman (1979)	Summer	Diurnal/nocturnal	4 (3:1)	Wyoming, USA	Field – telemetry
Zielinski et al. (1983)	Summer	Diurnal/crepuscular	5 (4:1)	California, USA	Field – telemetry
	Winter	Nocturnal/crepuscular	3 (1:2)	California, USA	Field – telemetry
Martin (1987)	Summer	Diurnal/nocturnal	6 (3:3)	California, USA	Field – telemetry
Thompson and Colgan (1994)	Summer	Arthythmic	20 (10:10)	Ontario, Canada	Field – telemetry
	Winter	Diurnal	10 (6:4)	Ontario, Canada	Field – telemetry
M. martes					
Pulliainen and Heikkinen (1980)	Winter	Nocturnal/crepuscular	–	Finland	Field – snow tracking
Clevenger (1993)	Fall/winter	Nocturnal	5 (2:3)	Spain	Field – telemetry
M. pennanti					
Zielinski (unpubl.)	Summer	Crepuscular	10 (2:8)	California, USA	Field – telemetry
Kelly (1977)	Annual	Crepuscular	10 (7:3)	New Hampshire, USA	Field – telemetry
	Summer	Crepuscular/diurnal	10 (7:3)	New Hampshire, USA	Field – telemetry
Powell (1979)	Winter	Crepuscular/diurnal	3 (2:1)	Michigan, USA	Field – telemetry
Johnson (1984)	Annual	Crepuscular	–	Wisconsin, USA	Field – telemetry
Arthur and Krohn (1991)	Annual	Crepuscular	43 (15:28)	Maine, USA	Field – telemetry

[a] Two activity types are listed when a significant amount of activity is reported during both periods. If a predominant type was mentioned, it is underlined.
[b] Number of individual animals that contributed to the activity data (males : females).
[c] Location of field study, or in the case of captive studies, the location where the study animals originated.
[d] Nocturnal when light/dark transitions are gradual and arhythmic when transitions are abrupt.
[e] Primary diurnal during anoestrus and increased nocturnal activity during oestrus.

does not necessarily precede the consumption of each meal and the motivation to eat (hunger) can be different from the motivation to find food. Daan (1981) distinguishes two optimisation problems related to feeding: optimisation with respect to the animal's metabolic requirements and optimisation with respect to the fluctuations in food availability. Here I review the data pertinent to the first problem.

Hungry animals actively seek food and sated animals usually rest (de Ruiter 1967). In classic meal pattern studies, large meals are followed by a post-prandial dip in activity proportional to the size of the meal; a phenomenon first demonstrated in captive rats, cats, and dogs (LeMagnen and Devos 1970; Kanarek 1974; Ardisson et al. 1981) but noted in wild and captive mustelids as well (e.g. Müller 1970; Kavanau and Ramos 1975; Thompson and Colgan 1994). Small carnivores that have secured a large prey item may be inactive for a day or two; the movements of American martens are greatest when food is least available (Thompson and Colgan 1990), and fishers are more active when they are provisioning young (Paragi et al. 1994). Laboratory evidence also suggests that hunger and locomotor activity are related (Müller 1970; Price 1971; Kavanau and Ramos 1975; Zielinski 1986b, 1988). Price (1971) discovered that total activity exhibited by least weasels was not affected by 8 h of food deprivation, but was nearly doubled in response to 24 h of deprivation. Similarly, Zielinski (1988) demonstrated that the total diel activity of a number of small carnivore species increased significantly when food availability was restricted. Locomotor and feeding behaviour appear closely linked in weasels (Zielinski 1986b, 1988; Sandell 1989; Figala and Tester 1992) although they can be easily dissociated in other species (Daan and Aschoff 1982).

Food intake can be extremely limited by digestive constraints in small carnivores. When allowed to feed freely after 16 h of deprivation least weasels could consume no more than an average of 3.1 g of food in an hour and 11.4 g (in 3–4 meals) in 8 h (Gillingham 1984). Meals were consumed about 3 h apart (near the lower limit of food passage time; Short 1961). This occurs despite the fact that least weasels must consume, on average, about 40% of their body weight in food each day (Gillingham 1984). This physiological constraint indicates why optimal meal timing can be very different from optimal foraging. Weasels do not have the option of consuming their daily requirement in 1 or 2 meals, but instead must distribute their feeding among 5–10 meals per day. The rapid processing of food by weasels means they must either stay near cached food and limit long-distance activities or, if cached food is not recovered, to forage every few hours. Martens apparently are not under similar constraints because of significantly slower gut passage times than the weasels (More 1978; Harlow 1994). The least weasel's frequent need to eat may explain the observation by some that they are arrhythmic or exhibit ultradian cycles of activity (Erdakov 1981; Zielinski 1986b), a pattern that also characterises the activity of another energy-demanding taxon, the shrews (Crowcroft 1954). The discontinuous pattern of activity (multiple short bouts) reported from radiotelemetry studies of some mustelids (Gerell 1969; Erlinge 1979; Powell 1979) could be influenced as much by frequent bouts of eating as by bouts of foraging, behaviours that cannot be distinguished using remote telemetry.

Not all locomotor activity is regulated by the need to replenish an energy deficit. Captive carnivores on ad libitum diets continue to exhibit spontaneous, rhythmic locomotor activity that usually bears some relationship to the LD cycle (Kavanau 1969; Kavanau and Ramos 1975; Zielinski 1986b, 1988; Robitaille and Baron 1987) and is presumed to be under endogenous control. Thus, activity is affected by the history of recent food intake (and other short-term activating or inhibiting events) *and* by cyclic circadian thresholds for activity that are in phase with the light cycle (Kavanau and Ramos 1975; Daan and Aschoff 1982).

In some experiments the circadian variation in food availability and light have been manipulated to understand their relative effects on locomotor activity. When placed in constant dark or constant light, weasels ran for prolonged periods on wheels to anticipate a meal provided with 24-h periodicity (Zielinski 1986b). This response suggested that periodic availability of food could entrain the endogenous rhythm. In a second experiment, weasels and mink were kept in LD 12 : 12 and required to run increasingly longer periods on a running wheel for food during the phase of the LD cycle they preferred, usually the dark (Zielinski 1988). This circumstance tested the animal's 'willingness' to abandon the 12 h of the day favoured for activity for the opposing period during which time food was available at a significantly lower energetic cost. Most of the weasels retained their preference for activity during the dark and merely ran the additional distance on the wheel to achieve sufficient food. A modest increase in diurnal activity, and reinforcement, indicated that each animal was aware of the lower cost to foraging during the light phase. Sometimes an order of magnitude of additional effort was expended to acquire food in the preferred phase rather than shift activity to the opposing phase of the LD cycle. This evidence suggests that there is an underlying circadian rhythm of locomotor activation that controls the magnitude of hunger-induced activity. This is similar to the observation that rats deprived of food for equivalent periods of time were more active, and ate more, when deprivation ended during the night (their normal activity time) than during the day (Bellinger and Mendel 1975).

Locomotor activity directed toward foraging is a part of the homeostatic control of food intake. However, animals appear more apt to respond to hunger with activity when the hunger occurs in phase with the underlying endogenous phase of activity. Very little is known about the interaction between the factors that regulate spontaneous activity versus hunger-driven activity in small mustelines, or any other mammal for that matter.

5.3.2 The Visual System

The visual system of most mammalian carnivores is adapted to permit activity at all times of day; many carnivores possess what has been referred to as 'the 24-h eye' (Walls 1963; Kavanau and Ramos 1975). Even the eye of primarily nocturnal mammals contains a duplex retina (rod and cone receptors) and adequate pupillary control to permit vision in daylight (Walls 1963). In fact, many carnivores that have eyes generally adapted to function

Table 5.2. Percent of various time periods that mustelids are active

Author (year)	Percent active[a]	Period of analysis	Season
Mustela nivalis			
Moors (1977)	21.1[b]	24 h	Fall/winter/spring
Buckingham (1979)	20.0	24 h	–
M. erminea			
Erlinge and Widen (1975)	19.0	24 h	Fall
Erlinge (1979)	25.0	24 h	Winter
	18.0	Day	Winter
	32.0	Night	Winter
	38.0	Day	Spring
	29.0	Night	Spring
Robitaille and Baron (1987)	17.8	24 h	Annual
	6.8	24 h	December
	14.4	24 h	September
M. frenata			
Kavanau and Ramos (1975)	70.6	Night	–
	10.0	Day	–
	9.2	24 h	–
M. vison			
Melquist et al. (1981)	50.0	24 h	Fall/winter
M. putorius			
Lode (1995)	31.0	24 h	Annual
M. nigripes			
Powell et al. (1985)	8.5	24 h	Winter
Martes americana			
Thompson and Colgan (1994)	59.2	24 h	Summer
	16.3	24 h	Late winter
M. martes			
Clevenger (1993)	53.0	Night	Fall/winter
	19.0	Day	Fall/winter
	34.0	24 h	Fall/winter
M. pennanti			
Powell and Leonard (1983)	27.0	24 h	Spring

[a] Mean for the period reported, except where otherwise specified.
[b] Percent of average daily metabolic rate, not time.

in darkness will choose the brightest setting when allowed to select among different ambient light levels (Kavanau et al. 1973). Because of the adaptability of the mustelid eye, it is unlikely that the amount of light available is a serious constraint on the choice of activity time.

Kavanau and Ramos (1975) recognised the dual influences of physiology and ecology on the temporal pattern of carnivore activity by contrasting the 'visual activity type', the genetically-influenced activity pattern exhibited in the laboratory when food was provided ad libitum, and the 'ecological activity type' exhibited in the field. The ecological type was considered a reflection of an animal's response to the times of food availability and threat of predation. The long-tailed weasel's eye appears to be adapted best for dim-light vision, suitable for daytime vision, and least well adapted for vision in very dim light even though the animal was reported to be primarily nocturnal in the field (Kavanau and Ramos 1975). The least weasel had a nocturnal visual type and a reported arrhythmic ecological activity (Kavanau 1969). Both cases suggest that evolution had not yet adjusted the endogenous rhythm to the selective environment.

'Aschoff's Rule' (as described by Pittendrigh 1960) can also be used to assess the genetic, or visual, activity pattern. In Aschoff's original formulation nocturnal animals are characterised by a circadian rhythm with (1) a period (cycle length) that is less than or equal to 24 h in constant dark, (2) a period that is shorter in constant dark than it is in constant light, and (3) an active portion that is greater in constant dark than in constant light. Mink, the only mustelid with adequate data to test the rule, conform to conditions 1 and 2 (Zielinski 1986b), and are nocturnal in the field (Table 5.1), suggesting that their genetic and ecological activity patterns are equivalent.

5.3.3 Temperature and Season

Reduced activity in winter by mammalian carnivores is considered an adaptation to cold stress and has been noted in ursids, procyonids, and mustelids (Ewer 1973). Most mustelids living in northern latitudes are less active during winter than summer and also reduce their activity during extremely cold winter weather (Pulliainen and Heikkinen 1980; Zielinski et al. 1983; Clark et al. 1984; Robitaille and Baron 1987; Buskirk et al. 1988; Arthur and Krohn 1991; Thompson and Colgan 1994; Table 5.2). High winds probably also contribute to reduced activity (Taylor 1993). Thompson and Colgan (1994) observed several martens that were continuously inactive for more than 30 h when the temperature was -25 °C. Given the high energy demands of weasels, especially during winter, reducing activity during winter can yield considerable energy savings. Activity can increase energy consumption up to six times that of the basal metabolic rate in some mustelids (Powell 1979; Karasov 1992). Chappell (1980) found that most Arctic mammals (the least weasel included) could save an average of 30% of thermoregulatory costs if they were diurnal during winter. In contrast, a number of studies on mustelines have reported that nocturnal activity actually increases in winter (Bäumler 1973; Zielinski et al. 1983; Debrot et al. 1985), a result that may have more to do with prey availability (see below) than physical factors. The energetic cost of winter nocturnality may not be as great when an animal forages under the snow (Formozov 1946), as is the case for most northern weasels (King 1989) and martens (Buskirk et al. 1988).

Small carnivores that are less active in the winter than the summer must omit some activities during winter that they are normally engaged in during the summer. Obviously the subtraction of activity associated with reproductive behaviour will reduce activity, but some mustelids may exhibit less territorial behaviour as well (Thompson and Colgan 1994). If foraging time is reduced during winter, either the basal metabolic rate is lower in winter than summer (for which there is no evidence), weasels are more efficient at finding the same prey they consume during the summer, or they consume prey of larger average size. With less time available, foraging should be particularly efficient during the winter (Sandell 1985). There is some evidence that this efficiency occurs, at least for martens, by hunting and eating larger-sized prey (Zielinski et al. 1983; Bull and Heater, in press). Anytime large prey are cached near a rest site (Henry et al. 1990), or when the predator resides in the dens of the prey it kills (e.g. Fitzgerald 1977; Powell and Brander 1977; Buskirk et al. 1989; King 1990) the component of activity stimulated by hunger will be reduced. Deep snow also hinders the movements of some mustelids, most notably fishers (Raine 1983; Krohn et al. 1997), but this is unlikely to account entirely for the significant reductions in activity noted.

5.3.4 Competition

Because most mammalian carnivores, mustelids included, are either nocturnal or crepuscular (Table 5.1; Gittleman 1986), the range of activity types does not exist to suggest that these species are either using resources at different times or disassociating because of interference competition. There is no mammalian carnivore equivalent to the extreme partitioning of the temporal niche axis observed in owls and hawks (Jaksic 1982). Considerable overlap is typical among mammalian species in primarily nocturnal and primarily diurnal communities, and between these groups (Daan 1981), though there are some exceptions (e.g. Ziv and Smallwood, Chap. 9). If competition has any role in the evolution of activity pattern it is more likely via interference than exploitation. This is because time is not an independent niche axis, but is dependent on partitioning of either the habitat or food axes (Schoener 1974; Carothers and Jaksic 1984). The same foods can be exploited by many species, no matter how different the activity type.

Larger carnivores are typically dominant over smaller ones (Rosenzweig 1966) and within the Mustelidae there are examples of larger species killing smaller ones (e.g. *Mustela frenata* and *M. nivalis*, Polderboer et al. 1941; *Martes pennanti* and *M. americana*, de Vos 1952; *Martes americana* and

Mustela erminea, Thompson and Colgan 1987). In the southern Sierra Nevada, sympatric fishers and martens exhibit similar crepuscular activity patterns (Zielinski, unpubl. data). There is no evidence that the activity pattern can be influenced by the presence of a potential competitor, unless that competitor is also an important predator (e.g. King 1975).

5.3.5 Predator Avoidance

Predation threats that occur at a particular phase of the LD cycle can influence daily activity patterns in any species where predation is a significant mortality factor (Curio 1976). Predation is considered the reason that most small mammals are nocturnal or crepuscular (Park 1940; Daan and Aschoff 1982). Most of the carnivores considered here are relatively small and as a result are susceptible to predation by avian and larger mammalian predators. If darkness provides some cover from predators, or if predation pressure at night is less than that during the day, nocturnalism may be favoured by selection.

Other small, non-mustelid carnivores (*Genetta genetta*, *Potos flavus*, *Bassariscus astutus*) become active very consistently shortly after sunset, a pattern that is attributed in part to avoidance of predators (Kavanau and Ramos 1972). Although least weasels are most typically described as nocturnal (Table 5.1) they are more active by day than by night in Marley Wood, England (King 1975). Although the weasel's prey are nocturnally active, so is a major predator of weasels, the tawny owl (*Strix aluco*). The threat of predation at night is considered the primary reason for the weasel's diurnal activity. Zielinski (1988) found that weasels (*Mustela* spp.) were more reluctant than American mink to shift activity from the nocturnal period, when food was available at a high energetic cost, to the diurnal period were food was available at significantly lower cost. Selection in the significantly smaller weasels may have favoured a stronger association with the nocturnal phase, which provides some cover from predators, than has occurred in the mink. Incidentally, a number of species of carnivores appear to increase their nocturnal activity in the presence of human disturbance, including some mustelids (Löhrl 1972; Rosevear 1974; Curio 1976; Mason and Macdonald 1986). If humans are perceived as a threat the increased nocturnal activity may be an anti-predatory response.

5.3.6 Social and Reproductive Behaviour

Most mustelines are solitary except during the breeding season. Regardless of the season during which reproduction occurs, male activity and movements generally increase when they are searching for mates (Gerell 1970; King 1975; Debrot and Mermod 1983; Sandell 1986; Arthur and Krohn 1991). Presumably the polygynous mating system that is thought to characterise most mustelids demands that males engage in extensive movements in search of females. Robitaille and Baron (1987) suggest that reproductive behaviour is the dominant activity-generating behaviour of male ermine in midsummer. Female activity is lowest during pregnancy (Gerell 1969; Leonard 1980) and appears to increase, at least in fishers, after parturition (Paragi et al. 1994) and with the age of the kits (Leonard 1980). Amount or intensity of motor activity is not affected by estrous condition or by the administration of estradiol in domestic ferrets (Stockman et al. 1985; Donovan 1987), contrary to the stimulating effect of estrogen on activity in female rats (Blizard 1983). The late winter activity in captive and wild *Martes martes*, previously referred to as 'false heat', may instead represent late dispersal of overwintering juveniles (Helldin and Lindström 1995).

The lack of complex social structure in the Mustelinae, compared with some of the viverrids for example, may account in part for the fact that no small mustelids are diurnal. The most social species of viverrids, *Mungo mungo*, *Helogale parvula*, and *Suricatta suricatta*, are all diurnal (Hinton and Dunn 1967; Rood 1975; Waser 1981; Palomares and Delibes, Chap. 6). Whether diurnalism facilitates the evolution of complex social behaviour, is a consequence of it, or is unrelated, is unknown. However, in an analysis of the correlation between activity pattern and type of social system, 14 of 19 (73.7%) of the social species in the Carnivora were primarily diurnal (Zielinski, unpubl. data).

5.3.7 Prey Availability

Foods are distributed patchily in space and time, but the analysis of predator response to spatial variation in food has received considerably more attention than analysis of daily temporal variance in food availability. Predators experience a foraging problem that herbivores do not; their foods are differentially active, and thus differentially available throughout the 24-h period. It is a logical extension of foraging theory to suggest that predators that hunt at the time of day when the probability of prey capture is highest should enjoy greater prey capture success, and at lower cost, than individuals that initiate foraging at random. Although this area has received only modest theoretical consideration (Schoener 1974; Caraco 1980; Williams and Nichols 1984; Zielinski 1986a; Belovsky et al. 1989), empirical evidence of the phenomenon is growing.

In theory, predators should synchronise their activity with the time of day when their prey are most vulnerable. This may be when prey are active, and are producing visual and auditory cues that betray their location, or when prey are resting when they may be more difficult to locate but are easier to catch and subdue (Zielinski 1986a). Two prey species can have the same activity patterns but if they have different detectability and capturability functions (e.g. one rests in inaccessible, and the other in accessible, refugia) each can be most vulnerable to a predator at a different time of day. Whether carnivores should forage when prey are active or inactive will depend on the trade-off among these factors. A graphical illustration of the interaction between activity time, detectability, capturability and vulnerability is presented in Fig. 5.2. Prey populations that have the same activity pattern (diurnal in Fig. 5.2) can have sufficient variety in their detectability and capturability functions that they manifest very different vulnerability functions over the 24-h day.

The most thorough studies of this problem have focused on the activity of raptors and their microtine prey (Rijnsdorp et al. 1981; Raptor Group 1982). Small mammals are only available to raptors when they are active and two species of raptors synchronised their flight time with the ultradian pattern of above-ground activity of the voles. Rijnsdorp et al. (1981) estimated that kestrels (*Falco tinnunculus*) that were active in synchrony with the ultradian rhythm of voles could experience sufficient flight-time savings (10–20 kJ per day) to contribute significantly to the number of offspring produced.

Mammalian predators and their prey have not been the subjects of similar studies but we can evaluate their foraging mode and review related literature to help understand the importance of predator-prey synchrony. Because

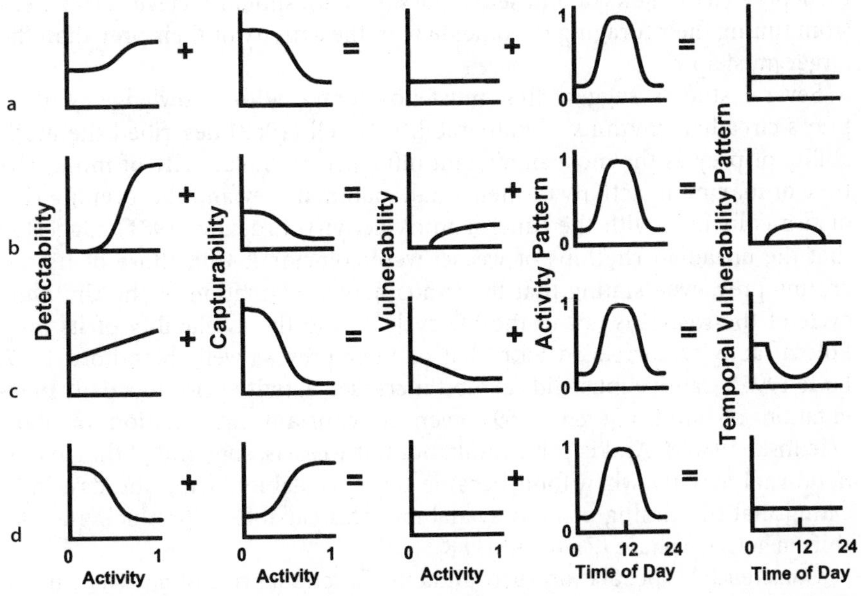

Fig. 5.2. Relationships between detectability, capturability, vulnerability, and time of day for four hypothetical prey species (a–d). **a** Detectability and capturability are increasing and decreasing functions of prey activity, respectively. Relative effects cancel and vulnerability does not vary with time of day. **b** Detectability when active is greater than capturability when inactive, thus vulnerability is greatest during prey activity. **c** Detectability when active is less than capturability when inactive, thus vulnerability is greatest during prey inactivity. **d** Detectability and capturability are decreasing and increasing functions of prey activity, respectively. Relative effects cancel and vulnerability does not vary with time of day

moving prey are easier to detect than inactive prey (Curio 1976; Norrdahl and Korpimäki 1998) predators that hunt primarily using sight and sound should forage during the active phase of their prey; so too should predators that cannot physically enter the resting refugia of their prey. In the least weasel and polecat, vision and olfaction are of equivalent importance in hunting with audition somewhat less important (Murphy 1985), and movement detection is more important than visual acuity in mink and ferrets (Apfelbach and Wester 1977; Dunstone and Sinclair 1978). Audition probably complements the other senses, especially because some small mustelids can hear ultrasound in the range of small mammal vocalisations (Heffner and Heffner 1985; Powell and Zielinski 1989). Thus the sensory modalities of small mustelids suggest that they might benefit from synchrony with the active phase of their prey. However, the smallest predators considered here can enter the burrows of their prey and, unlike raptors or larger predators, are able to locate prey when they are resting. Nams (1981) carefully examined the activity of radio-collared ermine in large enclosures and concluded that prey selection was independent of prey activity. Weasels killed inactive voles more frequently than active voles. Weasels, which can enter burrows and kill rodents when they are inactive, and which can change the activity patterns of their prey (Jedrzejewska and Jedrzejewski 1990), should receive less benefit from timing their foraging to coincide with the activity of their prey than the larger mustelines.

Several studies suggest that mustelids forage with knowledge of their prey's circadian rhythms of vulnerability. Gerell (1969) described the availability of prey as the most significant influence on the activity of mink. The time of maximum activity of their small mammal prey and the 'evening rise' of fish coincided with the time of mink activity. Erdakov (1981) suggested that the ultradian rhythms of weasel were coincident with those of its microtine prey, even stating that the fundamental adaptation of the rhythmic cycle of the weasel is not to the LD cycle, but to the availability of its prey. Polecat activity correlated with that of their prey as well (Blandford 1987, Lode 1995). Captive mustelids exhibit increased activity prior to a daily presentation of food (Gerell 1969) even in constant illumination or dark (Zielinski 1986b). And captive mink, but not weasels, substituted diurnal for nocturnal activity when food became more available during the day, indicating that the timing of food availability was capable of producing a 180° shift in activity phase (Zielinski 1988).

Considerable speculation surrounds the functional role of activity time in American martens. In California, martens are primarily diurnal in summer and exhibit a significant increase in nocturnal and crepuscular activity during winter (Zielinski et al. 1983; Martin 1987). This shift at first appears maladaptive because temperatures during the time of peak activity during the summer and winter are above and below thermoneutrality, respectively (Worthen and Kilgore 1981; Buskirk et al. 1988). Zielinski et al. (1983) accounted for the activity change by proposing that the time of maximum vulnerability of summer prey was different than that for winter prey, a phenomenon also described for polecats (Lode 1995). Ground-dwelling sciurids (i.e., *Tamias*, *Spermophilus*) are available only during the summer and then only during the day; martens cannot access them in their burrows at night during the summer nor at all during the winter when they are hibernating. During late fall and winter the crepuscular and nocturnally-active lagomorphs, flying squirrels (*Glaucomys sabrinus*), and voles comprised most of the diet. The energetic benefits accrued to martens by tracking the diel availability of seasonally available prey apparently exceed the thermoregulatory costs for doing so. Hauptman (1979) also found that the activity of American martens in Wyoming was congruent with the activity of their primary prey.

Thompson and Colgan (1994) monitored the activity of American martens in Ontario but did not find evidence for synchrony with prey activity. Martens were arhythmic during summer and diurnal during winter; nearly opposite to that described in California. If synchrony was important martens should have been nocturnal in winter when their snowshoe hare (*Lepus americanus*) and ruffed grouse (*Bonasa umbellus*) prey were active. The authors reasoned that because winter temperatures are so much colder in Ontario than in California that the thermoregulatory benefits of winter diurnality exceeded the foraging advantage of synchrony with prey. More (1978) also believed that winter diurnality was an energy saving behaviour for American martens in Alberta.

Because meal patterning and foraging patterns are driven by different incentives we do not expect that predators will be active every day at the peak of prey activity. Time since the last meal, availability of cached food, and other short-term phenomena will produce only a general synchrony even when environmental conditions favour it. However, this does not discount the potentially large energy savings that can be gained by foraging when probability of success is greatest. Whether these gains can be realised, however, depends on the magnitude of other costs incurred, particularly thermoregulatory costs in northern climates.

5.4 Conclusions

Body size is key to understanding many behaviours (Peters 1983), including the factors that influence the amount and distribution of daily activity. Smaller herbivores spend proportionally more time active and are more nocturnal than large herbivores (Belovsky and Slade 1986); a relationship that may apply to mammalian carnivores as well. In a review of 146 species in the Carnivora, nocturnal species were significantly smaller than those that were not nocturnal, a pattern that was exhibited in the Mustelidae as well (Zielinski, unpubl. data).

The activity patterns of small mustelids, like many other aspects of their ecology, are influenced by their body size and shape and the consequences these have on energetics. Small carnivores must take small prey, need energy-rich and frequent meals, are potentially more vulnerable to predators themselves and are poorer interspecific competitors than large carnivores. The smallest mustelids should be active primarily at night because their prey are usually nocturnal and darkness provides more cover from predators and competitors than daylight. Field and laboratory evidence largely support this prediction. However, their demanding metabolism and limited gut capacity make frequent meals a necessity and some diurnal eating and foraging is necessary. Consequently, weasels have a visual system that permits adequate vison at all natural light levels.

There are fewer incentives for nocturnalism in the larger mustelines (minks, polecats, martens and fishers) and they appear to have a greater diversity of activity patterns. Their diet is not comprised exclusively of small nocturnal rodents and fewer other predators can interfere with their activities or can kill them. Although their larger sizes may make them somewhat less vulnerable to predators than the weasels, they probably experience a sufficient predator threat to find some protection by restricting most of their activity to darkness. However, because their diets change seasonally more than the diets of weasels, we should expect more flexibility in activity patterning in these species than in the smaller, microtine-dependent weasels. The limited data on marten and mink suggest that a more diverse diet and larger size free these larger mustelids from some of the constraints acting on weasels and result in more flexibility in their activity times.

Two factors appear to universally affect the activity of all mustelines; cold weather and reproductive behaviour. Cold suppresses the amount of activity during the winter, especially at night. Martens at the southern limit of their range may be an exception because milder nocturnal temperatures permit them to hunt nocturnally-active prey. During the breeding season the activity of all male mustelines appears to increase. Polygynous mating systems favour multiple matings by males and the increased activity is probably attributable to mate search and the additional foraging needed to support this behaviour.

An understanding of the factors that affect small carnivore activity patterns will require more field and laboratory studies directed toward the study of physical and biotic factors that affect activity. The knowledge of a carnivore's size may explain a considerable amount of variation in its' activity time (Bunnell and Harestad 1990) but more fieldwork on the diel regime of predator and prey behaviours (e.g. Raptor Group 1982) and laboratory work that evaluates the effect of circadian variation in food availability on activity pattern (e.g. Zielinski 1988) are both required before we will understand the roles of exogenous and endogenous factors in the expression of activity patterns. It has been a handicap to the development of the field that activity data are usually collected only incidentally to the primary objective of a particular research study. These data are usually improperly sampled and uncritically analysed. Using the data available today it is difficult to know, for example, whether the activity patterns of two species differ in some fundamental way or whether methodological differences or unequal sample sizes account for the difference.

The metabolic costs of being small, long and thin are high. Elevated basal metabolic rate, high locomotion costs (especially in winter), and high reproductive costs are difficult to maintain without the ability to store much body fat. Weasels, martens and other small northern carnivores require other adaptations to help balance their energy equation. Reductions in activity during winter and during extreme cold and a generalised visual system that permits foraging at any time of day, are important behavioural and physiological traits that help conserve energy in small predators. The ability to identify and forage during the most vulnerable phase of their prey's activity pattern may be yet another important adaptation that reduces weasel and marten metabolic costs in northern climates.

References

Ables ED (1969) Activity studies of red foxes in southern Wisconsin. J Wildl Manage 33:145–153

Anderson E (1970) Quaternary evolution of the genus *Martes* Carnivora, Mustelidae. Acta Zool Fenn 130:1–133

Anderson E (1994) Evolution, prehistoric distribution, and systematics of *Martes*. In: Buskirk SW, Harestad AS, Raphael MG, Powell RA (eds) Martens, sables, and fishers biology and conservation. Cornell University Press, Ithaca, New York, pp 13–25

Apfelbach R, Wester V (1977) The quantitative effect of visual and tactile stimuli on the prey-catching behavior of ferrets *Putorius furo* L. Behav Proc 2:187–200

Ardisson JL, Dolisi C, Ozon C, Crenesse D (1981) Caracteristiques de preses d'eau et d'aliments spontanees chez des chiens en situation ad lib. Physiol Behav 26:361–371

Arthur SM, Krohn WB (1991) Activity patterns, movements, and reproductive ecology of fishers in southcentral Maine. J Mammal 72:379–385

Aschoff J (1964) Survival value of diurnal rhythms. Symp Zool Soc Lond 13:79–98

Bäumler W (1973) Circadian activity-rhythm of the polecat *Mustela putorius* and the ermine *Mustela erminea* and its influence on pelage cycle of the ermine. Säugetierkdl Mitt 21:31–36

Bellinger LL, Mendel VE (1975) Effect of deprivation and time of refeeding on food intake. Physiol Behav 14:43–46

Belovsky GE, Slade JB (1986) Time budgets of grassland herbivores: body size similarities. Oecologia 70:53–62

Belovsky GE, Ritchie ME, Moorehead J (1989) Foraging in complex environments: when prey availability varies over time and space. Theor Pop Biol 36:144–160

Beltrán JF, Delibes M (1994) Environmental determinants of circadian activity of free-ranging Iberian lynxes. J Mammal 75:382–393

Blandford PRS (1987) Biology of the polecat *Mustela putorius*: a literature review. Mammal Rev 17:155–198

Blizard DA (1983) Sex differences in running-wheel behaviour in the rat: the inductive and activational effects of gonadal hormones. Anim Behav 31:378–384

Brown JH, Lasiewski RC (1972) Metabolism of weasels: the cost of being long and thin. Ecology 53:939–943

Buckingham CJ (1979) The activity and exploratory behaviour of the weasel, *Mustela nivalis*. PhD thesis, Exeter University

Bull EL, Heater TW (in press) Ecology of American marten in northwestern Oregon. Wildl Monogr

Bunnell FL, Harestad AS (1990) Activity budgets and body weight in mammals: how sloppy can mammals be? Curr Mammal 2:245–305

Buskirk SW, Harlow HJ (1989) Body-fat dynamics of the American marten *Martes americana* in winter. J Mammal 70:191–193

Buskirk SW, Harlow HJ, Forrest SC (1988) Temperature regulation in American marten *Martes americana* in winter. Natl Geogr Res 4:208–218

Buskirk SW, Forrest SC, Raphael MG, Harlow HJ (1989) Winter resting site ecology of marten in the central Rocky Mountains. J Wildl Manage 53:191–196

Caraco T (1980) On foraging time allocation in a stochastic environment. Ecology 61:119–128

Carothers JH, Jaksic FM (1984) Time as a niche difference: the role of interference competition. Oikos 42:403–406

Casey TM, Casey KK (1979) Thermoregulation of arctic weasels. Physiol Zool 52:153–164

Chappell MA (1980) Thermal energetics and thermoregulatory costs of small arctic mammals. J Mammal 61:278–291

Charles-Dominique P (1978) Nocturnality and diurnality: an ecological interpretation of these two modes of life by an analysis of the higher vertebrate fauna in tropical forest ecosystems. In: Luckett WP, Szalay FS (eds) Phylogeny of the primates. Plenum Press, New York, pp 69–88

Clark TW, Richardson L, Casey D, Campbell TM III, Forrest SC (1984) Seasonality of black-footed ferret diggings and prairie-dog burrow pluggings. J Wildl Manage 48:1441–1444

Clevenger AP (1993) Pine marten (*Martes martes*) home ranges and activity patterns on the island of Minorca, Spain. Z Säugetierkd 58:137–143

Crawford SC (1934) The habits and characteristics of nocturnal animals. Q Rev Biol 9:201–214

Crowcroft P (1954) The daily cycle of activity in British shrews. Proc Zool Soc Lond 123:715–729

Curio E (1976) The ethology of predation. Springer, Berlin Heidelberg New York, pp 250

Daan S (1981) Adaptive daily strategies in behavior. In: Aschoff J (ed) Biological rhythms. Handbook of behavioral neurobiology, vol 4. Plenum Press, New York, pp 275–298

Daan S, Aschoff J (1982) Circadian contributions to survival. In: Aschoff J, Daan S, Groos GA (eds) Vertebrate circadian systems. Springer, Berlin Heidelberg New York, pp 305–321

Danilov PI, Rusakov OS (1969) Peculiarities of the ecology of *Mustela putorius* in northwest districts of the European part of the USSR (in Russian). Zool 48:1383–1394

Davidson RP (1975) The efficiency of food utilization and energy requirements of captive fishers. MS thesis, University of New Hampshire, Concord

Debrot S, Mermod C (1983) The spatial and temporal distribution pattern of the stoat (*Mustela erminea* L). Oecologia 59:69–73

Debrot S, Weber JM, Marchesi P, Mermond C (1985) The day and night activity pattern of the stoat *Mustela erminea* L. Mammalia 49:13–17

de Ruiter L (1967) Feeding behavior of vertebrates in their natural environment, chap 7. In: Handbook of physiology, sect 6, alimentary canal, vol 1: food and water intake. Am Physiol Soc, Washington DC

de Vos A (1952) Ecology and management of fisher and marten in Ontario. Tech Bull, Ontario Dept Lands and Forests

Donovan BT (1987) Reproductive state and motor activity in female ferrets. Physiol Behav 40: 717–724

Dunstone N, Sinclair W (1978) Comparative aerial and underwater visual acuity of the mink, *Mustela vison* Schreber, as a function of discrimination distance and stimulus luminance. Anim Behav 26:6–13

Erdakov LN (1981) Adaptability of sleep and waking rhythms in the weasel *Mustela*, Mustelidae. Izv Sib Otd Akad Nauk SSR Biol Nauk 3:140–144

Erlinge S (1979) Movements and daily activity patterns of radio-tracked male stoats, *Mustela erminea*. In: Amlaner CJ, Macdonald DW (eds) A handbook on biotelemetry and radio tracking. Wheaton and Company, Exeter, UK, pp 703–715

Erlinge S, Widen P (1975) Activity patterns of stoat. Fauna Och Flora 70:137–142

Ewer RF (1973) The carnivores. Cornell University Press, Ithaca, New York

Falls JB (1968) Activity. In: King JA (ed) Biology of *Peromyscus*. Am Soc Mammal Spec Publ 2, Provo, pp 543–570

Figala J, Tester JR (1992) Seasonal changes in locomotor and feeding activity patterns in stoat *Mustela erminea*, Carnivora in captivity. Acta Soc Zool Bohemoslov 56:7–13

Fitzgerald BM (1977) Weasel predation on a cyclic population of the montane vole *Microtus montanus* in California. J Anim Ecol 46:367–397

Fitzgerald BM (1981) Predatory birds and mammals. In: Bliss LC, Cragg JB, Heal DW, Moore JJ (eds) Tundra ecosystems: a comparative analysis. Cambridge University Press, Cambridge, pp 485–508

Formosov AN (1946) Snow cover as an intergral factor in the environment and its importance in the ecology of mammals and birds. Translated from Russian by Prychodko W, Pruitt WO. Edmonton, Boreal Institute for Northern Studies, University of Alberta, Occasional Publ no 1 (1973)

Gerell R (1969) Activity patterns of the mink, *Mustela vison*, Schreber, in southern Sweden. Oikos 20:451–460

Gerell R (1970) Home range and movements of the mink *Mustela vison* Schreber in southern Sweden. Oikos 21:160–173

Gillingham BJ (1984) Meal size and feeding rate in the least weasel *Mustela nivalis*. J Mammal 65:517–519

Gittleman JL (1986) Carnivore brain size, behavioral ecology, and phylogeny. J Mammal 67:23–36

Harestad AS, Bunnell FL (1979) Home range and body weight - a reevaluation. Ecology 60:389–404

Harlow HJ (1994) Trade-offs associated with the size and shape of American martens. In: Buskirk SW, Harestad AS, Raphael MG, Powell RA (eds) Martens, sables, and fishers biology and conservation. Cornell University Press, Ithaca, New York, pp 391–403

Hauptman TN (1979) Spatial and temporal distribution and feeding ecology of the pine marten. MS thesis, Idaho State University, Boise

Heffner RS, Heffner HE (1985) Hearing in mammals: the least weasel. J Mammal 66:745–755

Helldin JO, Lindström ER (1995) Late winter social activity in pine marten (*Martes martes*) false heat or dispersal? Ann Zool Fenn 32:145–149

Henderson PF, Springer PF, Adrian R (1974) The black footed ferret in South Dakota. Tech Bull no 4, South Dakota Dept Game, Fish, and Parks, Pierre, South Dakota

Henry SE, Raphael MG, Ruggiero LF (1990) Food caching and handling by marten. Great Basin Nat 50:381–383

Hinton HE, Dunn AMS (1967) Mongooses: their natural history and behaviour. Oliver and Boyd, Edinburgh

Iverson JA (1972) Basal energy metabolism of mustelids. J Comp Physiol 81:341–344

Jaksic FM (1982) Inadequacy of activity time as a niche difference: the case of diurnal and nocturnal raptors. Oecologia 52:171–175

Jedrzejewska B, Jedrzejewski W (1990) Antipredator behaviour of bank voles and prey choice of weasels - enclosure experiments. Ann Zool Fennici 27:321–328

Johnson SA (1984) Home range, movements, and habitat use of fishers in Wisconsin. MS thesis, University of Wisconsin, Stevens Point

Kanarek RB (1974) The energetics of meal patterns. PhD thesis, Rutgers Unversity

Karasov WH (1992) Daily energy expenditure and the cost of activity in mammals. Am Zool 32:238–248

Kavanau JL (1969) Influences of light on activity of the small mammals, *Peromyscus* spp., *Tamias striatus*, and *Mustela rixosa*. Experientia 25:208–209

Kavanau JL (1971) Locomotion and activity phasing of some medium sized mammals. J Mammal 52:386–403

Kavanau J, Ramos J (1972) Twilights and onset and cessation of carnivore activity. J Wildl Manage 36:653–657
Kavanau J, Ramos J (1975) Influence of light on activity and phasing of carnivores. Am Nat 109:391–418
Kavanau JL, Ramos J, Havenhill RM (1973) Compulsory regime and control of environment in animal behavior, II: light level preferences of carnivores. Behaviour 46:279–299
Kelly GM (1977) Fisher (*Martes pennanti*) biology in the White Mountain National Forest and adjacent areas. PhD thesis, University of Massachusetts, Amherst
Kemp TS (1982) Mammal-like reptiles and the origin of mammals. Academic Press, London
King CM (1975) The home range of the weasel *Mustela nivalis* in an English woodland. J Anim Ecol 44:639–668
King CM (1989) The advantages and disadvantages of small size to weasels, *Mustela* species. In: Gittleman JL (ed) Carnivore behavior, ecology, and evolution. Cornell University Press, Ithaca, New York, pp 302–334
King CM (1990) The natural history of weasels and stoats. Cornell University Press, Ithaca, New York
Kraft VA (1966) Influence of temperature on the activity of the ermine in winter. Zool Zhurnal 45:148–150. Translated in: King CM (ed) (1975) Biology of mustelids: some Soviet research. British Library, Boston Spa, Yorkshire, pp 104–107
Krohn WB, Zielinski WJ, Boone RB (1997) Relations among fishers, snow, and martens in California: results from small-scale spatial comparisons. In: Proulx G, Bryant HN, Woodard PM (eds) Martes: taxonomy, ecology, techniques, and management. The Provincial Museum of Alberta, Edmonton, Canada, pp 211–232
Leonard RD (1980) Winter activity and movements, winter diet, and breeding biology of the fisher in southeastern Manitoba. MS thesis, University of Manitoba, Winnipeg
LeMagnen J, Devos M (1970) Metabolic correlates of the meal onset in the free food intake of rats. Physiol Behav 5:805–814
Lode T (1995) Activity patterns of polecats *Mustela putorius* L in relation to food habits and prey activity. Ethology 100:295–308
Löhrl H (1972) *Martes foina* (Mustelidae) Sichern. Encycl Cinemat 1:6–12
Martin LD (1989) Fossil history of the terrestrial carnivora. In: Gittleman JL (ed) Carnivore behavior, ecology, and evolution. Cornell University Press, Ithaca, New York, pp 536–568
Martin SK (1987) The ecology of the pine marten *Martes americana* at Sagehen Creek, California. PhD thesis, University of California, Berkeley, California
Mason CE, Macdonald SM (1976) Otters: ecology and conservation. Cambridge University Press, Cambridge
McNab BK (1978) The evolution of endothermy in the phylogeny of mammals. Am Nat 112:1–21
Melquist WE, Whitman JS, Hornocker MG (1981) Resource partitioning and coexistence of sympatric mink and river otter populations. In: Chapman JA, Pursley D (eds) Worldwide furbearer conference proceedings, vol I. University of Maryland, Frostberg, Maryland, pp 187–220
Moors PJ (1977) Studies of the metabolism, food consumption and assimilation efficiency of a small carnivore, the weasel *Mustela nivalis* L. Oecologia 27:185–202
More G (1978) Ecological aspects of food selection in pine marten *Martes americana*. MS thesis, University of Alberta, Edmonton, Alberta
Murphy MJ (1985) Behavioural and sensory aspects of predation in mustelids. PhD thesis, University of Durham
Müller H (1970) Beiträge zur Biologie des Hermelins *Mustela erminea* L. Säugetierkdl Mitt 18: 293–380
Nams V (1981) Prey selection mechanisms of the ermine *Mustela erminea*. In: Chapman JA, Pursley D (eds) Worldwide furbearer conference proceedings, vol II. University of Maryland, Frostberg, Maryland, pp 861–882
Norrdahl K, Korpimäki E (1998) Does mobility or sex of voles affect risk of predation by mammalian predators? Ecology 79:226–232

Oksanen T, Oksanen L, Fretwell SD (1985) Surplus killing in the hunting strategy of small predators. Am Nat 126:328–346

Paragi TF, Arthur SM, Krohn WB (1994) Seasonal and circadian activity patterns of female fishers, *Martes pennanti*, with kits. Can Field Nat 100:52–57

Park O (1940) Nocturnalism – the development of a problem. Ecol Monogr 10:485–536

Peters RH (1983) The ecological implications of body size. Cambridge University Press, Cambridge

Pittendrigh CS (1960) Circadian rhythms and the circadian organization of living systems. Cold Spring Harbor Symp Quant Biol 25:159–184

Polderboer EB, Kuhn WL, Hendrickson GO (1941) Winter and spring habits of weasels in central Iowa. J Wildl Manage 5:115–119

Powell RA (1979) Ecological energetics and foraging strategies of the fisher *Martes pennanti*. J Anim Ecol 48:195–212

Powell RA, Brander RB (1977) Adaptations of fishers and porcupines to their predator prey system. In: Phillips RL, Jonkel C (eds) Proceedings of the 1975 predator symposium. Montana Forest and Conservation Experiment Station, University of Montana, Missoula, Montana, pp 46–53

Powell RA, Leonard RD (1983) Sexual dimorphism and energy expenditure for reproduction in female fisher *Martes pennanti*. Oikos 40:166–174

Powell RA, Clark TW, Richardson L, Forrest SC (1985) Black-footed ferret *Mustela nigripes* energy expenditure and prey requirements. Biol Conser 34:1–15

Powell RA, Zielinski WJ (1989) Mink response to ultrasound in the range emitted by prey. J Mammal 70:637–638

Price EO (1971) Effect of food deprivation on activity of the least weasel. J Mammal 52:636–640

Pulliainen E, Heikkinen H (1980) Behaviour of the pine marten (*Martes martes*) in a Finnish forest lapland in winter. Soumen Riista 28:30–36

Raine RM (1983) Winter habitat use and responses to snow of fisher *Martes pennanti* and marten *Martes americana* in southeastern Manitoba. Can J Zool 61:25–34

Raptor Group (1982) Timing of vole hunting in aerial predators. Mammal Rev 12:160–181

Rijnsdorp A, Daan S, Dijkstra C (1981) Hunting in the kestrel, *Falco tinnunculus*, and the adaptive significance of daily habits. Oecologia 50:391–406

Robitaille JF, Baron G (1987) Seasonal changes in the activity budget of captive ermine, *Mustela erminea* L. Can J Zool 65:2864–2871

Rood JP (1975) Population dynamics and food habits of the banded mongoose. E Africa Wildl J 13:89–111

Rosenzweig ML (1966) Community structure in sympatric carnivora. J Mammal 47:602–612

Rosevear PR (1974) The carnivores of West Africa. Br Mus Nat Hist, London

Sandell M (1985) Ecological energetics and optimum body size in male and female stoats *Mustela erminea*: predictions and test. In: Ecology and behaviour of the stoat *Mustela erminea*, a theory on delayed implantation. PhD thesis, University of Lund, Sweden

Sandell M (1986) Movement patterns of male stoats *Mustela erminea* during the mating season: differences in relation to social status. Oikos 47:63–70

Sandell M (1989) The mating tactics and spacing patterns of solitary carnivores. In: Gittleman JL (ed) Carnivore behavior, ecology, and evolution. Cornell University Press, Ithaca, New York, pp 164–182

Scholander PF, Walters V, Hock WR, Irving L (1950) Body insulation of some arctic and tropical mammals and birds. Biol Bull 99:225–236

Schoener T (1971) Theory of feeding strategies. Annu Rev Ecol Syst 2:369–403

Schoener T (1974) The compression hypothesis and temporal resource partitioning. Proc Natl Acad Sci USA 71:4169–4172

Schmidt-Nielsen K (1983) Animal physiology: adaptation and environment, 3rd edn. Cambridge University Press, Cambridge

Short HL (1961) Food habits of a captive least weasel. J Mammal 42:273–274

Stockman ER, Albers HE, Baum MJ (1985) Activity in the ferret: oestradiol effects and circadian rhythms. Anim Behav 33:150–154

Taylor SL (1993) Thermodynamics and energetics of resting site use by the American marten *Martes americana*. MS thesis, University of Wyoming, Laramie

Thompson ID, Colgan PW (1987) Numerical responses of martens to a food shortage in north-central Ontario. J Wildl Manage 51:824–835

Thompson ID, Colgan PW (1990) Prey choice by marten during a decline in prey abundance. Oecologia 83:443–451

Thompson ID, Colgan PW (1994) Marten activity in uncut and logged boreal forests in Ontario. J Wildl Manage 58:280–288

Walls GL (1963) The vertebrate eye and its adaptive radiation. Hafner, New York

Waser PM (1981) Sociality or territorial defense? The influence of resource renewal. Behav Ecol Sociobiol 8:231–237

Webb T III, Cushing EJ, Wright HE Jr (1983) Holocene changes in the vegetation of the midwest. In: Wright HE Jr (ed) Late-quaternary environments of the United States, the Holocene, vol 2. University of Minnesota Press, Minneapolis, pp 142–165

Williams BK, Nichols JD (1984) Optimal timing in biological processes. Am Nat 123:1–19

Worthen GL, Kilgore DL Jr (1981) Metabolic rate of pine marten in relation to air temperature. J Mammal 62:628–630

Wozencraft WC (1989) The phylogeny of the recent carnivora. In: Gittleman JL (ed) Carnivore behavior, ecology and evolution. Cornell University Press, Ithaca, New York, pp 495–535

Zielinski WJ (1986a) The effect of diel variation in food availability on the circadian activity of small carnivores. PhD thesis, North Carolina State University, Raleigh

Zielinski WJ (1986b) Circadian rhythms of small carnivores and the effect of restricted feeding on daily activity. Physiol Behav 38:613–620

Zielinski WJ (1988) The influence of daily variation in foraging cost on the activity of small carnivores. Anim Behav 36:239–249

Zielinski WJ, Spencer WD, Barrett RH (1983) Relationship between food habits and activity patterns of pine martens. J Mammal 64:387–396

6 Mongooses, Civets and Genets – Carnivores in Southern Latitudes

Francisco Palomares and Miguel Delibes

6.1 Introduction

Mongooses, civets and genets are small and medium-sized carnivores of Africa, the Middle East and southern Asia, mainly inhabiting tropical and subtropical regions. Their body weights range from 0.45 to 14 kg, although most of them are smaller than 3–4 kg. Two species (one mongoose and one genet) have well-established natural populations on the European continent. These populations probably result from escapes of tame individuals brought into Europe in historical times by humans. One mongoose species has also been introduced to Hawaii and the Antilles.

Traditionally, mongooses, civets and genets have all been included in one family, the Viverridae. However, some authorities now consider mongooses to represent a separate family, the Herpestidae. Here, we follow the recent classification using the list of species and the common and latin names proposed by Wozencraft (1989).

Morphologically, these carnivore species can be considered overall as the most diverse of the Carnivora, with civets being considered to represent most closely the ancestors of the modern Carnivora (Dücker 1965; Eisenberg 1981). There are arboreal (mostly civets) as well as terrestrial species (mostly mongooses), and genets present good adaptations for both terrestrial and arboreal life. Hence, ecologically, civets and genets show more diversification in trophic specialisation and substrate use than most of other carnivore families.

Information on activity patterns can include three related aspects: diel or circadian activity (i.e., time of the day when animals are active), daily activity time (i.e., percentage of the 24-h cycle that animals are active), and time budget (i.e., what types of activities animals perform throughout the day). Most of the available information for herpestids and viverrids relates to the first aspect, while the second one has been studied only for a few species, and there is hardly any information about the third aspect.

This review will be subdivided in two major parts: (1) we will review the current knowledge on diel activity of mongooses, civets and genets and its

Ecological Studies, Vol. 141
Halle/Stenseth (eds.) Activity Patterns in Small Mammals

relation to social structure, feeding habits, habitat use and body weight, and (2) we will then describe general activity patterns in these species with a special focus on the European genet (*Genetta genetta*), a nocturnal viverrid of 1.9 kg body weight, and the Egyptian mongoose (*Herpestes ichneumon*), a diurnal herpestid of 2.9 kg body weight. Both species were studied by us in the same Mediterranean study area using identical methods. Therefore, differences in activity patterns will reflect specific adaptations rather than different environmental contraints.

6.2 General Trends in Activity Patterns

6.2.1 Overview

Data on diel activity, social structure, diet, habitat, and body weight were obtained from previous reviews such as those of Gorman (1979), Smither (1983), Gittleman (1986), Rood (1986), Nowak (1991) and Rasa (1994), from published original papers, and from personal communications of some authors. Problems were encountered due to the lack of information for most species (see Rood 1986 for further comments regarding mongooses), so we made as few categories as possible within each ecological or social attribute.

For 14 herpestids and 13 viverrids no data on activity patterns were available. Two other viverrids known to be nocturnal (*Civettictis civetta* and *Viverra zibetha*) have large body weights (11 and 8 kg, respectively) and were not included in this review. The remaining 42 species were classified as primarily nocturnal or primarily diurnal (Table 6.1) although data for many species came from opportunistic observations and captures, and may, therefore, not accurately reflect the true diel activity. For instance, *Herpestes ichneumon* had been previously reported as diurnal, nocturnal or crepuscular throughout its range. However, data from two places where this species has been radio-tracked indicated that it is strictly diurnal (Delibes and Beltrán 1985; Palomares and Delibes 1992; Maddock and Perrin 1993). Detailed data on activity patterns only exist for a few species for which radio-tracking has been extensively used, and some that are gregarious and diurnal. Furthermore, data on time that animals spent active each day are lacking in some of these studies, although this information could have been reliably obtained.

Twenty-nine species were classified as solitary or gregarious (Table 6.1). This classification refers to the size of the most common foraging group. Solitary species include those classified by Rood (1986) or Rasa (1994) as living in pairs, living in a colony because several individuals rest together, and solitary. We adopted this definition because many of the species traditionally considered to be solitary are found in fact to have more complex social systems and inter-individual contacts than expected when studied thoroughly. Nevertheless, they are indeed better characterised as solitary than as

Table 6.1. Relation between social structure, diet, habitat selection, body weight and activity patterns in Herpestidae and Viverridae. Numbers refer to the number of species except for body weight and total time active. Note that information on social structure, diet, habitat, body weight and total time active is not available for all species where there is information on activity patterns

	Herpestidae		Viverridae	
	Diurnal (n=16)	Nocturnal (n=7)	Diurnal (n=0)	Nocturnal (n=19)
Social structure (n = 29)				
Solitary	11	6	0	8
Gregarious	4	0	0	0
Diet (n = 36)				
Insectivores	6	3	0	1
Small vertebrates	3	2	0	5
Omnivores	6	2	0	7
Frugivores	0	0	0	1
Habitat (n = 40)				
Dense	10	4	0	16
Open	5	3	0	0
Open/dense	1	0	0	1
Body weight (kg)				
Mean	1.21[a]	2.22[a]	–	2.15
SD	0.81	1.15	–	0.86
n	15	7	0	18
Total time active (percent)				
Mean	38.8[b]	45.0	–	46.3[b]
SD	5.76	–	–	9.40
n	5	1	0	6

[a] differences in body weight are significant ($Z = 2.21, p = 0.0267$, Mann-Witney U-test)
[b] differences in total active time are significant ($Z = 2.48, p = 0.0133$, Mann-Witney U-test)

gregarious (e.g. see Waser and Waser 1985 for *Ichneumia albicauda*; Palomares and Delibes 1993a for *Herpestes ichneumon*).

Food preferences were assigned to 36 species (Table 6.1) following the criteria proposed by Gittleman (1986). We distinguished between insectivores, frugivores, and small vertebrate/omnivore feeders. Most herpestids and viverrids are opportunistic feeders, thus for most of them the diet probably varies in different areas, making an accurate dietary classification difficult (e.g. see Smithers 1983 for different species and Palomares 1993, for *H. ichneumon*).

Habitat preferences were assigned to 40 species (Table 6.1) distinguishing between two broad categories, i.e., dense and open habitats. Habitat preference refers to the use of foraging habitat, even though it was not always possible to obtain these data with the exception of a few radio-tracked species and in easily observable species. For the remaining species the general characteristics of the habitat in which the species is observed or captured were used to classify it as preferently using dense or open habitats.

6.2.2 Relationships Between Activity Patterns and Ecological Attributes

Herpestids and viverrids exhibit clearly different general trends (Table 6.1). Diel activity is primarily nocturnal for all the viverrids for which data are available. Herpestids, however, are either primarily diurnal (16 species) or primarily nocturnal (7 species).

All viverrids are solitary, most of them are small vertebrate/omnivore feeders (86%) and inhabit dense habitats (94%). On the other hand, herpestids are mainly solitary (73%) when diurnal and exclusively solitary when nocturnal. Diurnal herpestids are either insectivore (40%) or small vertebrate/omnivore (60%) feeders, and similar feeding habits are found in the nocturnal species (insectivores 43%, small vertebrate/omnivores 57%). Both diurnal and nocturnal species inhabit dense vegetation more often (63% and 57% of the species, respectively) than open habitats.

Considering the possible combinations of the above ecological and behavioural factors, 54% of diurnal herpestids (7 out of 13 species) are mainly solitary, small vertebrate/omnivore feeders and inhabit dense habitats, three species are gregarious, insectivorous and inhabit open habitats, two species are solitary, insectivorous and inhabit dense habitats, and one species is solitary, insectivorous and inhabits open habitats (Fig. 6.1). Nocturnal herpestids show less variation. Three out of six species are solitary, insectivorous and inhabit open habitats, and the other three species are solitary, small vertebrate/omnivore feeders and inhabit dense habitats (Fig. 6.1). For viverrids, six out of seven species are solitary, small vertebrate/omnivore feeders and

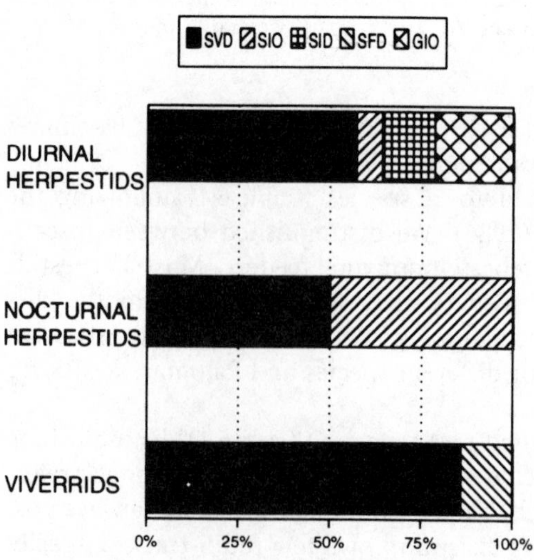

Fig. 6.1. Combinations of the ecological and behavioural attributes found in diurnal herpestids (13 species), nocturnal herpestids (6 species), and viverrids (7 species, all noctural). *S* solitary, *G* gregarious, *V* small vertebrate/omnivores, *I* insectivores, *F* fruguivores, *D* dense habitat, *O* open habitats. Diurnal herpestids: SVD: *Galidia elegans, Herpestes ichneumon, Herpestes pulverulentus, Herpestes sanguineus, Herpestes urva, Herpestes naso, Herpestes auropunctatus*; SIO: *Cynictis penicellata*; SID: *Mungotictis decemlineata, Salanoia concolor*; GIO: *Helogale parvula, Mungos mungo, Suricata suricata.* Nocturnal herpestids: SVD: *Galidictis fasciata, Galidictis grandidieri, Atilax palodinosus*; SIO: *Ichneumia albicauda, Rhynchogale melleri, Paracynictis selousi.* Viverrids: SVD: *Genetta genetta, Genetta tigrina, Genetta maculata, Prionodon lisang, Prionodon pardicolor, Paguma larvata*; SFD: *Nandinia binotata*

inhabit dense habitats, and one species is solitary, frugivorous and inhabits dense habitats (Fig. 6.1).

Interestingly, diurnal herpestids are significantly smaller than nocturnal ones (Table 6.1), which is unusual within the Carnivora (Zielinski, Chap. 5). Nocturnal herpestids and viverrids are similar in body weight (Table 6.1). Diurnal herpestids spend significantly less time per day in activity than nocturnal viverrids (Table 6.1). However, differences were not significant when we corrected for allometry in body weight (i.e., percent activity/$W^{0.75}$). The only nocturnal herpestid with available information on this aspect spends a similar time being active as nocturnal viverrids (Table 6.1). Food habits of the species seem to have no relation to time spent being active each day, since diurnal vertebrate/omnivore feeders and diurnal insectivores have similar activity times (mean = 27% and 30% for two and three species, respectively). Nocturnal species are either small vertebrate/omnivore feeders (all viverrids) or insectivores (the only nocturnal herpestid with data available).

6.3 Activity Patterns of Egyptian Mongooses and European Genets

6.3.1 Data Collection

Data on activity patterns of *G. genetta* and *H. ichneumon* were obtained in the Doñana National Park, southwestern Spain, between 1985 and 1989 during a total of 36 months following the methods described in Palomares and Delibes (1988, 1991a, 1992, 1993b, 1994). Thirty mongooses and 12 genets were equipped with radio-collars containing tip switches to monitor activity. Each individual was located once or twice every day. Additionally, mongooses and genets were radio-tracked intensively during the full period of activity for 82 days (3 adult males, 12 adult females, 3 young females and 1 young male) and 51 days (5 adult males, 2 adult females, 2 young females and 1 young male), respectively. Daily activity patterns were obtained by tabulating the percentage of activity observations during each 1-h interval, using data from the periods of intensive tracking. From these data, the time spent being active every day (percentage of active locations with regard to the 24-h period) was estimated. This last method to compute daily activity time was considered reliable because we also knew the actual time in minutes spent by mongooses in activity as recorded from the tip switch pulses of the radio-collars. Time budgets (eating/foraging and travelling activities) of 13 mongooses (one adult male, 9 adult females and 3 young females) could also be estimated by recording speed of movement during 15-min intervals in 40 intensive tracking periods (Palomares and Delibes 1993b).

Most mongooses and genets were studied simultaneously and in widely overlapping home ranges, thus enviromental contraints for activity were identical for both species. Social structure, habitat use and feeding habits of both species were also studied (Palomares 1993; Palomares and Delibes 1991b, 1993a, c, 1994).

6.3.2 Start and End of Activity

Egyptian mongooses were strictly diurnal, starting activity on average 129-min (SD = 65 min) after sunrise and ending activity on average 51 min (SD = 73 min) before sunset. On the other hand, with the exception of one young female, European genets were exclusively nocturnal. Activity started after sunset (most often in the second following hour, 70.5% of occasions) and was terminated before sunrise (most often one (31.8% of occasions) or two hours (31.8% of occasions) preceding sunrise). The one young female genet that showed some diurnal activity was the only individual inhabiting a continuous patch of dense shrub vegetation (100% ground cover).

Similar patterns of starting and ending the activity has been found for three diurnal (*Mungos mungo*, *Cynictis penicillata*, *Herpestes sanguineus*) and two nocturnal (*Atilax paludinosus*, *Ichneumia albicauda*) species studied (Rood 1975; Waser and Waser 1985; Cavallini 1993a; Maddock and Perrin 1993). In spite of exhibiting a restricted period of activity, both Egyptian mongooses and European genets may exceptionally be active out of the normal period of activity. This is also the case in other species, for instance *Herpestes auropunctatus* or *Genetta maculata* (Buskirk et al. 1990; Fuller et al. 1990).

Onset of activity showed a regular pattern throughout the year in both species as it was correlated with sunrise and sunset for mongooses and genets, respectively. For three intensively sampled mongooses, 25% of the variation in the onset of activity was explained by the combined effects of sunrise time and maximum and minimum daily temperatures; nevertheless, partial regression showed that only the effect of sunrise time was statistically significant (Palomares and Delibes 1992). Mongooses started activity later when sunrise was also later. Cavallini (1993a) found that the start of activity in *Cynictis penicillata* was positively correlated with sunrise time and mist hours (i.e., mist presence delayed start of activity).

The cessation of activity was less predictable in both species, and other factors than sunset and sunrise seemed to also be important. Again, for three intensively sampled mongooses, the combined effects of sunset time, maximum and minimum temperature, total daily activity time, and onset activity time explained 38% of the variation in the time of the end of activity, although only total activity time and onset of activity time were statistically significant (Palomares and Delibes 1992). Mongooses ended activity later when they started activity later and when they were active longer. Cavallini (1993a) found that cessation of activity in *Cynictis penicillata* was positively correlated with sunset time and maximum temperature, and negatively correlated with windspeed. Additionally, on seven occasions mongooses ended activity earlier than usual shortly after consuming large prey. On one occasion a genet ended activity earlier than ususal after capturing and feeding on a rabbit (*Oryctolagus cuniculus*).

6.3.3 Total Activity Time per 24-h Day

The time per day spent in activity was 26% for Egyptian mongooses (n = 18 individuals, SD = 6.5%) and 29% for genets (n = 10 individuals, SD = 5.0%). Young Egyptian mongooses spent significantly more time in activity than adults, but differences between age-classes were not significant in European genets. Both species had resting periods ('siestas') within their activity phases. In mongooses, siestas made up almost 3 h (mean = 161 min, SD = 89 min), distributed over five periods (mean = 5.4, SD = 2.3). Siestas were longer during the summer, probably due to heat avoidance. Temperatures in the study area exceed 40 °C in summer.

Mungos mungo also have siestas during the hottest time of the day (Rood 1975), and Cavallini (1993a) found a positive correlation between the duration of siestas and maximun tempature in *Cynictis penicillata*, which means that animals rest more during daylight when it is hotter. For this last species and *Helogale parvula*, Rasa (pers. comm.) also found that animals have a long siesta in the summer, a shorter one in spring and autumn, and no siesta in winter.

The number of individuals moving together (i.e., the size of the foraging group) affected the daily time spent in activity by Egyptian mongooses. It was shorter when individuals foraged by themselves (mean = 24%, SD = 4.9%, n = 14 individuals) than when mongooses foraged in groups of two or more individuals (mean = 29%, SD = 6.7%, n = 12; Z = 1.96, $p = 0.0499$, Mann-Whitney U-test). There was no evidence for genets ever moving in groups. Comparing the daily activity of single mongooses and genets, the former spent a significantly shorter time per day in activity than the latter (Z = 2.15, $p = 0.0318$, Mann-Whitney U-test).

6.3.4 Time Budget

Mongooses spent most of their activity time in foraging/eating activities (77%), whereas the major part of the remaining time represented travelling (or moving between foraging areas). Social interactions were relatively frequent only during mating. From several observations of the radio-tracked individuals other activities seemed to have little importance.

Foraging is also the predominant activity in other diurnal species studied, such as *Mungos mungo*, *Helogale parvula* and *Cynictis penicilata* (Rood 1975;

Rasa 1989; Cavallini 1993a). Grooming and play can take a few minutes after leaving and before entering the den (Rood 1975).

6.3.5 Relationships Between Predators and Prey Activity

The diel activity patterns of Egyptian mongooses and their main prey do not coincide, but, as a rule, they do for European genets. Rabbits and amphibians comprise more than 88% of the biomass ingested by Egyptian mongooses in southwestern Spain. Among the amphibians, *Pelobates cultripes* represents 96.3% of the prey items. Rabbit activity is mainly crepuscular (Kufner 1986), whereas *P. cultripes* is nocturnal (Díaz-Paniagua and Rivas 1987). Mongooses capture prey mainly in nest or dens. On the other hand, small mammals and birds are the main prey of European genets. The former prey (mainly *Apodemus sylvaticus*; 50.4% of prey items) represents more than 74% of the biomass ingested by genets, the latter (mainly small forest passerines; 71.7% of prey items) 12% of the biomass. Activity of *A. sylvaticus* is nocturnal (Kufner 1986); the small passerines are diurnal, but nest and roost in the habitats that are used by genets during foraging.

6.4 Discussion and Conclusions

Herpestids are both diurnal and nocturnal, whereas viverrids are only nocturnal. Furthermore, within any activity category herpestids present more diversity in the possible combinations of the ecological and social attributes than viverrids. The only apparently clear trends in herpestids are that gregarious, insectivore and open habitat species, as well as solitary, insectivore and dense habitat species are all diurnal. Nevertheless, there is one exception to be specifically mentioned: *Cynictis penicillata.* This species is diurnal, solitary for foraging, mainly insectivorous and uses open habitats (Du Toit 1980; Earlé 1981; Mills et al. 1984; MacDonald and Nel 1986; Cavallini 1993a, b; Rasa 1994). Furthermore, it is small (0.86 kg). It remains as an interesting question as to why this species is diurnal and forages alone since all other small species are gregarious if diurnal or solitary if nocturnal. Small diurnal species using open habitats have a high risk of predation that can be substantially decreased through grouping (Rasa 1989). Also, feeding mainly on insects that usually have high rates of renewal, favours gregariousness (Waser 1981). Although quantitative data are lacking, nocturnality might also decrease the risk of predation. The case of *C. penicillata* may respond to a situation of decreased foraging efficiency during the night, or to high intraspecific scramble competition when foraging in groups during the day. Egyptian mongooses foraging in a group also spend more time being active than single individuals.

Probably the most remarkable aspect of diel activity in the species reviewed here is that most of them seem to be either exclusively diurnal or exclusively nocturnal. The species studied start activity after sunrise or sunset for diurnal and nocturnal species, respectively. Cessation of activity is also before sunset and sunrise, and only occasionally is activity prolonged into the following period. Anatomical characteristics of the eye might decisively influence the diel activity of viverrids and herpestids, since the retinas of diurnal species are rich in cones which allow for colour vision whereas those of nocturnal species are mainly composed of rods which respond to lower illumination levels than cones (Dücker 1965).

Age, sex and season do not seem to influence the diel activity markedly, as shown by our data from Egyptian mongooses and European genets. Nevertheless, there are too few detailed data from other species to reach a reliable conclusion.

Low levels of activity observed out of the normal period in some species might well be due to disturbance by the observer near resting places as argued in Palomares and Delibes (1994), or to comfort movements in the resting places. Another explanation might be that some populations or individuals have a lower risk of predation due to cover provided by specific habitats (Palomares and Delibes 1988).

As a rule, the onset of activity seems to be triggered by sunrise or sunset for diurnal and noctural species respectively, and subsequently by maximum and minimum daily temperatures, as these co-vary with sunrise and sunset times throughout the year. Thus, onset of activity is quite predictable. However, cessation of activity is less precise, although it also often correlates with sunset or sunrise times. Additionally, other factors not clearly determined yet also seem to influence the end of the activity period. Foraging success, for example, might be important, but unfortunately this is difficult to assess in such secretive species. Diurnal and easily observable species might be used to test this hypothesis.

Like cessation of activity, the daily activity time can be quite variable between individuals of the same species, and between different days for the same individual. Young individuals seem to spend more time being active, probably due to a lower foraging efficiency than adults. Other aspects that might also influence the daily amount of activity are foraging success, grouping or territory marking and patrolling.

Southern lattitudes are characterised by hot and, in many places, dry weather. Hot and dry weather should promote behaviour to avoid water loss and high body temperatures in mammals (Kuntzsch and Nel 1990). Of course, this will be especially true for diurnal species. Diurnal species did not totally avoid foraging during daylight due to hot temperatures, but rather tend to increase the duration of the midday siestas when the weather is hotter. In *H. ichneumon* it has also been observed that animals more often use cooler resting places, such as underground dens, for siestas in the summer months than during other times of the year or than during the nocturnal continuous resting periods of the night (Palomares and Delibes 1993d).

Rain can also influence activity. Thus, it has been noted that rain can delay the start of activity or cease it. This response to rain might be to conserve body heat since most mongoose species lack an undercoat. Additionally, foraging efficiency might also be considerably decreased during rain.

Data on the daily times assigned to activity are scarce for the families reviewed. Nevertheless, the available information indicates that diurnal species spend shorter times being active than nocturnal ones, irrespective of food habits. This suggests that diurnal species might be more successful in acquiring food since most of the active time is spent foraging. However, diurnal species are also smaller, therefore the observed trend might be a result of lower metabolic needs rather than a higher foraging efficiency. In fact, when we corrected for allometry in body weight, the differences disappeared.

Nevertheless, data from *G. genetta* and *H. ichneumon* in southwestern Spain suggest that body weight might not be the reason for the observed pattern. *G. genetta* spend more time being active than *H. ichneumon* in spite of being smaller (note that correcting here for allometry will increase the differences because *G. genetta* is smaller than *H. ichneumon*). It may be important that these two species exhibit different hunting modes: detecting prey by hearing and then stalking in the genet, versus detecting prey by sight and odour and then harvesting in the mongoose.

These two general modes of acquiring prey can be extended to viverrids and herpestids in general (Dücker 1965). Thus, we propose that diurnal species can in fact have shorter daily activity times through a more efficient hunting mode adapted to daylight conditions. A clear prediction from this hypothesis would be that nocturnal herpestids with the 'typical' hunting mode of the family should spend more time being active than diurnal ones. The only nocturnal herpestid with some information (*Ichneumia albicauda*; Ikeda et al. 1982, 1983) indeed seemed to spend more time being active than the diurnal ones. However, the prediction cannot be conclusively tested with the currently available data.

Acknowledgements. Field work was supported by Dirección General de Investigación Científica y Técnica (projects 944, PB87-0405, PB90-1018 and PB94-0480). We are indebted to O.A.E. Rasa, P. Cavallini, P. Waser and M. Hwang who kindly reviewed a previous draft of this chapter and provided us with key references and unpublished information.

References

Buskirk SW, Delin W, Cleveland A (1990) Diel activity patterns of two female small indian mongooses (*Herpestes auropunctatus*) in relation to weather. Zool Res 11:355–357

Cavallini P (1993a) Activity of the yellow mongoose *Cynictis penicillata* in a coastal area. Z Säugetierkd 58:281–285

Cavallini P (1993b) Spatial organization of the yellow mongoose *Cynictis penicillata* in a coastal area. Ethol Ecol Evol 5:501–509

Delibes M, Beltrán JF (1985) Activity, daily movements and home range of an Ichneumon or Egyptian mongoose (*Herpestes ichneumon*) in southern Spain. J Zool Lond 207:610–613

Díaz-Paniagua C, Rivas R (1987) Datos sobre la actividad de anfibios y pequeños reptiles de Doñana (Huelva, España). Mediterranea 9:15-27

Dücker G (1965) Das Verhalten der Schleichkatzen (Viverridae). Handb Zool Berl 8(38):1–48

Du Toit CF (1980) The yellow mongoose *Cynictis penicillata* and other small carnivores in the Mountain Zebra National Park. Koedoe 23:179–184

Earlé RA (1981) Aspects of the social and feeding behaviour of the yellow mongoose *Cynictis penicillata* (G Cuvier). Mammalia 45:143–152

Eisenberg JF (1981) The mammalian radiations. An analysis of trends in evolution, adaptation, and behavior. University of Chicago Press, Chicago

Fuller TK, Biknevicius AR, Kat PW (1990) Movements and behavior of large spotted genets (*Genetta maculata* Gray 1830) near Elmenteita, Kenya (Mammalia, Viverridae). Trop Zool 3:13–19

Gittleman JL (1986) Carnivore brain size, behavioral ecology and phylogeny. J Mamm 67:23–36

Gorman ML (1979) Dispersion and foraging of small india mongooses, *Herpestes auropunctatus* (Carnivora: Viverridae) relative to the evolution of social viverrids. J Zool Lond 187:65–73

Ikeda H, Ono Y, Baba M, Doi T, Iwamoto T (1982) Ranging and activity patterns of three nocturnal viverrids in Omo National Park, Ethiopia. Afr J Ecol 20:179–186

Ikeda, H, Izawa M, Baba M,Takeishi M, Doi T, Ono Y (1983) Range size and activity pattern of three nocturnal carnivores in Ethiopia by radio-telemetry. J Ethol 1:109–111

Kufner MB (1986) Tamaño, actividad, densidad relativa, y preferencias de hábitat de los pequeños y medianos mamíferos de Doñana, como factores condicionantes de su tasa de predación. PhD thesis, University of Madrid

Kuntzsch V, Nel AJJ (1990) Possible thermoregulatory behaviour in *Giraffa camelopardalis*. Z Säugetierkd 55:60–62

MacDonald J, Nel JAJ (1986) Comparative diets of sympatric small carnivores. S Afr J Wildl Res 16:115–121

Maddock AH, Perrin MR (1993) Spatial and temporal ecology of an assemblage of viverrids in Natal, South Africa. J Zool Lond 229: 277–287

Mills MGL, Nel JAJ, Bothma JDP (1984) Notes on some smaller carnivores from the Kalahari Gemsbok National Park. Koedoe Suppl 1984:221–227

Nowak RM (1991) Walker's mammals of the world. Johns Hopkins University Press, Baltimore

Palomares F (1993) Opportunistic feeding of the Egyptian mongoose, *Herpestes ichneumon*, (L.) in southwestern Spain. Rev Ecol (Terre Vie) 48:295–304

Palomares F, Delibes M (1988) Time and space use by two common genets (*Genetta genetta*) in the Doñana National Park, Spain. J Mamm 69:635–637

Palomares F, Delibes M (1991a) Assessing three methods to estimate daily activity patterns in radio-tracked mongooses. J Wildl Manage 55:698–700

Palomares F, Delibes M (1991b) Alimentación del meloncillo *Herpestes ichneumon* y de la gineta *Genetta genetta* en la Reserva Biológica de Doñana, S.O. de la Península Ibérica. Doñana Acta Vert 18:5–20

Palomares F, Delibes M (1992) Circadian activity patterns of free-ranging large gray mongooses, *Herpestes ichneumon*, in southwestern Spain. J Mamm 73:173–177

Palomares F, Delibes M (1993a) Social organization in the Egyptian mongoose: group size, spatial behaviour and inter-individual contacts in adults. Anim Behav 45:917–925
Palomares F, Delibes M (1993b) Determining activity types and budgets from movement speed of radio-marked mongooses. J Wildl Manage 57:164–167
Palomares F, Delibes M (1993c) Key habitats for Egyptian mongooses in Doñana National Park, south-western Spain. J Appl Ecol 30:752–758
Palomares F, Delibes M (1993d) Resting ecology and behaviour of Egyptian mongooses (*Herpestes ichneumon*) in southwestern Spain. J Zool Lond 230:557–566
Palomares F, Delibes M (1994) Spatio-temporal ecology and behavior of European genets in southwestern Spain. J Mamm 75:714–724
Rasa OAE (1989) The costs and effectiveness of vigilance behaviour in the Dwarf Mongoose: implications for fitness and optimal group size. Ethol Ecol Evol 1:265–282
Rasa OAE (1994) Altruistic infant care or infanticide: the dwarf mongooses' dilemma. In: Parmigiani S, vom Saal FS (eds) Infanticide and parental care. Harwood Academic Publishers, Chur, Switzerland, pp 301–320
Rood JP (1975) Population dynamics and food habits of the banded mongoose. E Afr Wildl J 13:89–111
Rood JP (1986) Ecology and social evolution in the mongooses. In: Rubenstein DI, Wrangham RW (eds) Ecological aspects of social evolution. Birds and mammals. Princeton University Press, Princeton, pp 131–152
Smithers RHN (1983) The mammals of the Southern African subregion. University of Pretoria, Pretoria
Waser PM (1981) Sociality or territorial defense? The influence of resource renewal. Behav Ecol Sociobiol 8:231–237
Waser PM, Waser MS (1985) *Ichneumia albicauda* and the evolution of viverrid gregariousness. Z Tierpsychol 68:137–151
Wozencraft WC (1989) Classification of the recent Carnivora. In: Gittleman JL (ed) Carnivore ecology, behaviour, and evolution. Chapman and Hall, London, pp 569–593

7 Squirrels – Medium-Sized Granivores in Woodland Habitats

Luc A. Wauters

7.1 Introduction

All Holarctic tree squirrels (Sciuridae) are diurnal granivores (Reynolds 1985), offering the possibility of studying their activity pattern and behaviour through direct field observations. Some species, however, are not always easy to detect and the use of radio-tracking is necessary to obtain unbiased data of the activity pattern. Their diurnal activity is in contrast to some members of the nocturnal dormice family (Gliridae), which occupy similar habitats and have similar feeding habits (Schulze 1970; Bright and Morris 1991, 1993). Being diurnal might be related to the way squirrels move around in trees, often jumping between branches or from one tree to another, and to limiting heat loss for thermoregulation during winter, when cold-stress is highest at night (e.g. Pulliainen 1973). Dormice, in contrast, walk more through the trees and shrubs instead of jumping, and hibernate for up to 8 months per year (Schulze 1970).

Detailed information on the activity pattern from captive or natural populations was found for only three squirrel species: the Eurasian red squirrel (*Sciurus vulgaris*, Tonkin 1983; Wauters and Dhondt 1987; Wauters et al. 1992), the eastern grey squirrel (*Sciurus carolinensis*, Thompson 1977) and the pine squirrel (*Tamiasciurus hudsonicus*, Pauls 1978, 1979; Ferron et al. 1986).

7.2 The Daily Activity Pattern

7.2.1 General Pattern and Seasonal Variation

Unaided observations, studies in captivity and stomach contents analyses revealed a seasonal shift in the activity (or foraging) pattern, i.e., from a short, unimodal pattern in winter, to an intermediate pattern in spring and autumn, to a long, bimodal pattern in summer (Shorten 1962; Pulliainen

Ecological Studies, Vol. 141
Halle/Stenseth (eds.) Activity Patterns in Small Mammals

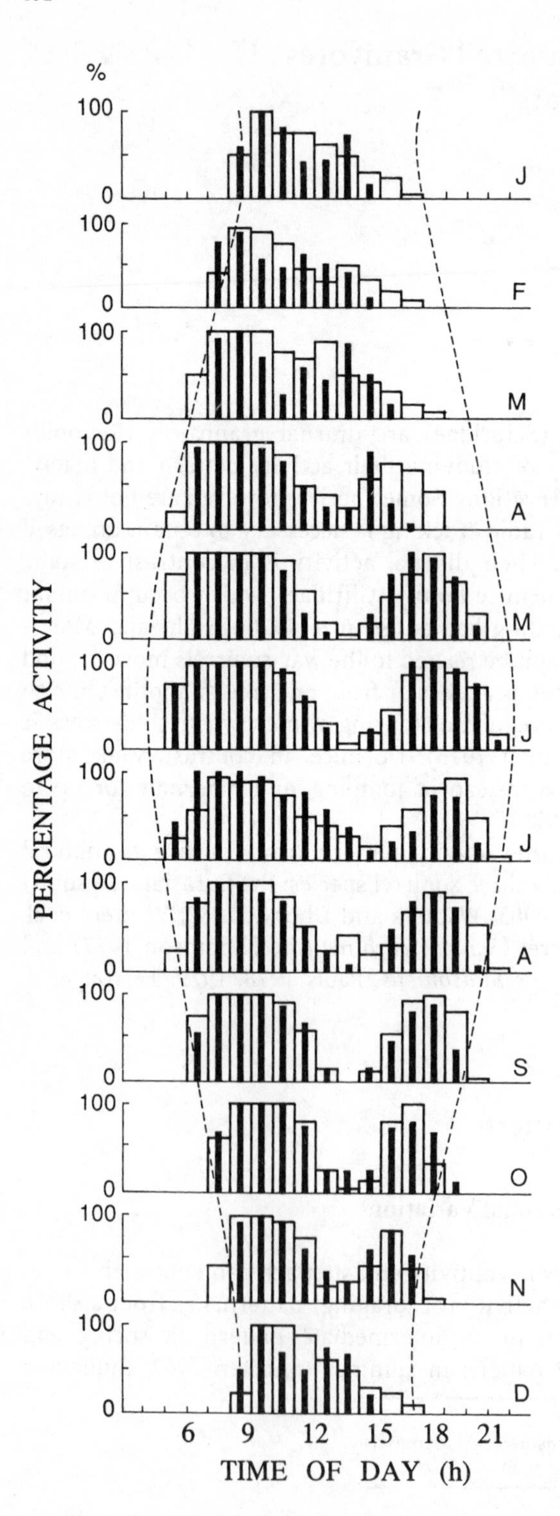

Fig. 7.1. The activity pattern of red squirrels (*Sciurus vulgaris*) over a whole year (1986). Coniferous area, *open bars*; deciduous area, *small black bars*. Times of sunrise and sunset are indicated by the *dashed lines*. Percentage activity per hour = (time observed active/time active + time not active) × 100. (From Wauters et al. 1992)

1973; Degn 1974; Zwahlen 1975; Thompson 1977; Tittensor 1977; Pauls 1978; Grönwall 1982). Intensive radio-surveillance of Eurasian red squirrels, which consisted of locating a radio-tagged squirrel and then following and observing it for about 1 h, carried out throughout 24-h cycles, confirmed that they were exclusively diurnal (Tonkin 1983; Wauters and Dhondt 1987; Wauters et al. 1992). For most species, the start of activity was closely related to sunrise (Zwahlen 1975; Pauls 1978; Wauters and Dhondt 1987) and the activity pattern varied considerably over the year (Fig. 7.1). In autumn/winter (October-April), when day length is shortest, red squirrels became active soon after first light, thus somewhat before sunrise. During the long summer months there was more than 1 h difference between daybreak and the onset of activity (Fig. 7.1). Free-ranging and captive pine squirrels shifted their activity in winter towards midday, probably as an adaptation to extremely low air temperatures at northern latitudes (Smith 1968; Pauls 1978). Most tree squirrels have short active periods in winter (ca. 4 to 6 h per day) and expand their afternoon activity when days become longer and temperature increases. Thus, the time spent being active is closely correlated with photoperiod and mean daily (or monthly) air temperature (Doebel and McGinnes 1974; Pauls 1978; Wauters et al. 1992).

The time spent in the nest, and the number of nests used by a red squirrel changed seasonally. Red squirrels used fewer dreys in autumn/winter (October–March) than in spring/summer (April-September), which coincided with an increase in home range size during the warmer seasons (Wauters and Dhondt 1990). Breeding behaviour also affected nest use. During lactation, females used fewer nests than outside the breeding season when they were not nursing young (Wauters and Dhondt 1990).

7.2.2 Geographical Variation

The activity pattern described above occurs in most tree squirrel populations throughout Europe and North America (Ognev 1940 in the Soviet Union; Saint-Girons 1966 in France; Degn 1974 in Denmark; Purroy and Rey 1974 in Spain; Pauls 1978, 1979 in Canada; Tonkin 1983 in Great Britain; Wauters and Dhondt 1987 in Belgium), although some geographical variation exists. In the west of France (Saint-Girons 1966) squirrels had long active periods with mild winter weather. On the other hand, in Canada, Pauls (1978, 1979) found reduced locomotion, increased time spent in the nest (accompanied by a decrease in body temperature), and a shift to warmer hours of the day under very cold winter conditions. He argued that when energy requirements for thermoregulation are extremely high, temperature modulated behaviour can (partly) uncouple a pine squirrel's daily energy requirements from air temperature. When temperatures become very high in summer, squirrels might be forced to reduce activity to avoid overheating, as suggested by the decrease in activity and strong increase in resting alert behaviour of pine squirrels at temperatures above 20 °C (Ferron et al. 1986).

Thus, seasonal changes in activity patterns of tree squirrels seem to result from a balance between resting and sleeping in the nest to conserve energy (winter), to avoid overheating (summer) or to digest food that is bulky and has a low energy content (spring/summer), and foraging and feeding to obtain sufficient energy (Tonkin 1983; Reynolds 1985; Ferron et al. 1986; Gurnell 1987; Wauters et al. 1992). Consequently, these seasonal changes in activity are accompanied by changes in the squirrel's diet: from late summer to early spring high-energy food resources (seeds of conifers and broadleaves) are their main food, while during the rest of the year a variety of vegetal and animal matter of poorer energy content is included in the diet (Möller 1983; Wauters 1986; Gurnell 1987; Wauters et al. 1992). The relation between the food situation, thermoregulation constraints and activity pattern are summarised in Table 7.1.

Hence, although the 24-h activity pattern of squirrels appears to have an endogenous basis linked to photoperiod, environmental factors, such as temperature, food availability and food quality strongly affect the annual variation of the time spent active per day (Pulliainen 1973; Purroy and Rey 1974; Pauls 1978; Reynolds 1985; Wauters and Dhondt 1987; Wauters et al. 1992).

7.3 Foraging Behaviour and the Activity Pattern

Apart from temperature-dependent variation in activity patterns of tree squirrels on a geographical scale, differences in habitat quality (food availability, population size, predator density, and others) might affect the intensity and timing of activity on a local scale. Comparative studies can reveal to what extent differences in habitat quality affect the squirrel's activity pattern and foraging behaviour.

The relationship, described for red squirrels in the following paragraph, between the seasonal shifts in foraging behaviour, caused by seasonal variation in the abundance of different food types, and the squirrel's activity pattern, is a general phenomenon among species of tree squirrels (see also Thompson 1977; Tamura and Miyashita 1984, for *Callosciurus erythraeus thaiwanensis* introduced to Japan; Ferron et al. 1986; States et al. 1988).

7.3.1 Foraging and Activity of Red Squirrels in Coniferous Woods

7.3.1.1 Winter (December-February)

During the short active period (4–5 h per day), most time was spent foraging on energy rich conifer seeds (> 90% of diet, Wauters and Dhondt 1987). Needing about 100 pine cones per day, at a feeding rate of 30 cones/h, the energy demands were fulfilled after only a few hours of feeding (Wauters et al. 1992). To avoid thermoregulatory heat loss with low temperatures, squirrels retreated into the drey shortly after midday, where the temperature (when the squirrel is in the drey) was about 10 to 20 °C above air temperature (Pulliainen 1973). Moreover, tree squirrels decreased body temperature while resting (Golightly and Ohmart 1978; Pauls 1979), making hiding in a nest an important behavioural response to decrease the energy costs of thermoregulation at low air temperatures.

7.3.1.2 Spring (March-May)

A decrease in the abundance of primary food items, as shown by an increase in the ratio of searching to handling time in Scots and Corsican pine (on average 0.10 in March, 0.29 in April and 0.85 in May; Wauters and Dhondt 1987), forced the squirrels: (1) to spend more time travelling between good feeding sites, and (2) to increase the diversity of their diet, feeding on buds, male flowers and shoots of pines, oak flowers, and cached cones. This shift towards food of lower energy content and the increased travelling time (Wauters and Dhondt 1987) decreased the rate of energy intake and increased energy expenditure, forcing the squirrels to spend more hours being active per day. Travelling mainly included movements between feeding sites or towards dreys. A negative correlation between the monthly percentages of foraging and travelling time suggested that the proportion of time available to squirrels for searching and handling food was influenced by the proportion of travelling time (Wauters et al. 1992). Squirrels gradually expanded their active period throughout the afternoon. Thus, by the end of March, the activity pattern became bimodal.

7.3.1.3 Summer (June–August)

All squirrels had a bimodal activity pattern: they were active in the morning, rested in a drey for about 3–4 h around noon, and had a second active bout later in the afternoon. The resting phase around noon is probably related to thermoregulation and feeding physiology. Squirrels clearly avoided activity during the warmest hours of the day, which they spent inactive and most of the time in a nest (Wauters et al. 1992). In Spain, where summers are much warmer, activity during hot weather is even more reduced, with squirrels being active only a few hours after sunrise and late in the evening, suggesting that summer resting is an important behavioural response to avoid overheating (J. Vilar, pers. comm.; Table 7.1). Squirrels also regularly interrupted foraging, remaining absolutely motionless, often spread out on a branch, to cool off. This so-called 'dozing' is considered another behavioural response to avoid overheating.

The end of morning activity in summer might also be related to the amount of food intake. After several hours of feeding, the contents of the stomach reach capacity, and a break in activity is necessary to digest the bulky summer food (Tonkin 1983; Wauters and Dhondt 1987; Wauters et al. 1992; see also Table 7.1). Squirrels groomed themselves more often than in winter/spring to get rid of the increasing number of skin parasites. Finally,

Table 7.1. The effects of the food situation (rate of energy-intake) and thermoregulation constraints in winter and spring-summer on the daily activity pattern of tree squirrels. In winter, thermoregulation constraints increase with decreasing temperature; in summer they increase with increasing temperature

Thermoregulation constraints	Food situation	
	High rate of energy intake	Low rate of energy intake
Winter		
Very cold	Short bimodal, delayed start of activity	Short bimodal, delayed start of activity
Cold	Short bimodal	Short unimodal
Warm	Longer bimodal	Longer bimodal, more afternoon activity
Summer		
Cool or temperate	Bimodal, less time active per day	Bimodal, more time active per day
Warm	Bimodal, long resting phase and shift of the active periods to coolest part of the day	

the time allocated to travelling between feeding sites increased: in June they spent some 108 min per day travelling against only 35 min in January (Wauters and Dhondt 1987; Wauters et al. 1992).

7.3.1.4 Autumn (September–November)

In autumn the afternoon activity peak became less important, but the absolute time spent active stayed around 6.5 h. During this time of the year high-energy food was most abundant, which was reflected in the decrease in travelling time and the increase in feeding time. From September onwards, beechnuts – when available – and to a lesser extent acorns formed an important part of the diet (Wauters and Dhondt 1987). The majority of time, however, was spent foraging on pine seeds. Feeding primarily on tree seeds, squirrels gained weight and put on some fat reserves. During this period they reached their maximum body mass (Wauters and Dhondt 1989).

Fungi, although a very nutritious and often preferred food of several tree squirrel species (Grönwall and Pehrson 1984; States et al. 1988), were not an important part of the diet in Belgian red squirrel populations (Wauters and Dhondt 1987). However, the occurrence of large numbers of edible fungi in conifer forests throughout the range of Holarctic tree squirrels is likely to have strong effects on their feeding behaviour, rate of energy intake and, hence, activity pattern (Smith 1968; Grönwall and Pehrson 1984; Sulkava and Nyholm 1987; States et al. 1988).

7.3.2 Habitat Effects

Between 68 and 81% of total activity in a coniferous woodland, and between 54 and 77% in a deciduous woodland was spent foraging (Wauters et al. 1992). In autumn/winter, when the diet differed most, squirrels spent relatively more time foraging in the coniferous than in the deciduous habitat, but not in spring/summer, when feeding on secondary (low-energy) food items in both habitats (Wauters et al. 1992). The higher proportion of time spent travelling in autumn/winter in the deciduous woodland showed that energy-rich food tended to be distributed in patches, while in the coniferous woodland, pine cones were more homogeneously distributed. In coniferous woodlands, squirrels also spent a higher proportion of their foraging time feeding on high-energy tree seeds than in deciduous woodlands, while in the latter, they needed to search longer when foraging on the ground for fallen or cached tree seeds in autumn/winter (Wauters et al. 1992).

During most of the year, active periods were about 0.5 h to 2 h longer in the coniferous than in the deciduous habitat (Wauters et al. 1992), due to squirrels in the coniferous woodland ending their activity later (Fig. 7.1). Moreover, in months when squirrels in both areas concentrated feeding on low-energy food items (June–July), or when deciduous seeds were important (October–November), the length of the active period did not differ between woodland types. Thus, differences in feeding behaviour, related to differences in food availability and food quality, resulted in longer active periods in coniferous than in deciduous woodlands.

Squirrels in deciduous woodlands had a bimodal activity pattern throughout the year, thus also in winter (December–February, Fig. 7.2), when animals in coniferous woodlands had only one active period per day. In deciduous woodlands, squirrels fed mainly on energy-rich, large seeds that had fallen on the ground or had been retrieved from food caches, filling the stomach after only a few hours of foraging (Wauters et al. 1992). They retreated into the nest to digest the food and reduce heat loss while resting. Since tree seeds are highly digestible, the animals could resume their activity after about 2 h. In the coniferous woodland, where squirrels fed mainly on small pine seeds which have relatively long handling times, no break in activity was possible. These differences in activity pattern and food choice re-

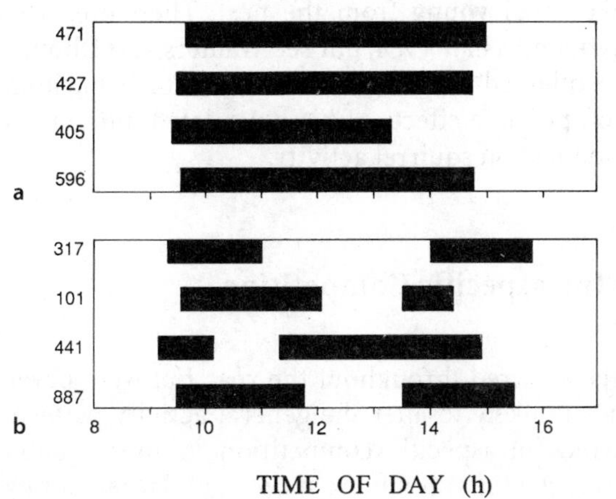

Fig. 7.2. Daily activity pattern from different red squirrels (numbers to the *left* stand for individuals). Each individual was monitored during 1 day in December 1986, between 8:00 and 17:00. **a** Coniferous woodland, **b** deciduous woodland. *Bars* indicate active periods. (From Wauters et al. 1992)

sulted in differences in daily energy intake according to habitat type. Both the estimated daily energy intake per squirrel and the average rate of energy intake (in kJ per h of activity) were higher in the deciduous (daily energy intake n = 4; mean ± s.d. = 419 ± 45 kJ) than in the coniferous area (n = 4; mean ± s.d. = 338 ± 31 kJ; Wauters et al. 1992). This calculated daily energy intake is confirmed by values of the metabolic rate during winter in related squirrel species (Knee 1983; Reynolds 1985).

Hence, despite a lower percentage of time spent feeding on tree seeds, the rate of energy intake was higher in the deciduous area, due to the larger size and relatively shorter handling times (Wauters et al. 1992) of the main food resources, at least in winter. In deciduous woods, the percentage of travelling time during winter months was higher and squirrels' home ranges were larger than in coniferous woods (Wauters and Dhondt 1992; Wauters et al. 1994), resulting in longer travelling distances per day per squirrel in broad-leaved woods. Therefore, energy expenditure was also likely to be higher, which was suggested by the between habitat variation in mean monthly body mass: adult males were significantly heavier in coniferous woods from January to April (Wauters and Dhondt 1989), when the difference in the proportion of tree seeds eaten was highest (Wauters et al. 1992). Hence squirrels in the deciduous habitat increased their daily rate of energy intake to compensate for higher energy losses caused by increased movemements and longer periods without food intake.

Little information is available on the effects predators (or predator densities) might have on squirrel activity. Scanning while foraging on the ground (every few seconds), i.e., when squirrels are most vulnerable to predator attacks, alarm-calls when a large bird of prey flies low over the trees, and sitting close to the trunk while foraging in trees (to avoid detection by avian predators) can all be considered as anti-predator behaviour. Goshawks (*Accipiter gentilis*) and to a lesser extent pine martens (*Martes martes*) regularly take squirrels, both on the ground and in trees (Münch 1994; K. Larsen, pers. comm.; pers. obs.). Other predators, such as corvids, cats, and stoats (*Mustela erminea*) might take small young from the nest. Therefore, nest choice (changing nest sites frequently, Lair 1984; but see Wauters and Dhondt 1990) could also be (partly) explained as being an anti-predator behaviour. No data are as yet available on possible effects of habitat-related differences in predator occurrence (abundance) on squirrel activity.

7.4 Activity Pattern and Intraspecific Competition

Interactions between squirrels occurred throughout the year, but were never numerous (Fig. 7.3) and hence unlikely to alter the general activity pattern. However, squirrels might avoid intraspecific competition in more subtle ways: (1) subordinate animals (subadults) might be more active when adults are in the nest or concentrate their foraging on other food items; (2) males and females, that have strongly overlapping home ranges (Wauters and Dhondt 1992), might also have different preferences, at least on seasonal foods. They might also avoid each other by using different parts of their home range at different periods of the day, as shown for grey squirrels that had a time-related pattern of home range use (Connolly 1979).

7.4.1 Sex Differences Between Adults

Adult male and female red squirrels had similar activity patterns in most months, except in May when the afternoon activity peak of females occurred later than that of males (Fig. 7.4). In July, the morning activity of females was much less synchronised and less intense than males, due to breeding activity (Fig. 7.4). The majority of radio tagged females were lactating in July, spending more time in the nest. During lactation, females have higher energy demands than males. However, lactating females were not active for more hours than males (Fig. 7.4), but spent relatively more time foraging during the breeding season (March–May and July–September, Fig. 7.3), thus increasing their energy intake, but not so during the non-breeding season.

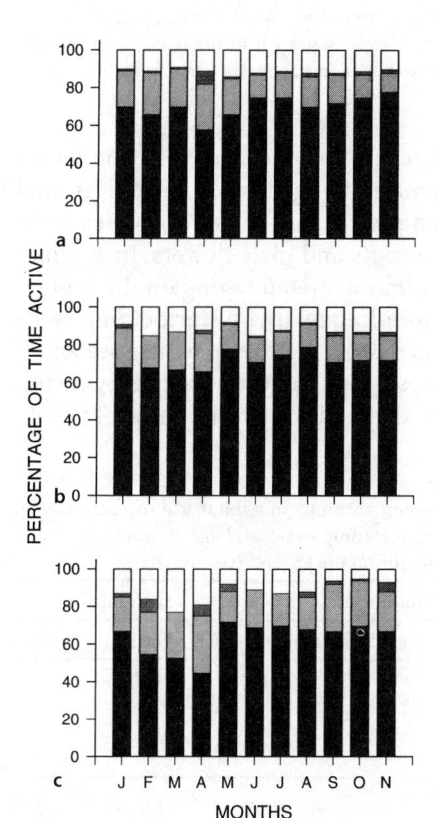

Fig. 7.3. Monthly activity budget of **a** adult male, **b** adult female, and **c** subadult red squirrels in a deciduous woodland in Belgium. The time allocated to each behavioural category is expressed as a percentage of total time active: foraging (*black bars*), travelling (*light grey bars*), interactions (*dark grey bars*), all other behaviour (*open bars*)

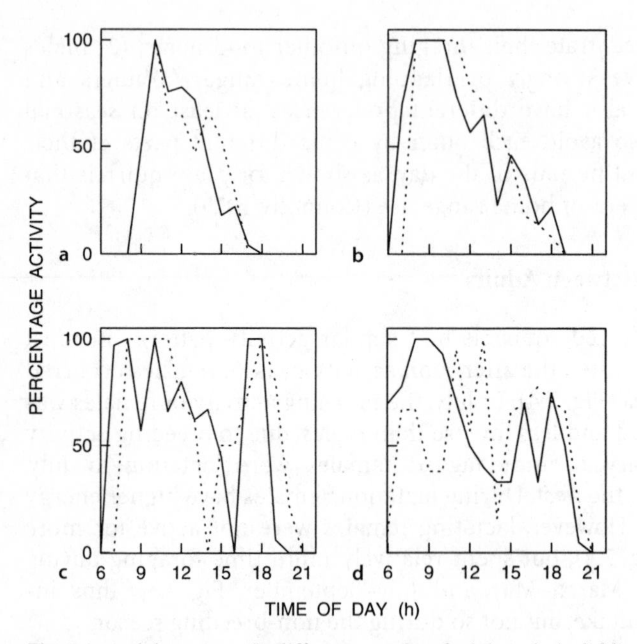

Fig. 7.4. Activity pattern in different months of adult male (*solid lines*) and adult female (*dotted lines*) red squirrels: **a** January, **b** March, **c** May, **d** July. Total time spent active per day for each sex: January: males 4 h 38 min, females 4 h 47 min; March: males 7 h 4 min, females 6 h 56 min; May: males 9 h 0 min, females 8 h 50 min; July: males 8 h 49 min, females 8 h 37 min

I also recorded some small sex differences in food choice in coniferous woodlands. In spring (March–May), females foraged more on the ground than males, feeding on cached and fallen pine cones (Table 7.2). Males spent more time in conifers, feeding on buds, shoots and male flowers. In summer (June–August), females were again found more often foraging on the ground (mainly for berries), while males fed more frequently on unripe pine seeds (Table 7.2). Since these differences occurred during the breeding season, it seems plausible they are related to females trying to maximise energy intake, probably as a response to higher energy demands due to lactation. They do

Table 7.2. Seasonal variation (winter, spring, summer, autumn) in habitat use of radio-tagged red squirrels in a coniferous woodland in Belgium, according to sex and age. Percentage of time active spent in pines (Pine), deciduous trees (Decid), or on the ground (Ground)

	Adult males			Adult females			Subadults		
Season	Pine	Decid	Ground	Pine	Decid	Ground	Pine	Decid	Ground
Winter	91	3	6	92	1.5	6.5	97	1	2
Spring	75	10	15	66	12	22	69	22	9
Summer	80	12	8	75	8	17	71	18	11
Autumn	68	16	16	70	18	12	54	29	17

this by spending more time foraging, feeding on energy-rich (cached) cones, on superabundant food items (flowers, insects), or on food that can be consumed without making long journeys away from the nest (low travelling time). In strictly territorial pine squirrels, on the other hand, reproductive constraints apparently did not affect the time budget of males and females differentially during spring breeding (April to early June, Ferron et al. 1986).

7.4.2 Age Differences

Data for subadult red squirrels were too few to investigate their activity pattern, but did allow a comparison of the behaviour of subadults and adults. Subadults foraged less intensively and spent more time travelling than adults (Fig. 7.3). They were quite often involved in interactions with adults during dispersal (April–May, October–November; Wauters and Dhondt 1993, and Fig. 7.3) and the mating period (January–February, Fig. 7.3). In the coniferous woodland, subadults spent more time in deciduous trees than adults, when oak and beech produced alternative food (flowers and insects in spring/summer, seeds in autumn; Table 7.2).

7.5 Conclusions

A strictly diurnal activity pattern, with seasonal changes from a short unimodal pattern in winter, an intermediate one in spring and autumn, to a long, bimodal one in summer was consistent for all tree squirrel species. However, both interspecific and geographical variation occurred in the fine-structure of the pattern. Moreover, tree squirrels adapted their activity to food availability and food quality by adjusting (1) the length of the active period, (2) the proportion of time spent foraging (autumn/winter) or spent feeding on primary food resources, and (3) the amount of time spent searching for food. Hence, differences in food availability (and rate of food intake) between woodland types caused habitat-related variation in the squirrel's activity pattern (Wauters et al. 1992). Little is yet known about how intraspecific competition affects the activity pattern of squirrels of different age and sex. It will need detailed studies of radio-tagged subadult and adult animals with overlapping home ranges to gain more insight into time-related home range use and behavioural mechanisms that might decrease the intensity of competition for space and food resources in non-territorial tree squirrels.

References

Bright PW, Morris PA (1991) Ranging and nesting behaviour of the dormouse, *Muscardinus avellanarius*, in diverse low-growing woodland. J Zool Lond 224:177–190

Bright PW, Morris PA (1993) Foraging behaviour of dormice, *Muscardinus avellanarius*, in two contrasting habitats. J Zool Lond 230:69–85

Connolly MS (1979) Time-tables in home range usage by gray squirrels (*Sciurus carolinensis*). J Mamm 60:814–817

Degn HJ (1974) Feeding activity in the red squirrel (*Sciurus vulgaris*) in Denmark. J Zool Lond 174:516–520

Doebel JH, McGinnes BS (1974) Home range and activity of a gray squirrel population. J Wildl Manage 38:860–867

Ferron J, Ouellet JP, Lemay Y (1986) Spring and summer time budgets and feeding behaviour of the red squirrel (*Tamiasciurus hudsonicus*). Can J Zool 64:385–391

Golightly RT Jr, Ohmart RD (1978) Heterothermy in free ranging Abert's squirrels (*Sciurus aberti*). Ecology 59:897–909

Grönwall O (1982) Aspects of the food of the red squirrel (*Sciurus vulgaris* L.). PhD thesis, University of Stockholm, Sweden

Grönwall O, Pehrson A (1984) Nutrient content in fungi as a primary food of the red squirrel *Sciurus vulgaris* L. Oecologia 64:230–231

Gurnell J (1987) The natural history of squirrels. Christopher Helm, London

Knee C (1983) Squirrel energetics. Mamm Rev 13:113–122

Lair H (1984) Adaptations de l'écureuil roux (*Tamiasciurus hudsonicus*) à la foret mixte conifères-feuillus: impact sur l'écologie et le comportement des femelles reproductrices. PhD thesis, University of Laval, Canada

Möller H (1983) Foods and foraging behaviour of Red (*Sciurus vulgaris*) and Grey (*Sciurus carolinensis*) squirrels. Mamm Rev 13:81–98

Münch S (1994) Wipfelakrobaten im Bergwald. Nationalpark 82:7–9

Ognev SI (1940) Animals of the U.S.S.R. and adjacent countries. 4. Rodents. Moscow, Leningrad. SSSR-Israel Programme for Scientific Translations 1966, Jerusalem

Pauls RW (1978) Behavioural strategies relevant to the energy economy of the red squirrel (*Tamiasciurus hudsonicus*). Can J Zool 56:1519–1525

Pauls RW (1979) Body temperature dynamics of the red squirrel (*Tamiasciurus hudsonicus*): adaptations for energy conservation. Can J Zool 57:1349–1354

Pulliainen E (1973) Winter ecology of the red squirrel (*Sciurus vulgaris* L.) in northeastern Lapland. Ann Zool Fennici 10:487–494

Purroy FJ, Rey JM (1974) Estudio ecologico y sistematico de la ardilla (*Sciurus vulgaris*) en navarra: distribution, densidad de poblaciones, alimentacion, actividad diana y anual. Bull Est Centr Ecol 3:71–82

Reynolds JC (1985) Autumn-winter energetics of Holarctic tree squirrels: a review. Mamm Rev 15:137–150

Saint-Girons MC (1966) Le rhythme circadien de l'activité chez les mammiféres holarctiques. Mém Mus Nat Hist Naturelle Paris Ser A40:101–187

Schulze W (1970) Beiträge zum Vorkommen und zur Biologie der Haselmaus (*Muscardinus avellanarius*) und des Siebenschläfers (*Glis glis*) im Südharz. Hercynia 7:355–371

Shorten S (1962) Squirrels, their biology and control. MAFF Bull 184:1–44

Smith CC (1968) The adaptive nature of social organization in the genus of tree squirrels *Tamiasciurus*. Ecol Monogr 38:31–63

States JS, Gaud WS, Allred WS, Austin WJ (1988) Foraging patterns of tassel-eared squirrels in selected Ponderosa pine stands. US Forest Service Publ Gen Techni Rep RM-166:425–431

Sulkava S, Nyholm ES (1987) Mushroom stores as winter food of the red squirrel, *Sciurus vulgaris*, in northern Finland. Aquilo Ser Zool 25:1–8

Tamura N, Miyashita K (1984) Diurnal activity of the Formosan squirrel, *Callosciurus erythraeus thaiwanensis*, and its seasonal change with feeding. J Mamm Soc Jpn 10:37–40

Thompson DC (1977) Diurnal and seasonal activity of the grey squirrel (*Sciurus carolinensis*). Can J Zool 55:1185–1189

Tittensor AM (1977) Red squirrel, *Sciurus vulgaris*. In: Corbet GB, Southern HN (eds) Handbook of British mammals. Blackwell, Oxford, pp 153–164

Tonkin JM (1983) Activity patterns of the red squirrel (*Sciurus vulgaris*). Mamm Rev 13:99–111

Wauters LA (1986) De eekhoorn en het bos. Bull Soc R For Belg 93:269–278

Wauters LA, Dhondt AA (1987) Activity budget and foraging behaviour of the red squirrel (*Sciurus vulgaris* Linnaeus, 1758) in a coniferous habitat. Z Säugetierkd 52:341–353

Wauters LA, Dhondt AA (1989) Variation in length and body weight of the red squirrel (*Sciurus vulgaris*) in two different habitats. J Zool Lond 217:93–106

Wauters LA, Dhondt AA (1990) Nest-use by red squirrels (*Sciurus vulgaris* Linnaeus, 1758). Mammalia 54:377–389

Wauters LA, Dhondt AA (1992) Spacing behaviour of red squirrels, *Sciurus vulgaris*: variation between habitats and the sexes. Anim Behav 43:297–311

Wauters LA, Dhondt AA (1993) Immigration pattern and success in red squirrels. Behav Ecol Sociobiol 33:159–167

Wauters LA, Swinnen C, Dhondt AA (1992) Activity budget and foraging behaviour of red squirrels (*Sciurus vulgaris*) in coniferous and deciduous habitats. J Zool Lond 227:71–86

Wauters LA, Casale P, Dhondt AA (1994) Space use and dispersal of red squirrels in fragmented habitats. Oikos 69:140–146

Zwahlen R (1975) Die lokomotorische Aktivität des Eichhörnchens (*Sciurus vulgaris*). Oecologia 22:79–98

8 Activity Patterns of Kangaroo Rats – Granivores in a Desert Habitat

Martin Daly, Philip R. Behrends, and Margo I. Wilson

8.1 Introduction

Kangaroo rats (*Dipodomys*, Heteromyidae) are nocturnal, facultatively bipedal, burrow-dwelling rodents, inhabiting arid and grassy habitats in western North America (Genoways and Brown 1993). The family Heteromyidae consists of six genera, of which pocket mice (*Chaetodipus* and *Perognathus*) and kangaroo mice (*Microdipodops*) are sympatric with kangaroo rats in deserts, and are to some degree their competitors. The approximately 20 *Dipodomys* species (Williams et al. 1993) are very similar in appearance, but vary about fourfold in body mass from the smallest species (*D. merriami* and *D. nitratoides*) to the largest (*D. ingens*). Notable features of all kangaroo rats include large hind feet (which they use for rapid, bipedal, saltatory locomotion), tail substantially longer than the body, and large heads housing enormous tympanic bullae.

Most of the *Dipodomys* species that have been studied are predominantly granivorous, although *D. microps* is a specialist folivore of halophytic shrubs, at least seasonally and in some parts of its range (Kenagy 1972; Csuti 1979). Even among the granivores, there are major differences in how kangaroo rat species make use of space and resources, as may be illustrated with reference to the two most thoroughly studied species. The small (ca. 35 g) Merriam's kangaroo rat (*D. merriami*) tolerates broad overlap (or at least interdigitation) of its large home ranges, changes day burrows frequently, and scatters the food it collects in numerous caches, often many meters apart (Behrends et al. 1986a, b; Daly et al. 1992b). The large (ca. 130 g) banner-tailed kangaroo rat (*D. spectabilis*) occupies small, non-overlapping territories, and accumulates all its hoarded food in a single larder in a large, complex, defended mound, which an individual may occupy for its entire life (Schroder 1979; Randall 1989; Waser and Jones 1991). These differences do not seem to reflect any dramatic difference in social systems: in both of these species – and in all *Dipodomys*, as far as we know – adults of both sexes dwell solitarily, parental care is exclusively maternal, and both sexes mate polygamously (Randall 1993).

Ecological Studies, Vol. 141
Halle/Stenseth (eds.) Activity Patterns in Small Mammals

The Heteromyidae and the closely related Geomyidae (pocket gophers) share the distinctive anatomical feature of external fur-lined cheek-pouches, in which the rodents gather seeds and other foodstuffs to be processed and consumed in safe refuges. Although kangaroo rats readily harvest seeds before they have fallen from grasses and herbaceous annual plants, they collect their seeds primarily - in most habitats, in most seasons - from the substrate, often far from cover. (*D. merriami* are poor and reluctant climbers, but we have observed them collecting mistletoe berries more than 1 m above the ground, after a prolonged drought and severe depletion of the soil seed bank had almost exterminated granivorous animals at our California study site.)

Kangaroo rats forage more in the open spaces between perennial shrubs than do sympatric heteromyid granivores, namely pocket mice (Price and Brown 1983). Kangaroo rats also differ from pocket mice in their bipedal locomotion and hypertrophied middle ear cavities, specialisations which are readily interpreted as adaptations to the higher level of predation risk in the kangaroo rat's preferred foraging environment (Webster and Webster 1980; Nikolai and Bramble 1983).

8.2 Nocturnality

Kangaroo rats are strongly nocturnal in their aboveground activities. Justice (1960) found that *D. merriami* and *D. spectabilis* confined all locomotory activity to the night, both in the field and in captivity. Kenagy (1973) studied the movements of *D. merriami* and *D. microps* implanted with radioactive gold wires, in a relatively cold part of *D. merriami*'s range, and likewise found them to be strictly nocturnal in their emergences throughout the year, although both species moved about frequently within their burrows during daylight hours. In captivity, *D. merriami* maintained under natural light confined wheel-running activity to the night (Kenagy 1976). We have observed *D. merriami* and *D. ordii* outside their burrows in daylight in the field, but only very rarely, primarily when the individuals in question were starving or otherwise near death.

Kenagy (1973, 1976) also claimed that the kangaroo rats on his site commonly spent less than 1 h per night away from their home burrows, a result that has been widely cited and even assumed to characterise desert heteromyids generally (e.g. Reichman 1983). Kenagy did not, however, present the evidence for this claim in detail, and it is certainly not generalisable. Our unpublished data on focal follows of radio-tagged *D. merriami* in southern California and *D. microps* and *D. ordii* in Utah indicate that kangaroo rats often spend several hours away from their day burrows on a given night, and may or may not spend most of this time in other burrows; observations of other species indicate the same (J.A. Randall, P. Kelly, pers. comm.). Seasonal and other determinants of the time spent aboveground require study.

In summer, nocturnal activity presumably protects desert-dwelling kangaroo rats from excessive heat loads and evaporative water losses (which are still a concern despite the heteromyid family's impressive physiological adaptations to these problems; see French 1993). In winter, excursions may be reduced to avoid excessive heat loss. At our study site, where temperatures below 0 °C are almost never recorded, comparisons that hold other factors constant indicate that even moderate cold suppresses *D. merriami* activity (Behrends 1984). For example, if we consider only those radio locations recorded within 1 h of midnight and within 3 days of the full moon, in Decembers in which females were not yet in reproductive condition, we find that *D. merriami* were out of their day burrows 44.7% of the time when the temperature was warmer than the median value of 8.5 °C, but only 29.7% of the time when it was colder. Nevertheless, Kenagy (1973) noted that both *D. merriami* and *D. microps* continued to emerge from their burrows even at –19 °C, the lowest temperature recorded in his 31-month study. Other heteromyid genera remain in their burrows and permit their body temperatures to fall in times of environmental stress, but such torpor is not part of the ordinary physiological repertoire of *Dipodomys* (French 1993).

Several writers have suggested that kangaroo rats' nocturnality is adaptively linked to their preference for foraging sites lacking protective cover. However, pocket mice, who forage mainly under and in shrubs and plant debris, are overwhelmingly nocturnal, too, as are most desert rodents other than sciurids. It is not yet known whether *Dipodomys* in dense, continuous grasslands are less strictly nocturnal than those in less vegetated deserts.

An interesting comparison group may be the old world gerbils, who are approximate ecological counterparts of the Heteromyidae (Mares 1993; Chap. 9). Typical granivorous gerbils are also mainly nocturnal, although they appear to be more often facultatively day-active than heteromyids. A notable exception is the purely diurnal gerbil, *Psammomys obesus*, whose folivorous diet allows it to forage in dense chenopod shrubs rather than in open ground (Daly and Daly 1974, 1975). *D. microps* is an ecological analogue of *Psammomys*, feeding on the hypersaline leaves of the same plant family (Kenagy 1972), but it is less anatomically divergent from other kangaroo rats than *Psammomys* is from other gerbils (perhaps because the gerbils are a much older radiation), and it remains strictly nocturnal.

Daylight observations of pocket mice were daily occurrences on our study site in December 1992, although extremely rare in 15 other winters. This was also the year of greatest abundance of kangaroo rats, which are aggressively dominant to pocket mice and may have forced them out of their preferred temporal niche into diurnality. The same idea – that subordinate species forage at non-preferred times to escape dominants – was proposed by Kenagy (1973) to explain why *D. microps* were maximally active soon after dusk on his site whereas *D. merriami* emerged mainly after the larger species' initial burst of activity subsided, and is discussed further by Ziv and Smallwood (Chap. 9).

On our study site, where it is the only *Dipodomys* species, *D. merriami* is maximally active soon after dusk (Daly et al. 1992a), like *D. microps* at Kenagy's site. Similarly, in a study in which activity was inferred from checking traps at 2-h intervals, O'Farrell (1974) found peak activity soon after dusk, and a decline over the night in *D. merriami*, *D. ordii*, and *D. panamintinus*, and the same pattern in kangaroo mice and pocket mice as well. Data for *D. microps* were noisier, but also showed the same declining pattern through the night when averaged over sexes and seasons.

This tendency for activity to decrease over the course of the night may involve additional factors besides just cold. Kenagy (1973) found such a pattern averaging across the entire year, but did not indicate whether it was seasonally variable. In the O'Farrell (1974) study, *D. merriami* exhibited steadier activity throughout the night in winter than in other seasons. On our study site, *D. merriami* apparently become more persistently active throughout the night in late winter and spring, but our midnight-to-dawn sampling has been limited, and effects of changing reproductive activity have not been separated analytically from effects of changing temperatures.

Kenagy (1976) reported on field trials in which traps were checked at 4-min intervals near dawn and dusk. Habituated kangaroo rats are quick to enter baited traps, and Kenagy found that both *D. merriami* and *D. microps* were regularly active in bright twilight: at 200 to 2 000 lx in the evening, and occasionally at up to 20 000 lx around sunrise (for comparison, light levels were just 0.5 lx on a clear night under a full moon). However, *D. merriami* in captivity were much more strictly nocturnal in their wheel running, and when the same trapping method was used to study a less habituated *D. ordii* population in the field, the animals did not emerge until at least 37 min after sunset, at light levels of 2 lx or less. These latter results suggest the possibility that emergence of highly trap-habituated *D. merriami* and *D. microps* in bright twilight was an artifact of the trapping regime.

In any event, regardless of whether kangaroo rats are ordinarily active in quite such bright twilight, it seems that some selection pressure(s) must favour early emergence, since crepuscular activity persists in the face of substantial predation from predominantly diurnal avian predators (see below). It has been suggested that aggregations of seeds in small wind shadows are replenished during daylight hours and are thus most profitable for the first animals to emerge after dusk (see Ziv and Smallwood, Chap. 9). However, Reichman and Price (1993) argue that available evidence does not support this idea. Kenagy (1976) proposed that there are socio-sexual advantages in emerging early in the night, but there have been no attempts to test this plausible idea.

8.3 Mating Effort and Aboveground Activity

Granivorous heteromyid rodents have been the objects of a large body of research in community ecology, focussed on the problem of how species of similar diet are able to coexist (Price and Brown 1983; Brown and Harney 1993). Perhaps for this reason, many ecologists studying these rodents (e.g. Reichman 1983) have unwittingly assumed that aboveground activity and 'foraging' for food are one and the same thing. The first published radiotelemetry study of *Dipodomys* movements (Schroder 1979) was entitled "Foraging behavior and home range utilisation of the bannertail kangaroo rat", for example, even though its data consisted solely of radio locations.

The trouble with this presumption is that aboveground activity must of course have social functions as well as foraging benefits. Since adult kangaroo rats do not nest together, emergence is necessary for mating. It is also of evident importance in defending a territory, and in the maintenance of a 'social presence' (Kenagy 1976). Repeated visits to a female when she is not yet in estrus, for example, can impart a mating advantage to a persistent and hence familiar male once the female is finally ready to mate (Daly 1977); this effect has been confirmed for *D. merriami* (Randall 1991) and *D. heermanni* (Thompson et al. 1995).

In many solitary rodents, males occupy much larger home ranges than females, not, presumably, because of greater energetic demands but because of their pursuit of polygamous mating opportunities. Indeed, sex differences in home range size characterise polygynous/promiscuous species but not monogamous congeners (Gaulin and FitzGerald 1988), and are often exaggerated in (or even confined to) mating seasons. In *D. merriami*, Behrends et al. (1986a, b) found no significant sex difference in home range size, even in the reproductive season, since male ranges were highly variable; however, reproductively active males travelled significantly farther between successive scheduled radio locations than either females or males with regressed testes. Females, too, exhibit activity increases that seem to reflect mating effort: there is often a substantial increase in female travels associated with the day of estrus (Behrends et al. 1986b). Whether this mobility at estrus reflects active mate choice on the part of females is not yet known.

8.4 Activity Levels and Predation Risk

It is often asserted that predators have the effect of 'weeding out' diseased, injured, parasite-ridden, senescent or otherwise enfeebled individuals of their prey species. In mammals, however, studies confirming such selective predation on the weak mainly involve large prey species (especially ungulates), whose principal anti-predator tactics consist of long-distance flight or active defence. In the case of small mammals, whose principal defence is crypticity or escape to a nearby refuge, it is questionable whether a similar pattern of selective predation prevails (Daly et al. 1990). Instead, we have hypothesised that the likeliest prey may often be those healthy, large, dominant adults who 'accept risk' because their phenotypes are such that the prospective fitness benefits of present reproductive efforts are sufficient to offset the predation risk costs (Daly et al. 1990). Our data on naturally occurring predation on radio-tracked *D. merriami* (Daly et al. 1990, 1992a, c) can be brought to bear on these ideas. In the following summary, we have extended previously published results of this study by the addition of subsequent years' data.

Between 1980 and 1992, we radio-tracked 207 *D. merriami* at a California site, for durations ranging from a single day (a male killed by a snake, within 24 h of his release after subcutaneous radio-implantation) to 280 days (a female tracked at intervals over a 39-month period). For further information about the methods and field site, see Behrends et al. (1986a) and Daly et al. (1990, 1992c). The kangaroo rats (104 females, 103 males) carried radio-transmitters for a total of 7 079 animal-days, during which time 38 were known victims of predators, and another 15 disappeared abruptly and were presumed predation victims. Identified predators included four species of snakes (*Masticophis flagellum*, *Crotalus cerastes*, *C. atrox*, and *C. ruber*), two birds (*Bubo virginianus* and *Lanius ludovicianus*), and one mammal (*Canis latrans*). These 53 known or presumed predations represent a rate of one predation death per 134 animal-days of radio-implantation.

Available data permit us to compare 44 predation victims' recent activity levels, over 1 to 3 previous nights, with the activity of other animals tracked simultaneously. In general, those killed had been out of their day burrows relatively often and had been travelling relatively far afield. For example, 14 of 20 female victims had travelled further between successive scheduled hourly radio locations than the contemporaneous median female and 5 had travelled less far ($p = 0.032$ by one-tailed sign test), while 1 value equalled the median. Among 24 males, 15 were more mobile than the median male and 5 less ($p = 0.021$), with 4 at the median. Thus, there was a significant association between mobility and predation risk within each sex considered separately.

Relatively mobile kangaroo rats might incur excess risk either because they are subordinates or because they are 'floaters' with no safe home. But neither of these possibilities appears to be correct. Being highly mobile is not associated with instability of day burrow sites or of home range, and at least during the reproductive season, predation victims of both sexes have tended to be somewhat heavier (and in the case of males to have somewhat larger testes) than average (Daly et al. 1990). Moreover, increased mobility is associated with mating effort in both sexes (Behrends et al. 1986b; Wilson et al. 1985). Thus, we interpret the association between mobility and predation as support for the proposition that those who incur excess predation risk are not the animals with relatively poor fitness prospects, but rather are those whose condition is good enough to make the prospective benefits of present reproductive effort cover the costs.

8.4.1 Moonlight Avoidance, Crepuscular Compensation, and Predation Risk

The hypothesis that highly active animals are relatively 'willing to accept risk' does not, of course, imply that they cease to respond adaptively to available cues of predation risk. The best-studied example of kangaroo rats' sensitivity to such a risk factor is moonlight avoidance. Like other nocturnal rodents that forage in relatively open habitats, kangaroo rats respond to moonlight by reducing aboveground activity, and by shifting such activity toward places with relatively dense cover (Lockard and Owings 1974; Price et al. 1984).

Rosenzweig (1974) proposed that increments in the amount of aboveground activity have diminishing marginal utility for kangaroo rats, as the most easily collected seeds are depleted and as information on the whereabouts and reproductive condition of conspecifics is updated. By contrast, he suggested that the costs of aboveground activity, including predation risk, cold stress, and evaporative losses, are likely to increase acceleratingly, because the animals can allocate activity preferentially to the lowest-cost times. It follows that there is an optimal amount of time allocated to such activity. With reasonable assumptions, that optimum is likely to decline if the cost curve rises while the benefit curve remains constant (see Fig. 1 in Rosenzweig 1974). Thus, if moonlight raises the predation risk cost of nocturnal activity while leaving the expected benefit curve (relatively) unaffected, one may predict that animals will reduce aboveground activity at the full moon compared with the new moon, and that they will preferentially allocate aboveground activity to moonless times of the night. The behaviour of *D. merriami* matches these predictions (Fig. 8.1). The left panels of Fig. 8.1 indicate that kangaroo rats were out of their burrows more and travelled farther at the new moon than at the full moon, especially near midnight, while the right panels indicate a tendency to shift activity toward the darker part of partially moonlit nights: before midnight moonrise during the waning moon phase and after midnight moonset during the waxing moon phase.

A more subtle issue is how animals should react to changing light levels at dawn and dusk (Lima 1988). Many nocturnal rodents emerge from their burrows when light is still adequate to support hunting by predominantly diurnal birds, and such emergences should vary with moon phase, not because of the moon's (minor) effect on light levels at twilight, but because of its influence on the availability of alternative low-risk emergence times. In particular, Daly et al. (1992a) argued that if moon phase has little effect on the predation risk costs of crepuscular activity but has a big effect on the predation risk costs of nocturnal activity, then reduced nocturnal activity at the full moon should be at least partially compensated by elevated crepuscular activity. This is because the incremental value of a given amount of crepuscular

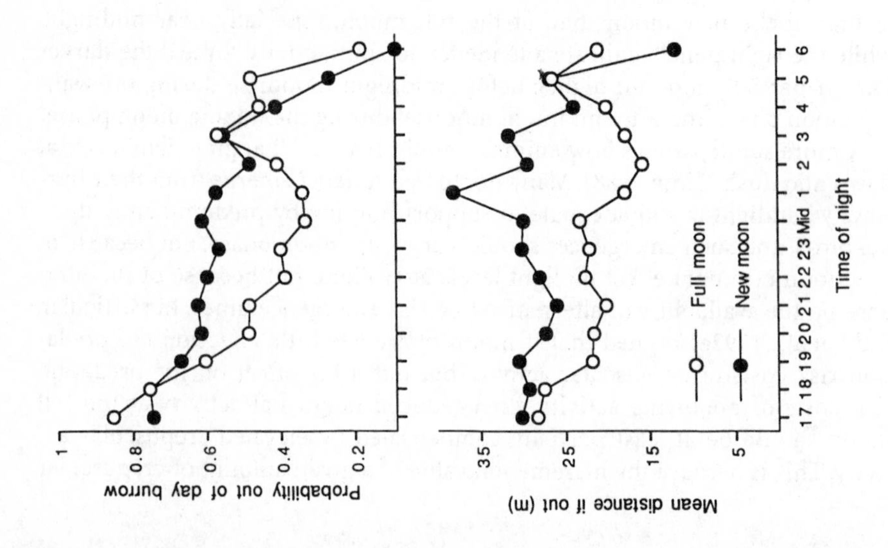

Fig. 8.1. Aboveground activity as revealed by animals' locations at scheduled hourly radio fixes, according to moon phase. Full and new moon phases refer to the 7-day periods surrounding those events; waxing and waning phases are the 7 to 8 day interims. During the waning moon phase, moonrise occurred between about 21:00 and 3:30; during the waxing moon phase, moonset occurred between about 22:00 and 4:30. Data are pooled for 156 individual animals tracked between 1980 and 1992, and are restricted to dates within 3 weeks of the winter solstice; sunset occurred within the hour before the 17:00 radio fix and sunrise within the hour after the 06:00 radio fix. The *upper panels* represent the probability of being outside one's day burrow at a scheduled radio fix (n = 16 901); the *lower panels* represent the mean distance from the day burrow for those animals who were away

Table 8.1. The distribution of predation on radio-implanted *D. merriami* by nocturnal versus diurnal predators, in relation to moon phase[a]

		Known diurnal predators		Known nocturnal predators		Unknown predators	
Moon phase	RAYs	N	Deaths per RAY	N	Deaths per RAY	N	Deaths per RAY
Full moon	4.81	4	0.83	0	0.00	7	1.46
Waxing or waning	10.13	3	0.30	5	0.49	20	1.97
New moon	4.44	0	0.00	6	1.35	8	1.80

[a] Moon phases are defined as in Fig. 8.1. *RAY* radio-animal-year (365.25 animal nights bearing a radio in the field). *N* number of deaths. Known diurnal predators are shrikes (*Lanius ludovicianus*); known nocturnal predators are coyotes (*Canis latrans*), owls (*Bubo virginianus*), and various snakes (*Crotalus cerastes*, *C. mitchelli*, *C. ruber* and *Masticophis flagellum*). The difference between diurnal and nocturnal predation distributions in relation to moon phase is significant at $p < 0.001$ by Moses's (1986) test of counts in ordered categories. See Daly et al. (1992a) for further details.

activity will be higher when nocturnal activity levels are lower. This prediction, too, is upheld by the data in Fig. 8.1: overall, animals stayed in their burrows more at the full moon than at the new moon, but as predicted, this difference was reversed at dusk and again at dawn.

If the utility of crepuscular activity is higher at the full moon because mid-nocturnal activity is reduced, then compensatory crepuscular activity should increase until its costs make it unprofitable. One such cost is elevated predation risk by predominantly diurnal predators. This reasoning led Daly et al. (1992a) to the further, seemingly paradoxical, prediction that any elevation of the predation rate, or components thereof, at the full moon would be disproportionately or solely imposed by day-active predators who do not actually exploit moonlight to hunt. This prediction, too, has been upheld (Table 8.1): predation by diurnal shrikes has occurred predominantly during the full moon phase, whereas predation by identified nocturnal predators has occurred predominantly at the new moon.

8.4.2 Does Predation Maintain Heterogeneity of Activity Profiles?

Predation on kangaroo rats at our study site tends to occur in a temporally clustered fashion, apparently as a result of the locally intensive activities of individual predators (Daly et al. 1990). Weeks often pass without any deaths among the radio-tagged animals, but when one is killed by a shrike or an owl, others are likely to be killed under nearly identical circumstances within a day or two.

It appears that scattered predators in desert habitats repeat successful perching and searching practices as long as encounter rates with potential prey remain high, and then move on. A possible result is that members of the prey species whose behavioural phenotypes are locally predominant – for example, those who are active at the times or in the microhabitats most favoured by their species – may be most at risk. In other words, local short-term learning by individual predators may result in frequency-dependent selection against locally prevalent activity profiles.

If this hypothesis is correct, one result could be the maintenance of genetically based polymorphism in the circadian timing of activity. Both the latency to emerge after dark and the distribution of activity through the night exhibit substantial individual differences in *D. merriami* in a simulated burrow-plus-open-arena environment in our laboratory, and we are presently investigating whether that variability is heritable. The issue is of some general interest since the reasons why heritable variation persists in nature remain obscure (e.g. Barton and Turelli 1989). Most decisions analysed by behavioural ecologists have a single optimum; well-understood polymorphisms typically involve two or three discrete alternatives (e.g. Austad and Howard 1984) rather than continuous variation. Frequency-dependent selection as a result of individual learning by predators could be one of the forces maintaining continuous heritable variation in various traits.

8.4.3 Sex Differential Predation Risk? A Methodological Caveat

The rate of predation on radio-tracked *D. merriami* males at our site (0.0106 deaths per radio-animal-day) has been almost twice that on females (0.0054). At first sight, this matches expectation: excess male mortality is the norm in non-monogamous mammals as a result of costly male mating effort (e.g. Trivers 1985). There is a complication, however. Year-to-year survivorship of females and males has not differed nearly as much as predation on the radio-implanted animals would lead one to expect, and several other studies indicate that female and male life tables are remarkably similar in *D. merriami* and other kangaroo rats (Behrends et al. 1986a; Zeng and Brown 1987).

It has seemed to us implausible that the sex difference in our predation data could be an artifact of the subcutaneously implanted radios, since radios should not have been more burdensome to the slightly larger males. Thus, Daly et al. (1990) proposed that male death rates normally exceed those of females before and early in the breeding season, when our radio-tracking efforts have been concentrated, and that sex differential mortality is reversed later in the summer. However, a subsequent analysis of predation risk and activity as a function of the time elapsed since radio-implantation suggests that the excess male mortality may indeed reflect a differential influence of radios after all, at least in part (Daly et al. 1992c). Shortly after radio-implantation, while the animals are presumably adjusting to the novelty of the implant, females exhibit reduced activity to a much greater degree than males (Fig. 8.2, bottom). And excess male mortality turns out to be concentrated in this early post-implantation period (Fig. 8.2, top).

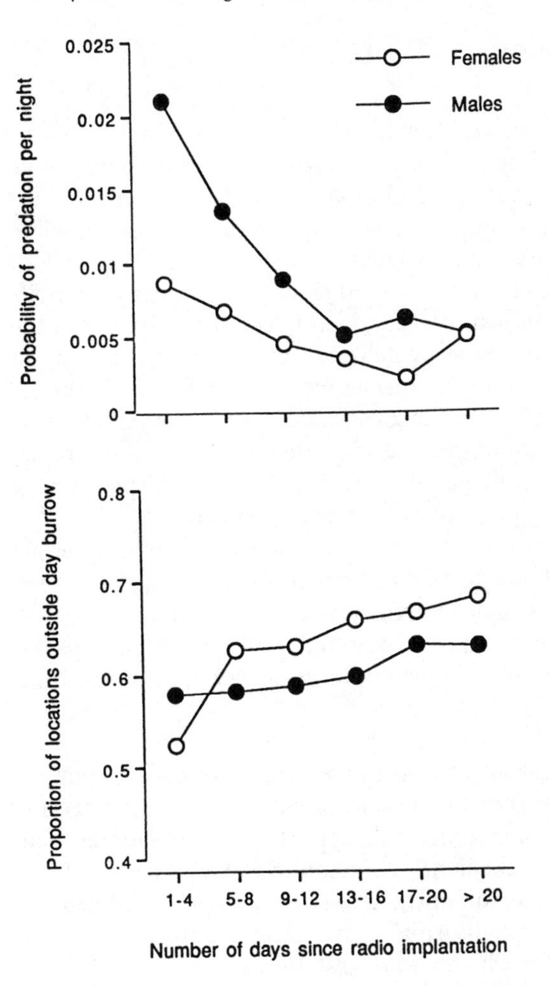

Fig. 8.2. Predation rates on radio-implanted kangaroo rats (*upper panel*) and probability that a radio-implanted kangaroo rat would be outside its day burrow at the time of a scheduled nocturnal radio fix (*lower panel*), as functions of sex and of the number of days elapsed since radio-implantation. The figure is based on 53 predation deaths, 27 919 nocturnal radio locations (scheduled no more than once per hour), and 357 radio-implantations between 1980 and 1992

To the best of our knowledge, this study is the first to have found sexually differentiated effects of radiotransmitter attachment. The implications are troublesome. If the sexually differentiated behavioural response portrayed in Fig. 8.2 is the source of the sexually differentiated mortality, as seems likely, a seemingly paradoxical conclusion follows. Sex differential boldness (or 'risk acceptance'), as manifested in more versus less cautious response to radio attachment, would indeed be a source of the observed sex difference in vulnerability to predators, and yet that sex difference might properly be deemed an artifact of the research methods.

8.5 An Unresolved Question

According to Reichman and Brown (1983), writing with reference to North American heteromyids, "Perhaps more is known about the comparative anatomy, physiology, behavior, ecology, and evolution of these rodents than is known about any comparable group of related organisms with the possible exceptions of Hawaiian Drosophila, Galapagos finches, and West Indian Anolis." A great deal more has been discovered since these words were written, but the sexual monomorphism of *Dipodomys* ranging activity and life tables, as noted above and as reviewed by Behrends et al. (1986b), remains an enigma. Studies of various kangaroo rat species consistently find male ranges to be scarcely or not at all larger than those of females, while solitary rodents in other families commonly exhibit massive sex differences, sometimes in the same field studies. Long-term field studies of several *Dipodomys* species have consistently failed to find even a hint of sex differential survival.

It appears that male kangaroo rats do not expend costly mating effort scrambling after females with the same reckless enthusiasm as gerbils, voles, lemmings, and various other solitary burrow-dwelling small rodents. It also appears that this generalisation holds across the genus (and perhaps for other heteromyids, too), despite ecological differences. We do not yet know why.

Acknowledgements. Our kangaroo rat studies have been supported continuously throughout the 16-year study by the Natural Sciences and Engineering Research Council of Canada, with additional support from the National Geographic Society and the National Science Foundation (USA). Al Muth, Lucia Jacobs, Jan Zabriskie, Tom Patterson and Suzi Patterson have facilitated our fieldwork at the University of California's Boyd Deep Canyon Desert Research Center in many ways, for which we are grateful.

References

Austad SN, Howard RD (eds) (1984) Symposium on alternative reproductive tactics. Am Zool 24:307–418

Barton NH, Turelli M (1989) Evolutionary quantitative genetics: how little do we know? Ann Rev Genet 23:337–370

Behrends PR (1984) Spatiotemporal activity patterns of Merriam kangaroo rats (*Dipodomys merriami*). PhD thesis, McMaster University, Hamilton

Behrends P, Daly M, Wilson MI (1986a) Range use patterns and spatial relationships of Merriam's kangaroo rats (*Dipodomys merriami*). Behaviour 96:187–209

Behrends P, Daly M, Wilson MI (1986b) Aboveground activity of Merriam's kangaroo rats (*Dipodomys merriami*) in relation to sex and reproduction. Behaviour 96:210–226

Brown JH, Harney BA (1993) Population and community ecology of heteromyid rodents in temperate habitats. In: Genoways HH, Brown JH (eds) Biology of the heteromyidae. Am Soc Mammal Spec Publ 10, Lawrence, pp 618–651

Csuti BA (1979) Patterns of adaptation and variation in the Great Basin kangaroo rat (*Dipodomys microps*). University of California Publ Zool 111:1–69

Daly M (1977) Some experimental tests of the functional significance of scent-marking in gerbils. J Comp Physiol Psychol 91:1082–1094

Daly M, Daly S (1974) Spatial distribution of a leaf-eating Saharan gerbil (*Psammomys obesus*) in relation to its food. Mammalia 38:591–603

Daly M, Daly S (1975) Behavior of *Psammomys obesus* (Rodentia: Gerbillinae) in the Algerian Sahara. Z Tierpsychol 37:298–321

Daly M, Wilson MI, Behrends PR, Jacobs LF (1990) Characteristics of kangaroo rats (*Dipodomys merriami*) associated with differential predation risk. Anim Behav 40:380–389

Daly M, Behrends PR, Wilson MI, Jacobs LF (1992a) Behavioural modulation of predation risk: moonlight avoidance and crepuscular compensation in a nocturnal desert rodent, *Dipodomys merriami*. Anim Behav 44:1–9

Daly M, Jacobs LF, Wilson MI, Behrends PR (1992b) Scatter-hoarding by kangaroo rats (*Dipodomys merriami*) and pilferage from their caches. Behav Ecol 3:102–111

Daly M, Wilson MI, Behrends PR, Jacobs LF (1992c) Sexually differentiated effects of radiotransmitters on predation risk and behaviour in kangaroo rats, *Dipodomys merriami*. Can J Zool 70:1851–1855

French AR (1993) Physiological ecology of the Heteromyidae: economics of energy and water utilization. In: Genoways HH, Brown JH (eds) Biology of the heteromyidae. Am Soc Mammal Spec Publ 10, Lawrence, pp 509–538

Gaulin SJC, FitzGerald RW (1988) Home-range size as a predictor of mating systems in *Microtus*. J Mammal 69:311–319

Genoways HH, Brown JH (eds) (1993) Biology of the heteromyidae. Am Soc Mammal Spec Publ 10, Lawrence

Justice KE (1960) Nocturnalism in three species of desert rodents. PhD thesis, University of Arizona

Kenagy GJ (1972) Adaptations for leaf eating in the Great Basin kangaroo rat, *Dipodomys microps*. Oecologia 12:383–412

Kenagy GJ (1973) Daily and seasonal patterns of activity and energetics in a heteromyid desert community. Ecology 54:120–1219

Kenagy GJ (1976) The periodicity of daily activity and its seasonal changes in free-ranging and captive kangaroo rats. Oecologia 24:105–140

Lima SL (1988) Initiation and termination of daily feeding in dark-eyed juncos: influences of predation risk and energy reserves. Oikos 53:3–11

Lockard RB, Owings DH (1974) Moon-related surface activity of bannertail (*Dipodomys spectabilis*) and Fresno (*D. nitratoides*) kangaroo rats. Anim Behav 22:262–273

Mares MA (1993) Heteromyids and their ecological counterparts: a pandesertic view of rodent ecology and evolution. In: Genoways HH, Brown JH (eds) Biology of the heteromyidae. Am Soc Mammal Spec Publ 10, Lawrence, pp 652–714

Moses LE (1986) Think and explain with statistics. Addison-Wesley, Reading, MA

Nikolai JC, Bramble DM (1983) Morphological structure and function in desert heteromyid rodents. Great Basin Naturalist Memoirs 7:44–64

O'Farrell MJ (1974) Seasonal activity patterns of rodents in a sagebrush community. J Mammal 55:809–823

Price MV, Brown JH (1983) Patterns of morphology and resource use in North American desert rodent communities. Great Basin Naturalist Memoirs 7: 117–134

Price MV, Waser NM, Bass TA (1984) Effects of moonlight on microhabitat use by desert rodents. J Mammal 65:353–356

Randall JA (1989) Territorial defense interactions with neighbors and strangers in banner-tailed kangaroo rats. J Mammal 70:308–315

Randall JA (1991) Sandbathing to establish familiarity in the Merriam's kangaroo rat, *Dipodomys merriami*. Anim Behav 41:267–275

Randall JA (1993) Behavioural adaptations of desert rodents (Heteromyidae). Anim Behav 45:263–287

Reichman OJ (1983) Behavior of desert heteromyids. Great Basin Naturalist Memoirs 7:77–90

Reichman OJ, Brown JH (1983) Introduction to symposium "the biology of desert rodents". Great Basin Naturalist Memoirs 7:1–2

Reichman OJ, Price MV (1993) Ecological aspects of heteromyid foraging. In: Genoways HH, Brown JH (eds) Biology of the heteromyidae. Am Soc Mammal Spec Publ 10, Lawrence, pp 539–574

Rosenzweig ML (1974) On the optimal aboveground activity of bannertail kangaroo rats. J Mammal 55:193–199

Schroder GD (1979) Foraging behavior and home range utilization of the bannertail kangaroo rat (*Dipodomys spectabilis*). Ecology 60:657–665

Thompson KV, Roberts M, Rall WF (1995) Factors affecting pair compatibility in captive kangaroo rats, *Dipodomys heermanni*. Zoo Biol 14:317–330

Trivers RL (1985) Social evolution. Benjamin/Cummings, Menlo Park, CA

Waser PM, Jones WT (1991) Survival and reproductive effort in banner-tailed kangaroo rats. Ecology 72:771–777

Webster DB, Webster M (1980) Morphological adaptations of the ear in the rodent family Heteromyidae. Am Zool 20:247–254

Williams DF, Genoways HH, Braun JK (1993) Taxonomy and systematics. In: Genoways HH, Brown JH (eds) Biology of the heteromyidae. Am Soc Mammal Spec Publ 10, Lawrence, pp 38–196

Wilson MI, Daly M, Behrends PR (1985) The estrous cycle of two species of kangaroo rats (*Dipodomys microps* and *D. merriami*). J Mammal 66:726–732

Zeng Z, Brown JH (1987) Population ecology of a desert rodent: *Dipodomys merriami* in the Chihuahuan Desert. Ecology 68:1328–1340

9 Gerbils and Heteromyids – Interspecific Competition and the Spatio-Temporal Niche

Yaron Ziv and Jeffrey A. Smallwood

9.1 Introduction

Competition theory has had a major influence on ecological thought for at least the last 60 years, with a special focus during the late 1950s through the 1970s (e.g. Hutchinson 1957, MacArthur and Levins 1964; Schoener 1974a). This led to the extensive development of niche theory (e.g. Hutchinson 1957; Pianka 1974; Whittaker and Levin 1975), linking competition pressures with observed resource use by competitors, both inter- and intraspecific (see Colwell and Fuentes 1975; Giller 1984 for reviews). As a result, terms such as niche release, niche shift and niche expansion are used to describe a species' response to its competitor's presence or absence. Dissimilarity between competing species in their foraging behaviour or habitat use is sometimes called 'niche differentiation'. Previous theoretical (e.g. May 1973) and experimental (e.g. Park 1962) studies on interspecific competition concluded that niche differentiation is necessary for competitive coexistence in interactive communities (i.e., communities that are mainly influenced by density-dependent processes such as competition and predation). As a result, the way in which ecologically-related species partition resources to avoid competitive exclusion has attracted the attention of ecologists for many years (e.g. MacArthur 1958; Schoener 1974a, b, 1986a).

Schoener (1974a) reviewed 81 studies on the relationship between competing species. He found that three major resource-partitioning dimensions may allow for niche differentiation between the species (see also Schoener 1986a for a more recent review): food, habitat (or space in general), and time (all corresponding to the 'separation components' suggested by MacArthur in a series of papers: e.g. MacArthur 1964; MacArthur and Levins 1964; MacArthur et al. 1966). He concluded that "habitat dimensions are important more often than food-type dimensions, which are important more often than temporal dimensions."

Among the three resource-partitioning dimensions, time has received the least attention (Schoener 1974a, 1986a). Theory seems to justify that because "no energetic gain can be derived from not feeding during most time peri-

Ecological Studies, Vol. 141
Halle/Stenseth (eds.) Activity Patterns in Small Mammals

ods", and "the relative rarity of experimentally produced shifts along time and diet dimensions is, in general, consistent with foraging theory" (Schoener 1986a). Some studies, however, did focus on temporal partitioning (Shkolnik 1971; Menge and Menge 1974; Kotler et al. 1993). These studies encouraged us to reevaluate the role of the temporal axis in niche partitioning. They also suggested that we should adopt a mechanistic approach that takes into consideration the complexity of species coexistence via the role of the limiting factor and environmental heterogeneity (e.g. Price 1986; Schoener 1986a; Rosenzweig 1991). Note that this does not change the fact that observing a pattern is usually the first step toward understanding ecological processes. Here we focus on two approaches: a 'theory of habitat selection' (Rosenzweig 1981) and 'mechanisms of coexistence on a single resource' (Brown 1986).

Rosenzweig (1981) developed a theory of habitat selection for two competing species, which he calls 'isoleg theory'. The theory builds on the intraspecific selection theory of Fretwell and Lucas (1970; Fretwell 1972), and optimal patch use theory (e.g. Rosenzweig 1974; Charnov 1976; Brown 1988). The theory assumes that individuals of each competing species choose the habitat set that maximises their fitness, and that the densities and distributions of each species affect this choice. In a practical sense, the theory of habitat selection allows one to map behaviours (i.e., different habitat preferences) onto a set of coordinate axes of population densities. As a result, it provides a way of incorporating niche relationships into a single two-dimensional picture (e.g. shared-preference versus distinct-preference models; see Rosenzweig 1985, 1991 for reviews). We can use this approach to move from the fundamental to the realised niche (Hutchinson 1957) of a given species by looking at the species behaviour with and without its competitor. It allows us to understand how the primary preference of each species (i.e., its resource preference in the absence of the competitor) for the competitively-limiting factor changes in a density-dependent manner. This may also allow us to hypothesise about the mechanism of coexistence between the competing species based on their community organisation.

Studying mechanisms of coexistence on a single resource, Brown (1986; 1989a, b) suggested that each mechanism of coexistence is composed of two essential features (see also Kotler and Brown 1988): (1) an environmentally heterogeneous resource axis (e.g. Levins 1979; Chesson and Warner 1981), and (2) an evolutionary trade-off between the abilities of the coexisting species to utilise various parts of this axis (e.g. Stewart and Levin 1973; Kotler and Brown 1990). These two features may provide a variety of mechanisms to give each species a relative advantage over its competitor under different sets of conditions. Hence, each may reproduce and maintain a non-decreasing population size (e.g. Brown et al. 1994). Furthermore, this approach may allow us to explain competitive coexistence even when only a single limiting resource is involved (Vance 1984; Brown 1986).

Many studies show that competitive relationships play an important role in organising rodent communities (e.g. Grant 1972; Frye 1983; Brown et al. 1986; Price 1986; Kotler and Brown 1988; Brown 1989b; Abramsky et al. 1991). In the following sections we review two studies on competing rodent species. These studies emphasise spatial and temporal partitioning. The first study focuses on the significance of the daily temporal axis between two gerbil species in Israel. The second study focuses on the contribution of the seasonal temporal axis between two species of pocket mice in Arizona. Both studies treat coexistence by examining how shared-preference habitat selection affects the competitive trade-off between the species. The gerbil study relies on manipulation experiments, while the pocket mouse study analyses patterns of distribution.

9.2 Gerbils and Daily Temporal Partitioning

The sandy habitats of the western Negev Desert in Israel are inhabited by up to five gerbil species (Zahavi and Wharman 1957). The most common species of this community are Allenby's gerbil (*Gerbillus allenbyi*; 26 g) and Egyptian sand gerbil (*G. pyramidum*; 40 g). These two species are nocturnal, inhabiting burrows during the day and consuming and collecting seeds and vegetative material while foraging during the night. Both species are mostly granivorous and have similar diets (Bar et al. 1984), and their densities are correlated with the productivity of seed-producing annual plants (Abramsky 1988). Habitat selection of *G. allenbyi* and *G. pyramidum* (Rosenzweig and Abramsky 1986; Abramsky and Pinshow 1989) and the amount of time they are active (Mitchell et al. 1990) are both affected directly by the other species' density. This competitive relationship stimulated an extensive study on the two species' coexistence mechanisms and the spatial and temporal niche partitioning between them (e.g. Abramsky et al. 1990, 1991; Kotler et al. 1993; Brown et al. 1994).

Two distinct activity patterns have been demonstrated experimentally for these species - the first relates to spatial partitioning and the second relates to temporal partitioning. Abramsky et al. (1990, 1991, 1992, 1994) have studied the habitat preference of the two gerbil species in an area with two habitat types: semistabilised dunes and stabilised sands. Both species primarily prefer the semistabilised dunes because each species enjoys a foraging advantage in this habitat type (Ziv et al. 1995). Hence, the two species reveal what is called 'shared-preferences habitat selection' (e.g. Pimm et al. 1985). However, habitat preference is both intra- and interspecifically density dependent. In the presence of a relatively low density of *G. pyramidum*, *G. allenbyi* uses both habitat types equally. In the presence of relatively moderate or high density of *G. pyramidum*, *G. allenbyi* shifts its habitat preference to the stabilised sands. In contrast, the preference of *G. pyramidum* for the semistabilised dunes increases in the presence of high densities of *G. allenbyi*. At natural densities (see Abramsky et al. 1991, 1992, 1994 for isocline analysis) *G. pyramidum* uses the semistabilised dunes while *G. allenbyi* mainly uses the stabilised sands ('apparent preference'; Rosenzweig 1991).

Regarding time, Kotler et al. (1993) showed, observationally, that each species is active during a different part of the night, thus suggesting a temporal-partitioning pattern. Additional studies showed that *G. allenbyi* is the more efficient forager at low resource abundances (Kotler and Brown 1990; Ziv 1991; Brown et al. 1994). That is, at resource densities when *G. pyramidum* cannot profit from foraging any further, *G. allenbyi* can still profit (for more details on optimal patch use see Brown 1986, 1988, 1989a; Kotler and Brown 1988, 1990; Kotler et al. 1993; Brown et al. 1994).

However, the main question regarding the two-species system remained unsolved: how do these two sympatric species coexist? Brown et al. (1994) found evidence against five mechanisms of coexistence that depend on habitat partitioning and annual temporal partitioning in resource abundance (see Kotler and Brown 1988 for 'mechanisms of coexistence' and Brown 1989b for testing some of these mechanisms). However, evidence of daily renewal of resource patches suggested that coexistence should depend on this scale of habitat heterogeneity (see Kotler et al. 1993 for detailed evidence of the daily renewal).

Following this information, Ziv et al. (1993) hypothesised that coexistence between *G. allenbyi* and *G. pyramidum* depends on one of two trade-offs: (1) interference (*G. pyramidum*) versus foraging efficiency (*G. allenbyi*), or (2) foraging efficiency in the early part of the night (*G. pyramidum*) versus foraging efficiency in the late part of the night (*G. allenbyi*; see Brown 1989a for similar a hypothesis on a seasonal scale). Interference versus foraging efficiency means that *G. pyramidum* monopolises rich resource patches at the beginning of the night. But, due to its higher foraging efficiency, *G. allenbyi* can and does exploit poorer resource patches available later at night after *G. pyramidum* quits foraging (Ziv et al. 1993). Foraging efficiency in different parts of the night means that each species can and does profit more from resource patches available at different times (Ziv et al. 1993; see also Kotler and Brown 1988 for a general review). If interference versus foraging efficiency is correct, then in the absence of its competitor, *G. allenbyi* will increase its activity in the first part of the night. However, if time of activity is determined only by foraging preferences without interference, then we should not expect any qualitative shift in *G. allenbyi*'s activity behaviour.

To test their hypotheses, Ziv et al. studied the two species in a sandy part of the Negev Desert, Israel. The area provides two main habitats: semistabilised dunes and stabilised sands (Danin 1978). Strong afternoon winds occur almost daily, and the average annual precipitation at the site is 108 mm. Ziv et al. used 1 ha enclosed and unenclosed grids to test the behaviour of *G. allenbyi* with and without *G. pyramidum* (see Abramsky et al. 1990, 1991 for detailed description of the enclosures). Each enclosure contained similar proportions of both habitats, and all fences were perforated with adjustable gates with small openings that only *G. allenbyi* could traverse to control the preferred densities (see Ziv et al. 1993). Forty 'sand-tracking' plots (0.4 × 0.4 m) in each enclosure (20 plots in each habitat) allowed scoring of plots for rodent activity by estimating percent coverage by tracks on a scale of 0 (no footprints at all) to 4 (100% coverage). Scores could be assigned to species on the basis of species-specific toe clips. The sum of scores of a species in a habitat in a grid divided by the number of individuals of that species in that grid gave the 'per-capita activity density' which represents an accurate measure of the species activity (e.g. Abramsky et al. 1990, 1991).

In the presence of both species, a temporal partitioning pattern exists (Fig. 9.1). *G. pyramidum* is active in the beginning of the night immediately after sunset with a sharp decrease in its activity towards midnight. From then on, almost no individuals of *G. pyramidum* are active. In contrast, *G. allenbyi* is hardly active until midnight. However, from midnight it increases its activity approximately threefold. Similar activity patterns were found in control grids, suggesting that the enclosures represent the natural situation accurately. This temporal-partitioning pattern is consistent with what has been shown by Kotler et al. (1993).

However, the activity pattern of *G. allenbyi* does depend on the density of *G. pyramidum*. In the absence of *G. pyramidum* (Fig. 9.1), the activity of *G. allenbyi* was highest immediately after sunset and declined moderately until the end of the night. During the earlier part of the night *G. allenbyi* was more active in the enclosed grids without *G. pyramidum* than on either the control grids or the enclosures where both species were present (Ziv et al. 1993).

The temporal niche shift of *G. allenbyi* as a result of *G. pyramidum*'s presence supports the hypothesis that the two species coexist due to a trade-off between interference competition and foraging efficiency (see also Ziv et al. 1993). This finding is consistent with previous studies regarding the relation-

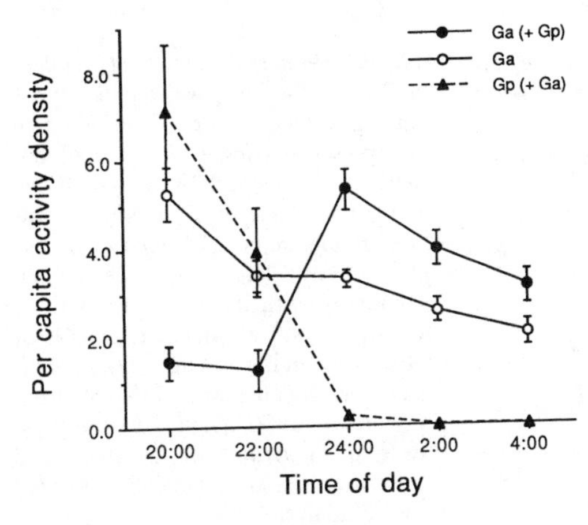

Fig. 9.1. Per capita activity density values (± SE) of *G. allenbyi* (*Ga*; *solid lines*) and *G. pyramidum* (*Gp*; *broken line*) during different hours of the night. *Closed* and *open circles* represent the per capita activity density of *G. allenbyi* in the presence and absence of *G. pyramidum*, respectively. Per capita activity density is the sum of scores of a species' sand tracking in a habitat in a grid divided by the number of individuals of that species in that grid (see text)

ship between ecologically-related species that differ slightly in body size (Vance 1984; Brown 1986). The bigger species should be dominant (e.g. Fyre 1983; Bowers et al. 1987) while the smaller species should forage more efficiently (Rosenzweig and Sterner 1970; Kotler and Brown 1990). This relationship leads to a dominant-subordinate community organisation (Rosenzweig 1991).

The temporal-partitioning pattern is only one pattern in the complex relationship between the species. Can we incorporate the spatial-partitioning pattern of the species to get a spatio-temporal understanding of their actvity? To do that, Ziv et al. (1993) went on to ask what proportion of *G. allenbyi*'s activity density in the semistabilised dunes occurs during different times of the night when *G. pyramidum* is present and absent. A proportion of 0.5 indicates no preference, while values above or below 0.5 indicate preference for the semistabilised dunes or for the stabilised sands, respectively. Figure 9.2 shows that in the presence of *G. pyramidum* during the first two time periods, *G. allenbyi* used mostly the stabilised sands. During later time periods when *G. pyramidum*'s activity ceased, *G. allenbyi* used both habitats equally. In the absence of *G. pyramidum*, *G. allenbyi* used both habitat types equally all night. The use of both habitats by *G. allenbyi* in the absence of *G. pyramidum* is expected from 'ideal free distribution' intraspecific habitat selection (see Fig. 1b in Abramsky et al. 1990).

With regard to space, Ziv et al. (1995) showed two ways that foraging advantages contribute to the preference of both species for the semistabilised dunes. First, both species have a higher foraging efficiency (Brown 1988) when foraging on the sandy substrate in the semistabilised dunes relative to that on the loess-based substrate of the stabilised sands. Second, the semistabilised dunes provide a better substrate for escape from predators than the stabilised sands, probably in part because a higher number of burrows exist in the semistablised dunes. Overall, foragers of both species should experience a higher net energy gain in the semistabilised dunes (see Ziv et al. 1995 for more details).

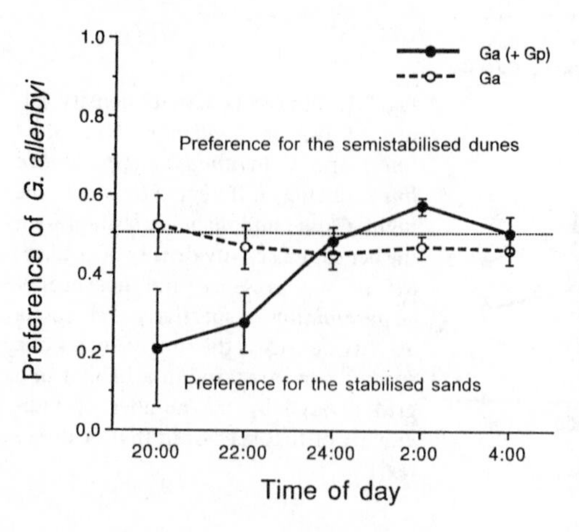

Fig. 9.2. Preference (± SE) of *G. allenbyi* (*Ga*) for the two habitat types during different hours of the night in the presence (*closed circles*) and absence (*open circles*) of *G. pyramidum* (*Gp*). Preference of *G. allenbyi* means the proportion of *G. allenbyi*'s per capita activity density in one habitat compared with its per capita activity density in both habitats. Preferences of *G. allenbyi* in the absence of *G. pyramidum* are significantly different from preferences of *G. allenbyi* in the presence of *G. pyramidum* at 20:00 and 22:00. Data were arcsine transformed for the analysis

With regard to time, the preference for the early part of the night by both species is consistent with the availability of resources in the area. The strong afternoon winds that occur almost daily are probably responsible for the daily renewal of seeds by uncovering seeds buried in the ground (Kotler et al. 1993). It is therefore advantageous to forage early in the night to get the newly exposed seeds before they are picked up by another forager (seed disinterment should be greatest in the semistabilised dunes because of their less stabilised substrate; Danin 1978). Additionally, harsher environmental conditions later at night, such as low temperature and high humidity, might increase foraging costs for gerbils.

We have other indirect evidence that competition for food resources may play a major role in the temporal-partitioning pattern. We conducted our experiment in October after a hot summer when resources should have been limited. However, a study conducted a few months later in late winter, when resources were abundant, revealed no temporal partitioning; both species were active from the beginning of the night (O. Ovadia and E. Vaginsky, pers. comm.). This suggests that resource availability is important in determining the temporal-partitioning pattern.

G. pyramidum prefers a particular habitat which is defined in both space and time, i.e., semistabilised dunes soon after sunset. It exploits this spatio-temporal habitat due to its dominance, while *G. allenbyi* gets the rest. So, the temporal and the spatial partitioning patterns exhibited by the species are not independent. Yet, the distribution of the seed resources in space and time provides more opportunities in the semistabilised dunes during the beginning of the night for foragers of both species. *G. allenbyi*'s restriction depends on *G. pyramidum*'s presence. Thus, the species demonstrate shared-preference spatio-temporal selection (Rosenzweig and Abramsky 1986; Abramsky et al. 1990; Kotler et al. 1993; Ziv et al. 1993).

G. allenbyi and *G. pyramidum* compete for food resources, mainly seeds (e.g. Bar et al. 1984; Abramsky 1988). Bar et al. (1984) showed that the two species have a high diet overlap and concluded that resource partitioning (i.e., food) cannot explain their coexistence. This conclusion is consistent with the known ecology of granivorous desert rodents in North America (see Kotler and Brown 1988 for review). In particular, the high temporal and spatial variance in desert ecosystems requires competing species to share diets (e.g. Reichman 1975, 1984; M'Closkey 1983; Price 1986). This precludes coexistence on the basis of resource partitioning. As a result, spatial and temporal habitat axes play major roles in species coexistence (Kotler and Brown 1988). Among the two species, *G. allenbyi* is the more efficient forager, partly due to its basal metabolic rate and thermoregulation (Linder 1988). However, *G. pyramidum*, the bigger species, dominates the shared-preferred parts of the temporal and spatial axes. Only after *G. pyramidum* stops foraging can *G. allenbyi* forage freely with no foraging costs of interference. Hence, the nature of the available resources, together with the differential abilities of the species, promote the existence of a dominant-subordinate community organisation. In turn, the community organisation affects the spatial and temporal partitioning patterns we observe.

9.3 Heteromyids and Seasonal Temporal Partitioning

The temporal niche also shows extensive variation on a seasonal or annual scale as well as the daily scale seen in the gerbil niche partitioning example. We therefore find it useful to review an example of niche shift which occurs at a seasonal scale. As with gerbilline systems, competition has been demonstrated in many heteromyid systems and resource partitioning is common (see Brown and Harney 1993; and Reichman and Price 1993 for recent reviews). Much research has shown partitioning of resources to occur by means of microhabitat specialisation, especially partitioning among the genera (Rosenzweig 1973; Price 1978; Kotler and Brown 1988; Brown and Harney 1993; Reichman and Price 1993). Despite extensive knowledge of activity patterns (Daly et al., Chap. 8), very little work has been done which demonstrates temporal partitioning of resources by heteromyids (Brown 1989b). Brown found that the two heteromyids which co-occur on his Southeast Arizona study site (*Dipodomys merriami* and *Perognathus amplus*) partition resources by means of seasonal variation in foraging costs; trade-offs between foraging efficiencies in different times of the year allow coexistence in that system.

The three most common species throughout the Sonoran Desert habitats of Southwest Arizona are Merriam's kangaroo rat (*Dipodomys merriami*; 36 g), desert pocket mouse (*Chaetodipus penicillatus*; 20 g) and rock pocket mouse (*C. intermedius*; 13 g), where they co-occur in two adjacent habitats, rocky slope and sandy flat (Schmidly et al. 1993). An extensive review of activity patterns and general ecology of kangaroo rats appears in Daly et al. (Chap. 8). The physiological differences (Hoover et al. 1977; French 1993) and size sequence represented by these three species provide an excellent opportunity to investigate competitive coexistence with respect to habitat use and habitat partitioning. In order to understand their coexistence, Smallwood and Swift (unpubl.) studied the effects of interspecific interactions on habitat use by utilising the differences in temporal activity patterns and their relationship to ambient temperatures and resource availability in a typical environment.

Variation in occurrence of important seed producing plants within these two habitats results in a generally more dense and continuous seed bank (richer resource for heteromyids) on the sandy flat compared with the rocky slope (Smallwood and Swift, unpubl.). Thus, resource availability is both spatially and temporally heterogenous in this location as has been demonstrated in other desert seed resource studies (Nelson and Chew 1977; Reichman 1984; Price and Reichman 1987). The sandy flat also apparently provides more favourable burrow locations with increased buffering of ambient temperatures (Hoover et al. 1977).

In the combined habitats, peak frequencies of *Dipodomys* and *Chaetodipus* occur at different times of the year (relationship between frequencies of the two genera: r = –0.896, $p < 0.001$, n = 12). Thus, the two genera partition the seed resource by varying the time of year when each genus is the numerically dominant forager. The out-of-phase pattern of the generic densities supports the idea that potentially competing heteromyid genera have seasonally distinct density peaks, as has been found to be the main mechanism of coexistence between the two co-occurring heteromyids in the study by Brown (1989b; see above).

Peak frequencies of the two *Chaetodipus* species in combined habitats are occasionally in phase (relationship between the two congeners: r = –0.274, ns, n = 12). In separate habitats, the relationships in either habitat between a *Chaetodipus* species and either of the other two common heteromyids are negative and significant where that particular *Chaetodipus* is the numerically dominant species through most of the year (Table 9.1; *C. penicillatus* most common on the sandy flat, *C. intermedius* most common on the rocky slope). The relationships between *D. merriami* and the *Chaetodipus* species that is not numerically dominant in either habitat are positive and either non-significant or weakly significant (Table 9.1; e.g. *D. merriami* with *C. penicillatus* on the slope, and with *C. intermedius* on the flat).

In addition to these general patterns, individuals move between habitats (Smallwood and Swift, unpubl.); the rare *C. intermedius* marked on the flat frequently move to the slope, particularly when *C. penicillatus* are increasing in density on the flat (i.e., in spring). The reverse movement by *C. penicillatus* is far less frequent, and most of those on the slope appear to be permanent residents there. Thus, while *C. intermedius* is usually more common on the rocky slope, their numbers of individuals using the sandy flat increases in the absence of *C. penicillatus*. *Chaetodipus penicillatus* prefers the sandy flat and is numerically and behaviourally dominant there most of the year; it seems able to preclude extensive use of the sandy flat habitat by *C. intermedius*

Table 9.1. Matrices of Pearson correlation coefficients (r) for the relationships of frequencies of all possible heteromyid species pairs in the rocky slope and sandy flat habitats. Correlations of frequencies on the slope are in the *upper right*, and those for the flat are in the *lower left* part of the matrix. Frequencies are proportions of capture rates averaged over the number of trap nights for each sampling date. Species abbreviations are as follows: Dm = *Dipodomys merriami*; Cp = *Chaetodipus penicillatus*; Ci = *C. intermedius*. In all cases n = 12 sampling dates. Significance levels: ns = not significant; * = $p < 0.05$; ** = $p < 0.01$; *** = $p < 0.001$

	Dm	Cp	Ci
		Rocky slope	
Dm	–	0.583 *	–0.816 ***
Cp	–0.872 ***	–	–0.742 **
Ci	0.553 ns	–0.773 **	–
	Sandy flat		

during all times when the former is active (Smallwood and Swift, unpubl.). *Chaetodipus penicillatus* can use the slope at will, but never in very high densities even if *C. intermedius* is absent (Fig. 9.3b: Dec 1994). Thus, *C. intermedius* can utilise the flat habitat only when *C. penicillatus* is at low densities there (Fig. 9.3a). Low densities are probably due to hibernation in cold weather (Hoover et al. 1977). The increase in the proportional presence of *C. penicillatus* on the slope at times of low *C. intermedius* densities is not the result of increased capture rates, rather it is an artifact of the use of proportions and the decrease of *C. intermedius* density. Alternatively, the increase of *C. intermedius* on the flat at low *C. penicillatus* densities is the direct result of increased trapping frequencies of both marked and unmarked individuals.

Movements and changing densities among habitats show that *C. intermedius* is able to expand its niche at times when *C. penicillatus* is absent. Trappings and markings show that individuals of *C. intermedius* are shifting

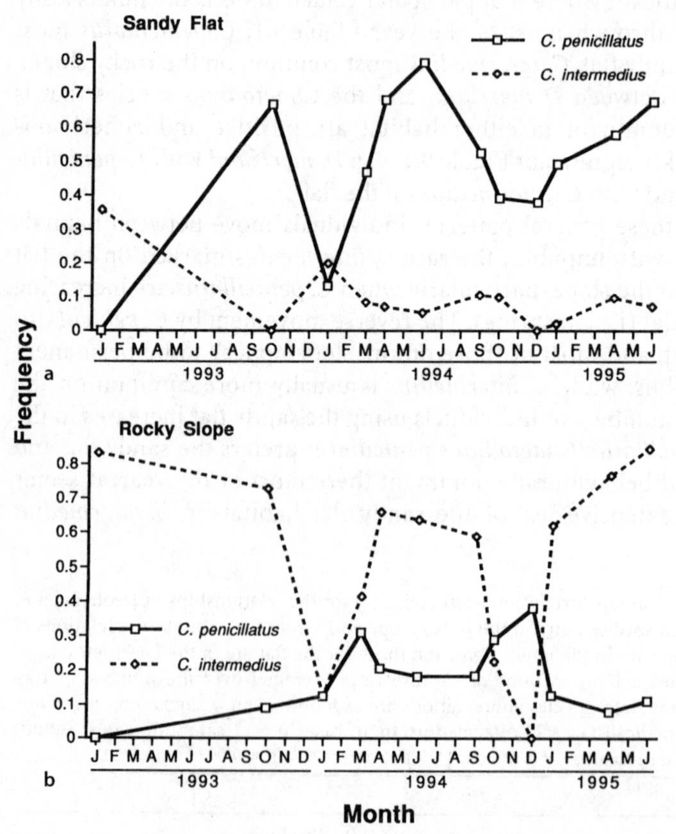

Fig. 9.3. Frequencies of *Chaetodipus* species in the two habitat types of the Sonoran Desert, Southwest Arizona. a Frequency of *Chaetodipus* species using the sandy flat habitat. b Frequency of *Chaetodipus* species using the rocky slope habitat. Frequencies are proportions of capture rates averaged over the number of trap nights for each sampling date

their niche in a temporal fashion in order to take advantage of the open niche in the flat habitat that is left by *C. penicillatus*' absence during periods of apparent inactivity or hibernation. The reverse is not true: *C. penicillatus* does not shift niches to utilise the slope habitat; certain individuals are permanent residents there, but additional individuals do not appear from the sandy flat marked population when *C. intermedius* is in low density on the slope. The use of both habitats by both species is evidence of extensively overlapping niches, therefore the use of habitats by these two species appears to fit a shared-preference model of habitat selection (Pimm et al. 1985; Rosenzweig 1991) as do the aforementioned gerbils.

Some evidence exists to suggest that closely related species differ in their foraging efficiencies (Kotler and Brown 1988), generally with the smaller species being more efficient foragers (Rosenzweig and Sterner 1970; Kotler and Brown 1990), and larger species being behaviourally dominant (Frye 1983; Bowers et al. 1987). Perhaps then, each *Chaetodipus* species has a range along the resource axis where resource availability is appropriate to enable each to tolerate the presence of competitors, and a range where each cannot tolerate the presence of competitors. Under this scenario, it may be possible that only one *Chaetodipus* species is able to compete for a specific resource density under certain cold conditions. Since larger species tend to be the less efficient foragers, *C. penicillatus* might be expected to require higher resource densities in order to out-compete *C. intermedius* in cold weather. At lower resource densities, *C. intermedius* may be able to persist while *C. penicillatus* cannot locate enough seeds to remain active. Thus, as with the gerbils, the temporal and spatial patterns exhibited by these two *Chaetodipus* species are not independent and they demonstrate shared-preferences habitat selection. The two pairs of congeners partition resources along a temporal axis, the scale being daily for the gerbils, and annual for the pocket mice.

Certainly other factors may influence these relationships such as finer scale dynamics of resource availability. Further analysis of the resource base is required to determine if this is the mechanism by which these two species partition the resource under stressful conditions, but this hypothesis of partitioning fits well with a shared-preference habitat selection model. While the occurrence of this phenomenon is observable without manipulation, the mechanism would appear to be highly complex, but at least involves variability in physiological tolerances, resource heterogeneity, differing foraging efficiencies, and competition.

9.4 Concluding Remarks and Synthesis

The concept of resource partitioning between sympatric species dominated studies of rodent communities (Rosenzweig and Winakur 1969; Grant 1972; Brown 1975; Rosenzweig 1977; Brown et al. 1979; Price 1986; Kotler and

Brown 1988; Brown and Harney 1993) that tried to explain how ecologically-related species partitioned their food, habitat or time of activity to allow co-existence (Schoener 1974a, 1986a). Here, we provide evidence for and discuss the contribution of temporal partitioning to the concept of resource-partitioning dimensions and species coexistence.

9.4.1 Temporal Partitioning and Its Relation to Species Coexistence

Both gerbils and pocket mice rely on the time axis to coexist. Gerbils partition the time of activity on a daily scale while pocket mice coexist due to environmental differences between different seasons on an annual scale. Between-year variability may also provide a temporal-partitioning dimension to allow the existence of competing species on a much larger scale. Although they represent different temporal scales, we can still attempt to generalise the contribution of time to coexistence and species diversity. In particular, we would like to emphasise three points emerging from our study systems that connect the time dimension to competition theory and community organisation.

The first point deals with the relationship between the competitively-limiting factor and the resource-partitioning dimensions. The temporal axis does not necessarily represent the limited factor on which species compete, but a dimension along which species can partition the limited factor and thereby coexist. For example, suppose a particular food type is a shared limiting factor for species. Competitive coexistence may occur when each species utilises a similar food type in different habitats or a similar food type at different times (as in cases 1 and 2 of Schoener 1974a, respectively). As a result, we may observe spatial or temporal partitioning between the species. However, in this case, food type is still the competitively-limiting factor that enhances interspecific competition; time or space are the dimensions along which species can partition it and thereby coexist. Hence, we should distinguish between the limiting factor on which species compete and the environmental axes on which each species can gain an advantage over the other. This may happen by consuming that limiting factor differently, and, thereby, generating the required niche differentiation for coexistence (e.g. Levins 1979; Chesson and Warner 1981; and see Kotler and Brown 1988; Cornell and Lawton 1992 for reviews). It is important to recognise that for a given limiting factor, several 'solutions' (i.e., mechanisms of coexistence) may be available. This is why we can group the above different temporal scales together to discuss the 'temporal-partitioning dimension'. The specific temporal scale may depend on the similarity between the species, on the range of the environmental axes in the particular environment, or even on the evolutionary scenario by which speciation events led to the evolution of the related species or groups of species. This also leads us to the conclusion that observing a given partitioning pattern between species does not necessarily tell us by what mechanism the species coexist.

The second point deals with the difference between a species' realised and fundamental niches. Very often, the resource-partitioning dimension (and in our case temporal partitioning) represents the species' realised niche, i.e., the observed resource use of species given the presence of other species in the community. However, to understand the contribution of resource partitioning, and more specifically temporal partitioning, to species coexistence, we should know the primary preference of each species (i.e., its resource preference in the absence of the competitor) for the competitively limiting factor as well as the primary preferences to other resource-partitioning dimensions. Although species' realised niches reflect the way species actually use resources, differences or similarities in species fundamental niches provide the basic understanding for the potential mechanisms that may allow for coexistence. Hence, the community organisation of the species play an important role in understanding the contribution of temporal partitioning to coexistence. For example, in the case of the gerbil species, both species prefer the first part of the night for their activity due to the better conditions available then. Temporal partitioning here is a product of the exclusion of *G. allenbyi* by *G. pyramidum* from the preferred part of the night. This emphasises the crucial effect of interference for the produced temporal-partitioning pattern and coexistence. It does not diminish the contribution of the temporal dimension for coexistence; without the differential ability of the species to consume resources during different parts of the night, coexistence might not have existed! Therefore, community organisation, through the fundamental niches of the species, is impoartant for understanding how temporal partitioning mediates and allows niche differentiation to promote coexistence.

The third point recognises the dependency between the different resource-partitioning dimensions. Temporal partitioning is not necessarily independent of spatial partitioning (for a general view of the tie between spatial and temporal scales see Holling 1992). For example, imagine that particular food types are available during specific times of the day in particular habitats (e.g. competitors specialising on prey types that are active during different hours of the day in different habitats will show an 'apparent spatio-temporal partitioning'). When interference competition occurs, the dominant species may be capable of monopolising a few preferred parts of different niche dimensions. This is true especially if these dimensions link together to provide the best resource gain. The subordinate species may be simultaneously excluded from preferred parts of the niche dimensions. Without looking carefully at the dependency between the different dimensions we could easily misinterpret the niche relationship of the species. Recognising one resource partitioning dimension should not suggest that we stop studying other potential resource-partitioning dimensions that may also be part of the picture.

We emphasise interference competition because it may be more likely to produce this dependency between temporal and other resource-partitioning dimensions. This is because a monopolisable limiting factor may be distributed in a specific habitat during a specific part of the day (or season). Given the likelihood of community organisation based on shared preferences of closely-related species, one, and maybe the most parsimonious outcome is a mechanism of coexistence based on a trade-off between dominance and foraging advantage on a linked spatio-temporal dimension. Therefore, temporal partitioning should at least be considered to promote coexistence.

9.4.2 Synthesis

We treated the three points above independently in order to focus on different important points regarding temporal partitioning and competition theory. However, the three of them represent a closely related sequence of processes and patterns. This does not mean that the sequence we found is the only general sequence expected from any two-species competitive system. Rather, different sequences may emerge depending on the relationship between the species, their primary resource utilisations, and the heterogeneity of the environment which they occupy. Temporal partitioning may provide a potential axis along which species can partition the limited factor and thereby coexist in many other sequences as well.

Closely-related species may have similar resource-utilisation curves due to their physiological, morphological or behavioural constraints. These, in turn, are likely to produce similar primary preferences (fundamental niches) for many resources and environmental dimensions. At this point we have to introduce another set of conditions: are the species' fundamental niches 'included' or 'reciprocal' (Colwell and Fuentes 1975)? Included niches mean that one species niche is included within the other species' niche. Reciprocal niches means that although species' niches overlap greatly, each species can use some resources that are not available to the other. Given included niches, the species with the smaller niche must better utilise the limiting resource, or must monopolise the limited resource by interference. As a result, one of the two following trade-offs may allow for coexistence: foraging advantage (e.g. higher consumption rate) versus tolerance (higher ratio of the variance of the fundamental niche to the best fitness possible), and interference (i.e., dominance) versus tolerance. A more tolerant species can exploit varied resources that cannot be used with the same efficiency by an intolerant species. However, in the case of reciprocal niches, an additional trade-off may exist: tolerance on one part of the niche dimension versus tolerance on the other part of the niche dimension. A specific example is the habitat selection of the gerbils in the presence of three habitat types (Rosenzweig and Abramsky 1986). The community organisation and the produced trade-off for coexistence determine whether species partition the limiting factor directly. Whenever this is impossible, other dimensions may help indirectly to partition the limiting factor. In the two systems presented in this chapter, both the spatial and the temporal dimensions allowed each species to have some advantage over the other by consuming the limiting factor better in different parts of those dimensions. The dependency between the spatial and temporal dimensions through the mechanism of coexistence suggests that we should examine them together and not necessarily separately.

The purpose of this chapter is not to dismiss or diminish the contribution of the concept of resource-partitioning dimensions. Rather, we wish to point out that the observation of such resource-partitioning dimensions is the first step on a journey whose roots are the primary preferences of the competing species and the trade-off(s) that allow for coexistence. We must distinguish between the processes and their products; in most cases we observe the products, but we really want to understand the processes. We need to order the ecological processes and patterns correctly. To do so, we should adopt mechanistic approaches to the study of species coexistence and community structure (e.g. Price 1986; Schoener 1986b). These should take into consideration the complexities mentioned above. For our studies we used two such approaches: 'a theory of habitat selection' (Rosenzweig 1981) and 'mechanisms of coexistence on a single resource' (Brown 1986, 1989a). By using these approaches we were able to distinguish between the processes emerging from the primary ecological needs of the species and the products, i.e., the resource-partitioning. We hope we were able to convince you that understanding competitive coexistence is more complex than has been previously thought, *but* that this complexity is not hopeless. We have the scientific tools to understand it.

Acknowledgements. We would like to thank Mike Rosenzweig for initialising the communication regarding this chapter, reviewing different versions of the manuscript, and supporting us during the entire writing. We would also like to thank Goggy Davidovitz and Elizabeth Sandlin for reviewing earlier versions of the manuscript. The gerbils study was done with the help of Zvika Abramsky and Burt Kotler from the Institute for Desert Research and Ben-Gurion University, Israel.

References

Abramsky Z (1988) The role of habitat and productivity in structuring desert rodent communities. Oikos 52:107–114

Abramsky Z, Pinshow B (1989) Changes in foraging efforts in two gerbil species correlate with habitat type and intra- and interspecific activities. Oikos 56:43–53

Abramsky Z, Rosenzweig ML, Pinshow B, Brown JS, Kotler BP, Mitchell WA (1990) Habitat selection: an experimental field test with two gerbil species. Ecology 71:2358–2369

Abramsky Z, Rosenzweig ML, Pinshow B (1991) The shape of a gerbil isocline measured using principles of optimal selection. Ecology 72:329–340

Abramsky Z, Rosenzweig ML, Subach A (1992) The shape of a gerbil isocline: an experimental field study. Oikos 63:193–199

Abramsky Z, Ovadia O, Rosenzweig ML (1994) The shape of a *Gerbillus pyramidum* (Rodentia: Gerbillinae) isocline: an experimental field study. Oikos 69:318–326

Bar Y, Abramsky Z, Gutterman Y (1984) Diet of Gerbilline rodents in the Israeli desert. J Arid Environ 7:371-376

Bowers MA, Thompson DB, Brown JH (1987) The spatial organization of a desert rodent community: food addition and species removal. Oecologia 72:77-82

Brown JH (1975) Geographical ecology of desert rodents. In: Cody ML, Diamond JM (eds) Ecology and evolution of communities. Harvard University Press, Cambridge, pp 315-341

Brown JH, Harney BA (1993) Population and community ecology of heteromyid rodents in temperate habitats. In: Genoways HH, Brown JH (eds) Biology of the heteromyidae. Am Soc Mammal Spec Publ 10, Lawrence, pp 618-651

Brown JH, Reichman OJ, Davidson DW (1979) Granivory in desert ecosystems. Annu Rev Ecol Syst 10:201-227

Brown JH, Davidson DW, Munger JC, Inouye RS (1986) Experimental community ecology: the desert granivory system. In: Diamond JM, Case TJ (eds) Community ecology. Harper and Row, New York, pp 41-61

Brown JS (1986) Coexistence on a resource whose abundance varies: a test with desert rodents. PhD thesis, University of Arizona, Tucson

Brown JS (1988) Patch use as an indicator of habitat preference, predation risk, and competition. Behav Ecol Sociobiol 22:37-47

Brown JS (1989a) Coexistence on a seasonal resource. Am Nat 133:168-182

Brown JS (1989b) Desert rodent community structure: a test of four mechanisms of coexistence. Ecol Monogr 59:1-20

Brown JS, Kotler BP, Mitchell WA (1994) Foraging theory, patch use and the structure of a Negev desert granivore community. Ecology 75:2286-2300

Charnov EL (1976) Optimal foraging: the marginal value theorem. Theor Pop Biol 9:129-136

Chesson PL, Warner RR (1981) Environmental variability promotes coexistence in lottery competitive system. Am Nat 117:923-943

Colwell RK, Fuentes ER (1975) Experimental studies of the niche. Annu Rev Ecol Syst 6:281-310

Cornell HV, Lawton JH (1992) Species interactions, local and regional processes, and limits to the richness of ecological communities: a theoretical perspective. J Anim Ecol 61:1-12

Danin A (1978) Plant species diversity and plant succession in a sandy area in the Northern Negev. Flora 167:400-422

French AR (1993) Physiological ecology of the Heteromyidae: economics of energy and water utilization. In: Genoways HH, Brown JH (eds) Biology of the heteromyidae. Am Soc Mammal Spec Publ 10, Lawrence, pp 509-538

Fretwell SD (1972) Populations in a seasonal environment. Princeton University Press, Princeton

Fretwell SD, Lucas HLJ (1970) On territorial behavior and other factors influencing habitat distribution in birds. Acta Biotheor 14:16-36

Frye R (1983) Experimental field evidence of interspecific aggression between two species of kangaroo rat (*Dipodomys*). Oecologia 59:74-78

Giller PS (1984) Community structure and the niche. Chapman and Hall, London

Grant PR (1972) Interspecific competition among rodents. Annu Rev Ecol Syst 3:79-106

Holling CS (1992) Cross-scale morphology, geometry, and dynamics of ecosystems. Ecol Monogr 62:447-502

Hoover KD, Whitford WG, Flavill P (1977) Factors influencing the distribution of two species of *Perognathus*. Ecology 58:877-884

Hutchinson GE (1957) Concluding remarks. Cold Spring Harbor Symp Quant Biol 22:415-427

Kotler BP, Brown JS (1988) Environmental heterogeneity and the coexistence of desert rodents. Annu Rev Ecol Syst 19:281-307

Kotler BP, Brown JS (1990) Rates of seed harvest by two species of gerbilline rodents. J Mammal 71:591-596

Kotler BP, Brown JS, Subach A (1993) Mechanisms of species coexistence of optimal foragers: temporal partitioning by two species of sand dune gerbils. Oikos 67: 548-556

Levins R (1979) Coexistence in a variable environment. Am Nat 114:765-783

Linder Y (1988) Seasonal differences in thermoregulation in *G. allenbyi* and *G. pyramidum* and their contribution to energy budget (in Hebrew with an English abstract). MS thesis, Ben-Gurion University Negev, Beer-Sheva
MacArthur RH (1958) Population ecology of some warblers of northeastern coniferous forests. Ecology 39:599–619
MacArthur RH (1964) Environmental factors affecting bird species diversity. Am Nat 98: 387–397
MacArthur RH, Levins R (1964) Competition, habitat selection, and character displacement in a patchy environment. Proc Natl Acad Sci USA 51:1207–1210
MacArthur RH, Recher H, Cody M (1966) On the relation between habitat selection and species diversity. Am Nat 100:319–332
May RM (1973) Stability and complexity in model ecosystems. Princeton University Press, Princeton
M'Closkey RT (1983) Desert rodent activity: response to seed production by two perennial plant species. Oikos 41:233–238
Menge JL, Menge BA (1974) Role of resource allocation, aggression, and spatial heterogeneity in coexistence in two competing intertidal starfish. Ecol Monogr 44:189–209
Mitchell WA, Abramsky Z, Kotler BP, Pinshow B, Brown JS (1990) The effect of competition on foraging activity in desert rodents: theory and experiments. Ecology 71:844–854
Nelson J, Chew R (1977) Factors affecting seed reserves in the soil of a Mojave Desert ecosystem, Rock Valley, Nye County, Nevada. Am Midl Nat 97:300–320
Park T (1962) Beetles, competition and populations. Science 138:1369–1375
Pianka ER (1974) Niche overlap and diffuse competition. Proc Natl Acad Sci USA 71:2141–2145
Pimm SL, Rosenzweig ML, Mitchell WA (1985) Competition and food selection: field tests of a theory. Ecology 66:798–807
Price MV (1978) The role of microhabitat specialization in structuring desert rodent communities. Ecology 58:1393–1399
Price MV (1986) Structure of desert rodent communities: a critical review of questions and approaches. Am Zool 26:39–49
Price MV, Reichman OJ (1987) Distribution of seeds in Sonoran Desert soils: implications for heteromyid rodent foraging. Ecology 68:1797–1811
Reichman OJ (1975) Relation of desert rodent diets to available resources. J Mammal 56:731–751
Reichman OJ (1984) Spatial and temporal variation in seed distributions in desert soils. J Biogeogr 11:1–11
Reichman OJ, Price MV (1993) Ecological aspects of heteromyid foraging. In: Genoways HH, Brown JH (eds) Biology of the heteromyidae. Am Soc Mammal Spec Publ 10, Lawrence, pp 539–574
Rosenzweig ML (1973) Habitat selection experiments with a pair of coexisting heteromyid rodent species. Ecology 54:111–117
Rosenzweig ML (1974) On the optimal aboveground activity of bannertail kangeroo rats. J Mammal 55:193–199
Rosenzweig ML (1977) Coexistence and diversity in heteromyid rodents. In: Stonehouse B, Perrins C (eds) Evolutionary biology. Macmillan, London, pp 89–99
Rosenzweig ML (1981) A theory of habitat selection. Ecology 62:327–335
Rosenzweig ML (1985) Some theoretical aspects of habitat selection. In: Cody M (ed) Habitat selection in birds. Academic Press, New York, pp 517–540
Rosenzweig ML (1991) Habitat selection and population interactions: the search for mechanisms. Am Nat 137:S5–S28
Rosenzweig ML, Abramsky Z (1986) Centrifugal community organization. Oikos 46:339–348
Rosenzweig ML, Sterner P (1970) Population ecology of desert rodent communities: body size and seed husking as a basis for heteromyid coexistence. Ecology 51:217–224
Rosenzweig ML, Winakur J (1969) Population ecology of desert rodent communities: habitat and environmental complexity. Ecology 50:558–572

Schmidly DJ, Wilkins KT, Derr JN (1993) Biogeography. In: Genoways HH, Brown JH (eds) Biology of the heteromyidae. Am Soc Mammal Spec Publ 10, Lawrence, pp 319–356

Schoener TW (1974a) Resource partitioning in ecological communities. Science 185:27–39

Schoener TW (1974b) The comparison hypothesis and temporal resource partitioning. Proc Natl Acad Sci USA 71:4169–4172

Schoener TW (1986a) Resource partitioning. In: Anderson D, Kikkawa J (eds) Community ecology: pattern and process. Blackwell, Oxford, pp 91–126

Schoener TW (1986b) Mechanistic approaches to community ecology: a new reductionism? Am Zool 26:81–106

Shkolnik A (1971) Diurnal activity in a small desert rodent. Int J Biometeor 15:115–120

Stewart FM, Levin BR (1973) Partitioning of resources and the outcome of interspecific competition: a model and some general considerations. Am Nat 107:171–198

Vance RR (1984) Interference competition and the coexistence of two competitors on a single limiting resource. Ecology 65: 1349–1357

Whittaker RH, Levin SA (eds) (1975) Niche theory and application. Dowden, Hutchinson and Ross, New York

Zahavi A, Wharman J (1957) The cytotaxonomy, ecology, and evolution of the gerbils and jirds of Israel (Rodentia: Gerbillinae). Mammalia 21:341–380

Ziv Y (1991) Mechanisms of coexistence, based on theories of optimal foraging and habitat selection, between *Gerbilus allenbyi* and *G. pyramidum* in the sandy habitats of the Western Negev (in Hebrew with an English abstract). MS thesis, Ben-Gurion University Negev, Beer-Sheva

Ziv Y, Abramsky Z, Kotler BP, Subach A (1993) Interference competition, and temporal and habitat partitioning in two gerbil species. Oikos 66:237–246

Ziv Y, Kotler BP, Abramsky Z, Rosenzweig ML (1995) Foraging efficiencies of competing rodents: why do gerbils exhibit shared-preference habitat selection? Oikos 73:260–268

10 Wood Mice – Small Granivores/Insectivores with Seasonally Variable Patterns

John R. Flowerdew

10.1 Introduction

The wood mouse (*Apodemus sylvaticus*) is found in a variety of habitats from woodland to ploughed fields and sand dunes to moorland (Flowerdew 1991). It has catholic feeding habits, taking fruits, seeds, shoots and invertebrates (Hansson 1985) and is thus broadly classed as a 'granivore/insectivore'.

The locomotory activity patterns of the wood mouse have been studied in the laboratory (Elton et al. 1931; Ostermann 1956; Saint-Girons 1959; Gelmroth 1970; Bovet 1972; Gurnell 1975; Lodewijckx et al. 1985; Miller 1995), in artificial enclosures using passage counters (Gurnell 1975) and in the field by live-trapping (Elton et al. 1931; Brown 1956; Kikkawa 1964), by passage counters (Halle 1988), by direct observation (Holisova 1961; Greenwood 1978), by radioactive tagging (Kikkawa 1964; Karulin et al. 1976) and by radiotelemetry (Wolton 1983). In general, these studies show that wood mice have a nocturnal, predominantly monophasic, but sometimes biphasic, pattern of activity (briefly reviewed by Montgomery and Gurnell 1985) and that short-term rhythms can be detected within these circadian rhythms (Elton et al. 1931; Ashby 1972; Wolton 1983). Some daytime activity has also been observed under field and laboratory conditions (Miller 1955; Ostermann 1956; Bäumler 1975; Gurnell 1975; Wolton 1983; Halle 1988) although this may reflect activity normally taking place in burrows or because of disruption to normal activity by live-trapping (Montgomery and Gurnell 1985).

It should be noted, however, that the various techniques used to monitor activity provide differing information and may lead to varying interpretations (Tapp et al. 1968; see also Appendix). Live-trapping and release offers only an indication of when most captures are occurring; interest in traps may vary with season (Kikkawa 1964) and capture may alter subsequent behaviour (Jewell 1966). Radiotelemetry and radioactive tagging provide details of 'in nest' and 'out of nest' behaviour and may provide additional details of movements. Activity counters in the field only provide data on the passage of an unknown number of unidentified animals through the counters and assume that it is the species being studied which triggers the counter.

Ecological Studies, Vol. 141
Halle/Stenseth (eds.) Activity Patterns in Small Mammals

Direct observation of activity provides more details of the timing of activity and its composition but has not been possible in long-term studies.

Small mammal activity may be affected by maintenance activities, growth, reproduction and many social and environmental factors (Falls 1968; Madison 1985; Montgomery and Gurnell 1985). This review will describe the patterns of locomotory activity found in wood mice and then consider the internal and external constraints which affect them.

10.2 Seasonal Patterns of Activity

The wood mouse is predominantly nocturnal and its activity is usually constrained by the seasonal changes in day length. The monophasic pattern of locomotory activity, characteristic of movements out of the nest from spring to autumn, is modified to become more biphasic in spring (Gurnell 1975; Wolton 1983; Lodewijckx et al. 1984).

Examples of these seasonal changes in locomotory activity and individual variation are available from field studies of individual wood mice (Wolton 1983). In the late spring and summer (Fig. 10.1a–c) male and female activity is predominantly monophasic (breeding females especially may show additional diurnal activity; Fig. 10.1b). This essentially monophasic pattern then changes in both sexes to become biphasic from October to March (Fig. 10.1d–g). The monophasic pattern of summer changes with decreasing day length to become strongly biphasic in the autumn, similar to early spring (Fig. 10.1g), and then more weakly biphasic in winter (Fig. 10.1d); some individuals show reduced total activity (less time out of the nest) at this time (Fig. 10.1e). As day length increases into the spring the biphasic pattern becomes stronger again (Fig. 10.1f–g). The biphasic patterns of autumn and early spring show peaks of locomotory activity (the greatest proportion of each half-hour spent away from the nest) occurring 2–4 h after sunset and around 2–4 h before sunrise (Brown 1956; Gurnell 1975; Wolton 1983; Halle 1988). As shown in Fig. 10.1e, the length of time spent out of the nest is not constant through the season and decreases when nights become longer (Wolton 1983). This may seem unusual, but may be related to a number of physiological and environmental changes which take place in the autumn and winter. In winter the mean body weight decreases, reproduction commonly stops and food supplies are stored below ground (Flowerdew 1991); in addition external temperatures are low and cover is at a minimum. These factors may or may not cause activity above ground to be reduced (cf. discussion of the effects of temperature and reproduction below).

In midsummer much time is spent out of the nest making almost maximum use of the available time between sunset and sunrise; at this time return to the nest is usually closer to sunrise than at any other time of the year (Wolton 1983). In studies where actual movements have been recorded there

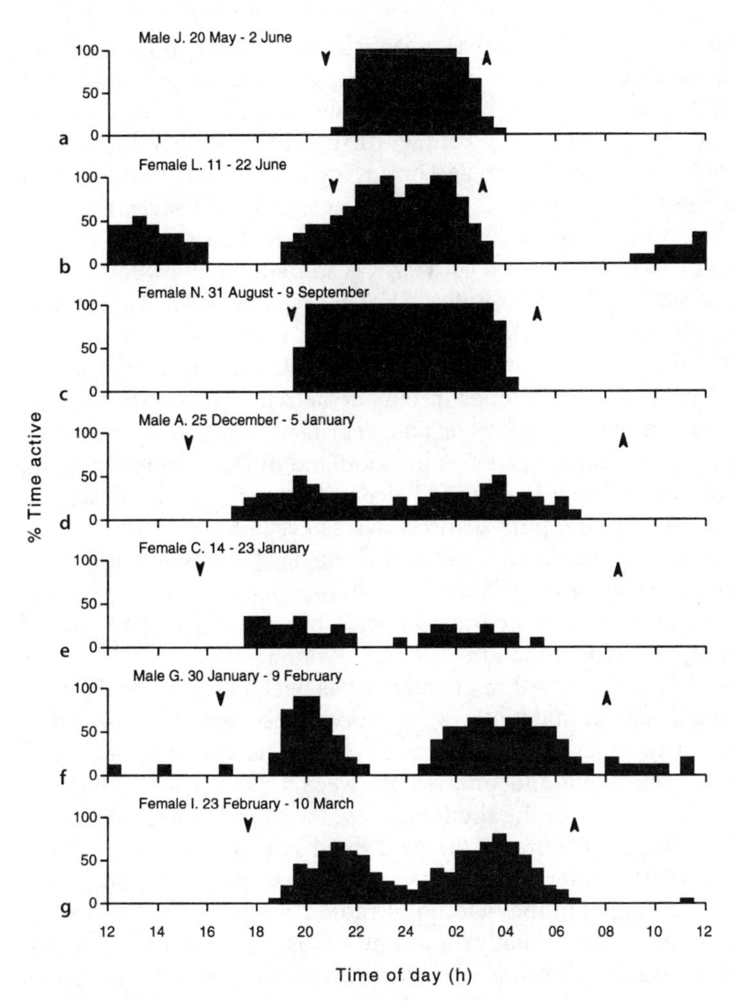

Fig. 10.1. Representative activity patterns of wood mice as indicated in the field by radio-tracking. For each 0.5-h interval the percent time active was computed from data collected over several days, assuming that periods of 15 min or more away from the nest in any 30-min period indicated 'activity'. The number of days in which the mouse was out of the nest for more than 15 min in the particular 0.5-h period was divided by the total number of days over which the mouse was monitored (expressed as a percent) to give percent time active. *Arrow heads* indicate times of sunset (*down*) and sunrise (*up*), respectively. (Adapted from Wolton 1983)

is a peak of locomotory activity around midnight or soon afterwards (Gurnell 1975; Wolton 1983; Lodewijckx et al. 1984); where trapping studies provide the data, there is a peak of captures soon after sunset (Brown 1956).

There is much variation between individuals in the precise pattern of activity shown each night and activity during daylight is not unusual (Ostermann 1956; Bäumler 1975; Gurnell 1975; Halle 1988; Corp et al. 1997). However, there is often no significant difference between sexes and different age groups studied in the laboratory (Lodewijckx et al. 1984). In the field the most pronounced difference in activity between the sexes is the additional bout of activity around midday in breeding females in June and July reported in Scottish woodland by Wolton (1983; Fig. 10.1b). However, this diurnal activity is not always restricted to females, occurring occasionally in both sexes in winter and early spring (Wolton 1983) and again in both sexes for circa 60 min each day throughout the breeding season in further observations in Scotland (Corp et al. 1997). Diurnal activity was sporadic in the wood mouse population studied by passage counters (Halle 1988) between August and December and absent in mid-October, but the sex of the animals was not known. The daytime activity described by Gurnell (see Fig. 3 in Gurnell 1975) occurred from July to October in outside activity cages with passage counters, but was absent in November and February; again either sex may have been responsible. Trapping studies in woodland in Derbyshire, England, indicate that diurnal activity in June is twice as common in males than in females (n = 26, from 40 trapping sessions over 20 years) and that juveniles will also be caught occasionally in traps during the day (Flowerdew, unpubl.). Diurnal activity was apparently absent from the populations studied in woodland in southern England by Brown (1956), possibly because the shortest nights in June were longer than in Scotland (Wolton 1983).

A short-term (ultradian) rhythm of activity has been observed in the field only during long nights in midwinter when wood mice leave their nests and return several times each night (Wolton 1983). Modal durations between these outings for one male and one female were 1.5–2.0 h and 3.0–3.5 h, respectively but the analysis of the rhythm for two other individuals was more ambiguous. It seems possible that in the field small sample sizes and individual variation prevent definite conclusions on ultradian presence and periodicity. During short nights in the field no ultradian rhythm was observed as the mice were out of the nest usually for a continuous period of time (Wolton 1983). Laboratory studies showing short-term rhythms give varying results and are thought to be unreliable (Wolton 1983). Thus it seems likely that ultradian rhythms in the longer biphasic activity periods are present in some individuals but further investigation is needed. It should be noted that Madison (1985), in reviewing activity in *Microtus* species, states that the period of the ultradian rhythm is very variable between and within individuals. In addition, there is a strong correlation between the frequency of the ultradian rhythm and body size (smaller body weight, shorter periodicity; Crowcroft 1953; Daan and Slopsema 1978); the relationship holds between and within species (Halle and Stenseth 1994). If these characteristics of *Microtus* are present in wood mice it may account for the variation observed in studies of wood mouse ultradian rhythms and render the detection of the ultradian rhythm difficult.

10.3 Environmental and Physiological Influences on Activity Patterns

Both environmental and physiological constraints impinge on the patterns of wood mouse locomotory activity. The information available on these constraints is varied, coming from laboratory and/or field studies and sometimes only from preliminary observation on one or a few individuals.

10.3.1 Environmental Influences

Photoperiodic changes, i.e., the times of sunset and sunrise, are generally thought to be important constraints on the activity of nocturnal mammals (Miller 1955; Falls 1968). However, studies of wood mice in the laboratory and in the field indicate that while sunset is an important influence on the start of activity (Fig. 10.2), sunrise is not a strong influence on the timing of return to the nest (Wolton 1983; Montgomery and Gurnell 1985). The nocturnal pattern breaks down in yellow-necked mice (*Apodemus flavicollis*) under high light intensity in the laboratory (Erkinaro 1973). Observations in the field have shown that wood mouse activity is reduced by moonlight (Kikkawa 1964). Radiotelemetry studies showed that the influence of moonlight was strongest at temperatures above 2–4 °C, when it tended to inhibit nocturnal activity (Wolton 1983). Further, the time spent out of the nest by male wood mice increases with ambient temperature and, in general, the influence of temperature is probably more important than moonlight below approximately 2–4 °C (Wolton 1983). This relationship may also be true for females but more data are needed. In a trapping study temperatures below freezing also appeared to cause reduced activity (Brown 1956).

Observations on the influence of snow cover on locomotory activity show possible seasonal and daily effects. Karulin et al. (1976) state that activity was of a shorter duration in winter under snow cover than in summer. This is similar to the seasonal change to a shorter activity time in longer (winter)

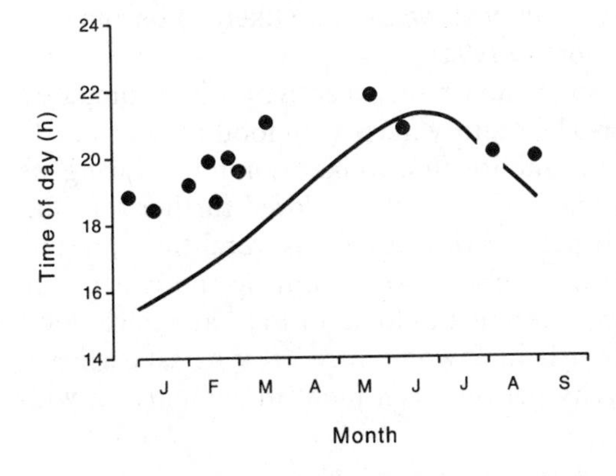

Fig. 10.2. Relationship between the time of first emergence from the nest each night (*dots*) and time of sunset (*solid line*). (Adapted from Wolton 1983)

nights discussed earlier. Wolton (1983) followed the activity of two male wood mice for a number of days in winter by radiotelemetry; his results suggest that snow cover lengthens the time spent out of the nest compared with ordinary winter nights. One male spent a mean of 9.2 h out of the nest during 4 days of snow but only 4.7 h out in 8 days without snow. The second male spent 7.2 h out and 7.9 h out in 2 days of snow compared with a mean of 5.0 h over 9 days without snow. Heavy rain during the night also appears to reduce activity as measured by catches in traps (Brown 1956; Gurnell 1976), but not in every case (Evans 1942).

Habitat characteristics also influence activity. In grasslands along field margins wood mice prefer tall vegetation and avoid activity in short vegetation, especially on moonlit nights (Plesner-Jensen and Hones 1995). Similarly, in cereal fields they emigrate or reduce activity in the stubbles after harvest (Tew and Macdonald 1993).

10.3.2 Physiological Influences

In the laboratory, breeding female wood mice are 30% more active than non-breeding females (Campbell 1974). Another laboratory study (Gelmroth 1970) shows more diurnal activity in lactating females, probably as a result of foraging for additional food (Grodzinski 1985; cf. variation in breeding female summer activity above). However, no difference in activity pattern could be found between sexually active and inactive wood mice captured in the field and placed immediately in activity recorders in the laboratory (Lodewijckx et al. 1985).

Parasitic infections can cause profound changes in the behaviour and activity patterns of their hosts (Healing and Nowell 1985) but detailed studies of laboratory or free-living wood mice are few. In the laboratory an individual infected with *Salmonella* showed reduced activity compared with a 'recovered' individual (Gelmroth 1970). Parasitism by the nematode *Heligmosomoides polygyrus* has been studied in the cereal ecosystem (Brown et al. 1994); infected wood mice travelled further and had higher movement parameters in a radiotelemetric study, but whether the infection caused the increase in activity or highly active individuals were more likely to become infected is debatable (Tew and Macdonald 1994).

Wood mice without food and water show reduced activity when compared with controls (Gelmroth 1970). In a laboratory maze with food removed early in the daylight (Gurnell 1975) the mouse first looked for food during its normal evening activity at dusk and subsequently explored further for food before returning to the nest. This pattern was repeated several times, resulting in a peak of activity before midnight; however, during the rest of the night and the next day the mouse continued to look for food at a lower level of activity. In a maritime sand dune habitat wood mice left the nest earlier in the evening and had longer activity periods each night in comparison with mice in deciduous woodland; the differences are attributed to lower food availability in the sand dunes (Gorman and Zubaid 1993).

Deprivation of food at low temperatures causes torpor in laboratory wood mice (Walton and Andrews 1981). Torpid mice were cold to the touch with their eyes fully or partly closed and some could not right themselves properly while others were quite active; the inactive mice were huddled together. In the field hypothermia has been found in three mice collected from a nest (Morris 1968); their body surface temperatures were at least 10 °C below normal, air temperature was about 3 °C and they showed behaviour similar to the active torpid individuals described above. Such changes in behaviour could have a marked effect on locomotory activity and further work is needed on this subject.

Wood mouse behaviour is also affected by organophosphate insecticide. Dell'Omo and Shore (1996a, b) have shown that in both the laboratory and in the field sublethal doses of dimethoate significantly reduce locomotory activity and induce unusual behaviour, at least on a temporary basis.

10.4 Intra- and Interspecific Influences on Activity

10.4.1 Intraspecific Influences

In summer, wood mice live in small social groups centred on single territorial females and their young. These breeding groups will interact with neighbouring territorial breeding females and with males and transients of both sexes in overlapping home ranges (Wolton 1985; Wolton and Flowerdew 1985). This changes to a less organised system with communal nesting in winter (Wilson et al. 1993). Behavioural interactions within field populations are difficult to study and little detail is known. In spring, movements to a baiting point have been observed to occur as a series of visits by mainly male and female associations in an ordered sequence during the night (Garson 1975). Thus social behaviour may influence the start and end of locomotory activity, the distribution of peaks of locomotory activity within the activity period and the total amount of activity shown by individuals.

Locomotory activity patterns are very variable between individuals (see above). Activity ranges over a larger area during the breeding season than in winter, especially in males (Attuquayefio et al. 1986; Tew and Macdonald 1994), although the distance travelled each night may not increase to the same extent as not all of the current home range is visited each night (Wolton 1985; Tew and Macdonald 1994). It is possible, however, that the total amount of locomotory activity increases during the short nights of much of the breeding season in parallel with the amount of time spent out of the nest (see above); individuals with the largest home ranges travel the furthest each night and move the quickest and most frequently (Tew and Macdonald 1994).

There is also some evidence in the laboratory of reduced activity (later capture after sunset) in juveniles compared with adults (Gurnell 1978). Other laboratory studies suggest that activity increases when unfamiliar individuals are placed in established groups of wood mice (Richard-Yris 1978). However, evidence of dynamic social interaction (radio fixes at the same place at the same time) is not common in the field (Tew and Macdonald 1994) and so verification of these laboratory results is needed. Groups of three or four wood mice in an enclosure (8.9 m^2) showed periods of one to several weeks of social behaviour with concurrent activity and frequent encounters interspersed with asocial periods when the subordinate individuals were active at times when the dominant individual was not (Bovet 1972). It is suggested, however, that avoidance of dominant individuals in the wild may take the form of dispersal and avoidance in space rather than in time (Bovet 1972).

10.4.2 Interspecific Influences

Wood mice have probably evolved the nocturnal habit in order to avoid predation (Wolton 1983). This is perhaps advantageous in not being so frequently preyed upon by mainly diurnal avian predators such as kestrels (*Falco tinnunculus*) and mammals such as weasels (*Mustela nivalis*) and stoats (*Mustela erminea*) which hunt by day and night and take many diurnal small mammals. The weasel, especially the female, however, can still take small mammals from their burrows and nests at any time of day (King 1989). Central European owl species show no main activity period in general and only a slight increase at dawn and dusk when all species are pooled (Halle 1993); more information is needed on the activity of individual predators and their wood mouse prey. The common peaks of activity observed once or twice during the night in population studies of activity (e.g. Elton et al. 1931; Brown 1956) indicate that there is some synchrony between the activity of different individuals and so possibly 'safety-in-numbers' works as an anti-predator strategy. Wood mice also show various behaviour patterns which can be interpreted as 'anti-predator' strategies; these include not coming to the surface as much on moonlit nights (see above) and sometimes preferring habitats with dense ground cover (Flowerdew 1991) or dense high-level cover (Montgomery et al. 1991).

When wood mice are confronted by yellow-necked mice in the laboratory they usually exhibit submissive behaviour (Montgomery 1978). Enclosure studies give similar results with wood mice avoiding yellow-necked mice, being dominated by them in a passive way so that the wood mice tend to hide and stay inactive (Hoffmeyer 1973). From this it is argued by Hoffmeyer (1973) that there may be some time segregation between the dominant and subordinate species as between bank voles and yellow-necked mice (Andrzejewski and Olszewski 1963) or spatial segregation by separating vertically, but evidence for this is weak (Montgomery 1980). If these behavioural characteristics are demonstrated in the field then the two species have the facility for interspecific competition through direct interference (Montgomery 1985). In populations of both species where one or the other was removed Montgomery (1985) found that trap points formerly visited mainly by the other species were being visited more frequently by the remaining species. However, the characteristic associations of each species with particular vegetation structures remained after the removals and so it is concluded that although interspecific competition may help maintain the spatial segregation of the two *Apodemus* species in shared habitats, their use of space is intrinsically different, possibly demonstrating 'the ghost of competition past'.

Interactions between wood mice and bank voles have been studied from the point of view of temporal segregation in woodland habitats (Greenwood 1978). The study indicates that it is probably the bank vole which changes its circadian pattern of activity to avoid contact with the wood mouse and not vice versa. This compares with the reduction of bank vole activity observed by Wójcik and Wołk (1985) at times of high density of yellow-necked mice. Interactions with short-tailed voles (*Microtus agrestis*) have been deduced from trapping studies by Brown (1956); she suggests that wood mice avoid grassland habitats at times of high *Microtus* density and colonise the grassland at times of low *Microtus* density. More information is needed on possible interactions with other small rodents and shrews.

10.5 Discussion

It is generally considered that the activity rhythm of an animal is composed of a short-term (ultradian) feeding rhythm over which is imposed a circadian rhythm (Southern 1954; Madison 1985). The circadian rhythm is usually strongly endogenous (Calhoun 1945; Miller 1955) and probably entrained by photoperiodic influences and temperature (Wolton 1983). The start of nocturnal activity is strongly linked to the time of sunset (Fig. 10.2) but the total length of time spent out of the nest is probably determined by a complex of factors such as weather, breeding and feeding (Wolton 1983). In wood mice it is only during the winter that the ultradian rhythm is easily detected. At this time the animals are least constrained by the demands of breeding and, often in the autumn, the need to forage widely for food, especially in woodland. Within the long dark period there is enough time to carry out essential activity and to return to the nest periodically (Wolton 1983). The short periodicity (1.5–2.0 h) of the ultradian rhythm in the field is considered to depend on metabolic demands such as hunger, thirst and excretion and may correspond to the time taken to empty a full stomach (Wolton 1983). However, Elton et al. (1931) suggest that the bimodality of the autumn and spring activity rhythm reflects the time taken to fill, digest and empty the stomach twice in the night, a much longer period, and so further work is needed on this problem. The persistence of the ultradian rhythm in the absence of food intake would negate the stomach-filling hypothesis explaining the short-term rhythm (Halle and Stenseth 1994).

The type of circadian rhythm observed may be related to the type of food ingested. Grodzinski (1962) found that *A. agrarius* had a monophasic rhythm with concentrated, high-calorific value food, and a triphasic rhythm with bulky low calorific food.

Clethrionomys and *Microtus* species feed on more bulky diets than wood mice and their activity patterns are biphasic or polyphasic (Miller 1955; Brown 1956; Bäumler 1975; Greenwood 1978; Rijnsdorp et al. 1981; Halle 1995). In addition, temporal changes in feeding preferences have been observed during the biphasic (autumn) activity period (Plesner-Jensen 1993); she found that wood mice show most variety in their choice of food at times corresponding to the peaks of activity at the start and end of the night and that both the amount of food eaten and the variation in it diminished from the first peak to the second peak of activity (whether similar changes occur under monophasic activity patterns is a matter for further study). The expected change from selection of protein-rich foods early in the night to carbohydrate-rich foods later in the night (Tempel et al. 1985) was not found. Only sugars were selected for early in the night (Plesner-Jensen 1993) and it is suggested that the mice may be selecting for a quick source of energy at the beginning of the first active period to help increase body temperature.

Feeding strategy may also explain the diurnal activity shown by breeding female wood mice (and some males; Fig. 10.1b). Wolton (1983) argues that in the summer most of the nocturnal activity is taken up with feeding to cope with the heavy energy demands of breeding, range and mate defence and that possibly food may be increasingly difficult to find. Under these circumstances, at least at northerly latitudes, the breeding females have to become active in broad daylight in order to find sufficient food. This argument is rejected by Halle (1988) because he found diurnal activity in passage counter studies in the field in autumn when breeding had stopped and food was abundant, explaining it by attributing great variability to individuals' activity patterns. Further work is obviously needed to fully understand the occurrence of, and reasons for, diurnal activity, especially as an unknown proportion of the population cause the readings on the passage counters in the day. Part of an explanation of individuality may come from woodland live-trapping studies in June and December (Flowerdew, unpubl.) where breeding females, juveniles and low weight individuals are often captured during the day. Small individuals have not been radio-tracked and they may need to feed in the day, despite abundant food, because of their small size or to avoid intraspecific encounters. Such individuals might still be present in the population in December, as found by Halle (1988), even if breeding had stopped earlier in the year. In addition, Haim et al. (1993) argue that in *A. hermonen-*

sis diurnal activity, which occurs only on cold days, is possibly an adaptation for survival in a cold environment.

It is clear that a wide range of environmental, physiological and social factors interact to produce the daily activity pattern of the wood mouse in the field and that much more information is needed on nearly all aspects of the subject.

References

Andrzejewski R, Olszewski J (1963) Social behaviour and interspecific relations in *Apodemus flavicollis* (Melchior 1834) and *Clethrionomys glareolus* (Schreber 1780). Acta Theriol 7:155-168

Ashby KR (1972) Patterns of daily activity in mammals. Mamm Rev 1:171–185

Attuquayefio DK, Gorman ML, Wolton RJ (1986) Home range sizes in the wood mouse *Apodemus sylvaticus*: habitat, sex and seasonal differences. J Zool Lond 210:45–53

Bäumler W (1975) Activity of some small mammals in the field. Acta Theriol 20:365–377

Bovet J (1972) On the social behaviour in a stable group of long-tailed field mice (*Apodemus sylvaticus*). II. Its relations with distribution of daily activity. Behaviour 49:55–67

Brown LE (1956) Field experiments on the activity of the small mammals *Apodemus*, *Clethrionomys* and *Microtus*. Proc Zool Soc Lond 126:549–564

Brown ED, Macdonald DW, Tew TE, Todd IA (1994) *Apodemus sylvaticus* infected with *Heligmosomoides polygyrus* (Nematoda) in an arable ecosystem: epidemiology and effects of infection on the movements of male mice. J Zool Lond 234:623–640

Calhoun JB (1945) Diel activity rhythms of the rodents, *Microtus ochrogaster* and *Sigmodon hispidus*. Ecology 26:251–273

Campbell I (1974) The bioenergetics of small mammals, particularly *Apodemus sylvaticus* (L.) in Wytham Woods, Oxfordshire. PhD thesis, University of Oxford

Corp N, Gorman ML, Speakman JR (1997) Ranging behaviour and time budgets of male wood mice *Apodemus sylvaticus* in different habitats and seasons. Oecologia 109:252–250

Crowcroft P (1953) The daily cycle of activity in British shrews. Proc Zool Soc Lond 123: 715–729

Daan S, Slopsema S (1978) Short-term rhythms in foraging behaviour of the common vole, *Microtus arvalis*. J Comp Physiol A 127:215–227

Dell'Omo G, Shore RF (1996a) Behavioral and physiological effects of acute sublethal exposure to dimethoate on wood mice, *Apodemus sylvaticus*. 1. Laboratory studies. Arch Env Cont Toxicol 31:91–97

Dell'Omo G, Shore RF (1996b) Behavioral effects of acute sublethal exposure to dimethoate on wood mice, *Apodemus sylvaticus*. 2. Field studies of radio-tagged mice in a cereal ecosystem. Arch Env Cont Toxicol 31:538–542

Elton C, Ford EB, Baker JR, Gardner AD (1931) The health and parasites of a wild mouse population. Proc Zool Soc Lond 1931:657–721

Erkinaro E (1973) Activity optimum in *Microtus agrestis*, *Arvicola terrestris* and *Apodemus flavicollis* (Rodentia) dependent on the intensity of illumination. Aquilo Ser Zool 14:89–92

Evans FC (1942) Studies of a small mammal population in Bagley Wood, Berkshire. J Anim Ecol 11:182–197

Falls JB (1968) Activity. In: King JA (ed) Biology of *Peromyscus* (Rodentia). Am Soc Mammal Spec Publ 2, Stillwater, pp 543–570

Flowerdew JR (1991) Wood mouse *Apodemus sylvaticus*. In: Corbet GB, Harris S (eds) The handbook of British mammals, 3rd edn. Blackwell, Oxford, pp 220–229

Garson PJ (1975) Social interactions of wood mice studied by direct observation in the wild. J Zool Lond 177:496–500

Gelmroth KG (1970) Über den Einfluß verschiedener äußerer und innerer Faktoren auf die lokomotorische Aktivität der Waldmaus (*Apodemus sylvaticus* L.). Z Wiss Zool 180:368–388

Gorman ML, Zubaid ABMA (1993) A comparative study of the woodmouse *Apodemus sylvaticus* in two contrasting habitats - deciduous woodland and maritime sand-dunes. J Zool Lond 229:385–396

Greenwood PJ (1978) Timing of activity of the bank vole (*Clethrionomys glareolus*) and the wood mouse (*Apodemus sylvaticus*) in a deciduous woodland. Oikos 31:123–127

Grodzinski W (1962) Influence of food upon the diurnal activity of small rodents. Symp Theriol Praha 1962:134–140

Grodzinski W (1985) Ecological energetics of bank voles and wood mice. Symp Zool Soc Lond 55:169–192

Gurnell J (1975) Notes on the activity of wild wood mice, *Apodemus sylvaticus*, in artificial enclosures. J Zool Lond 175:219–229

Gurnell J (1976) Studies on the effects of bait and sampling intensity on trapping and estimating wood mice, *Apodemus sylvaticus*. J Zool Lond 178:98–105

Gurnell J (1978) Observations on trap response in confined populations of wood mice, *Apodemus sylvaticus*. J Zool Lond 185:279–287

Haim A, Rubal A, Harari J (1993) Comparative thermoregulatory adaptations of field mice of the genus *Apodemus* to habitat challenges. J Comp Physiol B 163:602–607

Halle S (1988) Locomotory activity pattern of wood mice as measured in the field by automatic recording. Acta Theriol 33:305–312

Halle S (1993) Diel pattern of predation risk in microtine rodents. Oikos 68:510–518

Halle S (1995) Diel pattern of locomotor activity in populations of root voles, *Microtus oeconomus*. J Biol Rhythms 10:211–224

Halle S, Stenseth NC (1994) Microtine ultradian rhythm of activity: an evaluation of different hypotheses on the triggering mechanism. Mamm Rev 24:17–39

Hansson L (1985) The food of bank voles, wood mice and yellow-necked mice. Symp Zool Soc Lond 55:141–168

Healing TD, Nowell F (1985) Diseases and parasites of woodland rodent populations. Symp Zool Soc Lond 55:193–218

Hoffmeyer I (1973) Interaction and habitat selection in the mice *Apodemus flavicollis* and *A. sylvaticus*. Oikos 24:108–116

Holisova V (1961) Observation of long-tailed field mouse (*Apodemus sylvaticus*) under natural conditions. Folia Zool 10:85–86

Jewell PA (1966) The concept of home range in mammals. Symp Zool Soc Lond 18:85–109

Karulin BE, Nikitina NA, Khlyap LA, Litvin VY, Okhotskii YV, Al'Bov SA, Sushkin ND, Pavlovskii YS (1976) Diurnal activity and territory utilization of common field mice (*Apodemus sylvaticus*) according to observations of animals marked by Cobalt-60. Zool Zhur 55:112–121

Kikkawa J (1964) Movement, activity and distribution of the small rodents *Clethrionomys glareolus* and *Apodemus sylvaticus* in woodland. J Anim Ecol 33:259–299

King CM (1989) The natural history of weasels and stoats. Helm, London

Lodewijckx E, Verhagen R, Verheyen WN (1984) Activity patterns of wild wood mice, *Apodemus sylvaticus* (L.), from the Belgian northern Campine. Ann Soc R Zool Belg 114:291–301

Lodewijckx E, Verhagen R, Verheyen WN (1985) Influence of sex, sexual condition and age on the amount of activity of wild wood mice, *Apodemus sylvaticus* (L.). Ann Soc R Zool Belg 115:75–85

Madison DM (1985) Activity rhythms and spacing. In: Tamarin RH (ed) The biology of new world *Microtus*. Am Soc Mammal Spec Publ 8, Shippensburg, pp 373–419

Miller RS (1955) Activity rhythms in the wood mouse, *Apodemus sylvaticus*, and the bank vole, *Clethrionomys glareolus*. Proc Zool Soc Lond 125:505–519

Montgomery WI (1978) Intra- and interspecific interactions of *Apodemus sylvaticus* (L.) and *A. flavicollis* (Melchior) under laboratory conditions. Anim Behav 26:1247–1254

Montgomery WI (1980) The use of arboreal runways by the woodland rodents, *Apodemus sylvaticus* (L.), *A. flavicollis* (Melchior) and *Clethrionomys glareolus*. Mamm Rev 10:189–195

Montgomery WI (1985) Interspecific competition and the comparative ecology of *Apodemus* species. In: Cook LM (ed) Case studies in population biology. Manchester University Press, Manchester, pp 126–187

Montgomery WI, Gurnell J (1985) The behaviour of *Apodemus*. Symp Zool Soc Lond 55:89–115

Montgomery WI, Wilson WL, Hamilton R, McCartney P (1991) Dispersion of the wood mouse, *Apodemus sylvaticus*: variable resources in time and space. J Anim Ecol 60:179–192

Morris PA (1968) Apparent hypothermia in the wood mouse (*Apodemus sylvaticus*). J Zool Lond 155:235–236

Ostermann K (1956) Zur Aktivität heimischer Muriden und Gliriden. Zool Jahrb (Physiol) 66:355–388

Plesner-Jensen S (1993) Temporal changes in food preferences of wood mice (*Apodemus sylvaticus* L.). Oecologia 94:76–82

Plesner-Jensen S, Honess P (1995) The influence of moonlight on vegetation height preference and trappability of small mammals. Mammalia 59:35–42

Richard-Yris MA (1978) Experimental study of interindividual relationships in small groups of males of long-tailed field mice, *Apodemus sylvaticus*, murid rodent. Biol Behav 287–304

Rijnsdorp A, Daan S, Dijkstra C (1981) Hunting in the kestrel, *Falco tinnunculus*, and the adaptive significance of daily habits. Oecologia 50:391–406

Saint-Girons MC (1959) Les caractéristiques du rhythme nycthéméral d'activité chez quelques petites Mammifères. Mammalia 23:245–276

Southern HN (ed) (1954) Control of rats and mice. Vol 3: House mice. Clarendon Press, Oxford

Tapp JT, Zimmerman RS, D'Encarnacao PS (1968) Intercorrelational analysis of some common measures of rat activity. Psychol Rep 23:1047–1050

Tempel DL, Shor-Posner G, Dwyer D, Leibiwucz SF (1985) Diurnal patterns of macronutrient intake in freely feeding deprived rats. Am J Physiol 256:R541–R548

Tew TE, Macdonald DW (1993) The effects of harvest on arable wood mice *Apodemus sylvaticus*. Biol Cons 65:279–283

Tew TE, Macdonald DW (1994) Dynamics of space use and male vigour amongst wood mice, *Apodemus sylvaticus*, in the cereal ecosystem. Behav Ecol Sociobiol 34:337–345

Walton JB, Andrews JF (1981) Torpor induced by food deprivation in the wood mouse *Apodemus sylvaticus*. J Zool Lond 194:260–263

Wilson WL, Montgomery WI, Elwood RW (1993) Population regulation in the wood mouse *Apodemus sylvaticus* (L.). Mamm Rev 23:73–92

Wójcik JM, Wołk K (1985) The daily rhythm of two competitive rodents: *Clethrionomys glareolus* and *Apodemus flavicollis*. Acta Theriol 30:241–258

Wolton RJ (1983) The activity of free-ranging wood mice *Apodemus sylvaticus*. J Anim Ecol 52:781–794

Wolton RJ (1985) The ranging and nesting behaviour of wood mice, *Apodemus sylvaticus* (Rodentia: Muridae), as revealed by radio-tracking. J Zool Lond 206:203–224

Wolton RJ, Flowerdew JR (1985) Spatial distribution and movements of wood mice, yellow-necked mice and bank voles. Symp Zool Soc Lond 55:249–275

11 Voles – Small Graminivores with Polyphasic Patterns

Stefan Halle

11.1 Microtine Rodents, a Special Case of Diel Activity Patterns

Microtine rodents, i.e., voles and lemmings, are probably the group of small mammals for which most work on the ecological implications of diel activity patterns has been done. There are several reasons for a special interest in this group. First of all, microtines are in general a very popular system for ecological studies due to practical advantages like high trappability, easy handling, and a wide distribution with respect to geographical ranges and diversity of habitats. In some older studies information on the temporal pattern of activity was included to round up the description of the species' natural history. Second, microtines were thoroughly studied this century because of the fascinating phenomenon of cyclic density fluctuations, which provided solid knowledge about the general ecology of this group. Third, microtines typically appear in great numbers which makes them an important link in food chains of northern ecosystems. They contribute a lot to the flow of energy and nutrients, first from primary production to higher trophic levels as important prey for a wide range of predators (Halle 1993), second from primary production back to the decomposer system as small but numerous grazers (e.g. Grodzinski et al. 1977).

Finally, early research on microtines revealed that the pattern of diel activity differs substantially from other small mammal groups. Microtines are the most prominent representatives of a pattern type coined as 'polyphasic', although it was originally described for rats and mice (Szymanski 1920). In contrast to the 'normal' monophasic pattern with one long activity phase per 24-h day during either day or night, microtines exhibit a sequence of several short activity cycles with activity outside the nest or burrow during both day and night (cf. Ashby 1972). Hence, the prevailing feature that shapes microtine activity is not the 24-h component of the circadian rhythm, but rather a so-called ultradian rhythm (Daan and Aschoff 1981) with a period length of about 2–6 h.

Apparently, every microtine species surveyed so far follows this short-term rhythm in activity (for overviews see Madison 1985 and Halle and Stenseth

Ecological Studies, Vol. 141
Halle/Stenseth (eds.) Activity Patterns in Small Mammals

1994). For this feature body size is obviously not the crucial constraint. The range of microtine species with polyphasic patterns stretches from water voles (*Arvicola terrestris*) to pine voles (*Pitymys* spp.), while the activity patterns of murid rodents (references given in Halle and Stenseth 1994; see also Chap. 10), kangaroo rats (Chap. 8) and gerbils (Chap. 9) of about the same size are dominated by the circadian component. Also, the activity patterns of the much smaller bats are characterised by the circadian component (Chap. 14). Only shrews, which also show a distinct polyphasic pattern, seem to support a relationship with body size. Shrews are, however, a different case (Ashby 1972; Daan and Slopsema 1978), since there is clear evidence that the short-term rhythm of their activity simply reflects an exceptionally short-term hunger cycle due to small body size and high metabolic rates (Saarikko and Hanski 1990; Chap. 13). In contrast, the ultradian rhythm in microtines is obviously triggered by an endogenous pacemaker, as verified by the observation that the short-term rhythm is free-running in constant darkness (e.g. Daan and Slopsema 1978), and by experiments in which the short-term rhythm persisted under food, water or sleep deprivation (Daan and Slopsema 1978; Gerkema and van der Leest 1991). An endogenous clock mechanism that triggers the ultradian activity rhythm must be assumed when hunger, thirst or tiredness can be excluded as driving forces, since there is no environmental feature fluctuating with a frequency of 2–6 h that could possibly act as the zeitgeber.

So obviously, the activity pattern of microtine rodents represents a special case that has to be related to other peculiarities of microtine biology to find ultimate explanations for its evolutionary establishment. Unfortunately, there is more than one plausible approach. Microtines are highly specialised on bulky food with high cellulose content and low energy gain; hence, the ultradian pattern may be causally related to metabolism and digestion constraints. Microtines are also a prominent prey item for a large variety of aerial and ground predators; so possibly, the ultradian rhythm may be interpreted as an anti-predator strategy. Furthermore, population cycles are found in this group, which strongly hints at peculiarities with respect to population dynamics and demography; hence, the special behavioural feature of polyphasic activity may be somehow related to the density fluctuations. Finally, microtine populations are highly structured social communities; so the remarkable activity pattern may possibly be the result of peculiar social interactions. This coarse outline of ideas in fact comprises the main hypotheses that were put forward to explain ultradian rhythmicity in microtine rodents. It is interesting to realise that all of them involve ecological thinking, hence it is not surprising that attempts to understand diel activity patterns in terms of behavioural ecology occurred quite early in studies of this group.

11.2 A Historical Perspective

As early as 1920 a most remarkable comparative study on activity behaviour across a wide range of taxa was published by Szymanski from the University of Wien, in which, however, microtine rodents were not considered. The earliest quantitative study on microtine activity that I am aware of was done with *Microtus agrestis* by Davis (1933), working at Elton's Bureau of Animal Population at Oxford. Like Szymanski, Davis used specially designed cages named 'actographs' or 'kymographs'. They were quite popular in the early years of activity research with the common feature that sophisticated mechanical constructions translated animal movements into readable records (see Appendix). Davis' study revealed a distinct short-term rhythm with a period length of 2–4 h, which he interpreted as a consequence of small stomach capacity and low food quality. As part of this study he also did an early photoperiod manipulation experiment, in which activity patterns of voles living in constant darkness and normal light conditions were compared.

Four years later, a short but very interesting paper on *Microtus pennsylvanicus* was published by Hamilton (1937) in Ecology. He started off by reviewing the little information that was available at that time on *Microtus* activity and realised that even in the small data base of a mere handful of studies, a considerable confusion was prevalent regarding the apparently simple question as to whether *Microtus* is diurnal or nocturnal (cf. Erkinaro 1969). Hamilton's idea was that the discrepancies may be caused by the different methods employed to measure activity, especially in activity cages. So he monitored activity in the field by trapping with both snap- and live-traps, expecting data from the field to be "more truthworthy" than cages. He was, however, well aware that measuring activity in the natural habitat is technically much more challenging. In contrast to Davis (1933), who found caged voles to be more active at night, trapping success in Hamilton's study was highest during the day, and particularly so after dawn and before dusk. He suspected predation to be the determining factor for the different activity behaviour of free-ranging voles. Interestingly, the two early papers by Davis and Hamilton outlined some of the most persistent topics in the debate about *Microtus* activity during the following decades.

The next milestone was a study published by Calhoun in 1945, in which he presented a – for his times – high-tech device to record activity at individual as well as group levels. In my view, the most important contribution of this study was that 'activity' was for the first time not taken as a whole, but was instead broken down into several components that were compared directly. Furthermore, the results clearly indicated that the rhythm is of endogenous nature but may be affected by social interactions. Again, photoperiod manipulation was used as a tool to explore the features of the activity rhythm. The 1950s saw two methodologically similar studies by Miller (1955) and

Brown (1956), both dealing with ecological implications of diel activity patterns, i.e., predation, interspecific competition, and social interactions. Both studies included data on *Clethrionomys* activity, which became a second thoroughly studied vole genus besides *Microtus*. At the same time two studies with microtine rodents were published in Germany by Ostermann (1956) and Moyat (1957) which could be classified as the earliest modern chronobiological work in the sense that lab experiments with photoperiod manipulation were conducted to scrutinise rhythmicity.

Chronobiological experiments dominated the almost exploding literature on activity patterns in the 1960s and 1970s which cannot be reviewed here. This development was fostered by the outstanding pioneering work of Jürgen Aschoff at the Max-Planck-Institute at Erling-Andechs in Bavaria/Germany, who established general rules on diel activity rhythms and the underlying biological clocks. Although this research relied on birds at first, small mammals like microtine rodents and hamsters became a second widely used experimental system. To a considerable extent, physiological and metabolic aspects of activity patterns were the main focus during these decades, while the ecological meaning of the patterns was at best part of the discussion (with a few exceptions, e.g. Buchalczyk 1964).

An important turn was the work of Eino Erkinaro, partly in Finland and partly in the lab of Jürgen Aschoff (Erkinaro 1961, 1969, 1972). In long-term cage recordings of *Microtus agrestis* he found a very consistent seasonal change in the patterns with high nocturnality during summer, high diurnality during winter, and shifts of the main activity phase within a few weeks or even days in autumn and spring. This phenomenon, coined 'seasonal phase-shift', was later verified by numerous studies in various microtines as well as other small mammal species. Seasonal phase-shifts were particularly obvious in high latitudes, which was assumed to be a response to the dramatically changing day length and weather conditions in the course of the year.

The first to take an explicit ecological approach to microtine activity patterns was Serge Daan and his group at Groningen in the Netherlands. Being generally interested in the adaptive significance of daily activity patterns (Daan 1981; Daan and Aschoff 1982), they studied common voles (*Microtus arvalis*) with respect to predator-prey interactions with kestrels (*Falco tinnunculus*) and hen harriers (*Circus cyaneus*) as the main aerial predators in polder grasslands. During the most intense study period of the late 1970s and early 1980s, short-term trapping of voles with parallel observations of raptors was employed as the field method (Daan and Slopsema 1978; Rijnsdorp et al. 1981; Raptor Group 1982). The main objectives of these studies were to find out whether predators are able to adjust their hunting efforts temporally to the activity pattern of the main prey, and to test the hypothesis that the peculiar ultradian rhythm in microtines is an adaptive response to high predation risk. Later this was methodologically extended to experimental approaches to investigate the possible mechanisms of predator avoidance (Gerkema and Verhulst 1990). Based on the studies on common vole activity,

Menno Gerkema in Daan's group started extensive work on the triggering mechanism of the ultradian rhythm, searching for a short-term clock that works parallel to and in combination with the circadian pacemaker (Gerkema and Daan 1985; Gerkema et al. 1990, 1993; Gerkema and van der Leest 1991).

The latest important move in this field of research was initialised by new methodological developments which allowed measurement of vole activity directly in the natural habitat. Obviously, such approaches are of crucial importance for the study of the ecological dimension of temporal behaviour that can hardly be considered in cage experiments and even trapping (see Introduction, Chaps. 1, 4 and Appendix). Since the pioneering work of Banks et al. (1975) and Peterson et al. (1976) on brown lemmings (*Lemmus trimucronatus*), radio-tracking of a few individuals over short periods of time found its way into activity studies (e.g. Herman 1977; Madison 1981; Webster and Brooks 1981). With the still ongoing miniaturisation of transmitters and particularly batteries, this method is likely to be of increasing importance for studies of microtine activity in the near future.

Another trait was to develop the passage counter idea from a recording device for cages (Calhoun 1945) and enclosures (Mossing 1975) to a robust field technique for activity recording in free-ranging vole populations (Lehmann and Sommersberg 1980; Halle and Lehmann 1987). Although also burdened with specific problems (see Appendix) this method may be seen as a counterpart to radio-tracking because it covers the population level and allows for almost disturbance-free and continuous long-term recordings. Hence, a combination of radio-tracking and automatic recording is, at present, the most promising way to relate microtine activity behaviour to ecological factors like predation risk, population density, population cycle phase, competition, and habitat structure. This optimism is supported by the finding that the activity pattern of voles is affected very little by extrinsic factors (Halle 1995b). This robustness makes microtine rodents a most suitable system for experimental field studies on activity behaviour because changes in the temporal organisation, if they occur at all, can be reliably assumed as responses to treatments, and not to confounding environmental variables.

11.3 The Temporal Structure of Microtine Activity

In addition to the interesting ecological implications, the peculiar temporal structure of microtine activity as such also accounts for a specific interest in this group. In fact there are three – if not even four – different levels of temporal organisation, each worth considering on its own rights. On the shortest time-scale the principal ultradian rhythm structures the everyday live of individuals as well as of populations. The short-term rhythm, however, is clearly affected by the daily LD cycle, so the ultradian pacemaker interacts with the basic circadian clock. Since day length is not constant over time, the diel pattern has to be adjusted to the seasonally changing photoperiod. Additionally, field data hint at a possible long-term component of temporal organisation, i.e., a rhythm with a period length longer than 12 months. For an understanding of microtine activity behaviour this hierarchical structure of components has to be disentangled.

11.3.1 The Ultradian Component

11.3.1.1 General Features

Evidently, the ultradian rhythm is the feature that dominates the activity pattern of microtines. Although a variety of microtine species with different body sizes and ecological 'life-styles' was studied, the range of 2–6 h for the length of one activity cycle, i.e., the sequence of one activity bout and one rest period, was seldom exceeded to either end of the frame. Nevertheless, within this frame a clear dependency on body weight is prevalent, with the shortest cycle length in *Pitymys* spp. (1.5–3.0 h; Salvioni 1988) and the longest in *Arvicola terrestris* (4.0–6.0 h; Johst 1973; Airoldi 1979). Body weight dependency can be represented by the allometric equation

$$t = 1.00 \cdot W^{0.30},$$

with t being the period length and W being body mass. This relation holds true both among (Daan and Slopsema 1978) and within species (Gerkema and Daan 1985).

The ultradian rhythm is remarkably stable and robust. In continuous darkness the circadian 24-h rhythm disappears spontaneously in about 30% of the records, while the ultradian rhythm never fades away (Gerkema and Daan 1985; Gerkema et al. 1990). The short-term rhythm is even largely unaffected by deprivation of food, water and sleep (Daan and Slopsema 1978; Gerkema and van der Leest 1991). Deprivation experiments in particular are a very strong argument against homeostatic control of the ultradian rhythm and instead advocate for an ultradian clock, i.e., a short-term endogenous oscillator. However, robustness of the ultradian rhythmicity does not mean that period length is changeless. Rather, the bout-to-bout frequency is positively related to the metabolic rate and negatively correlated with body weight and food quality (Fig. 11.1). A modelling approach to evaluate the hypotheses on the triggering mechanism revealed that these features, together with the persistence of the rhythm under food deprivation, can only be explained by a pacemaker mechanism with some feedback loop to metabolism (Halle and Stenseth 1994). Although distinct landmarks for the characteristics of the triggering mechanism can today be derived from experimental findings, the very nature of the ultradian pacemaker itself still remains speculative. One study hints at a genetic component: cyclic *Microtus agrestis* from North Sweden showed a clear-cut ultradian rhythm in indoor

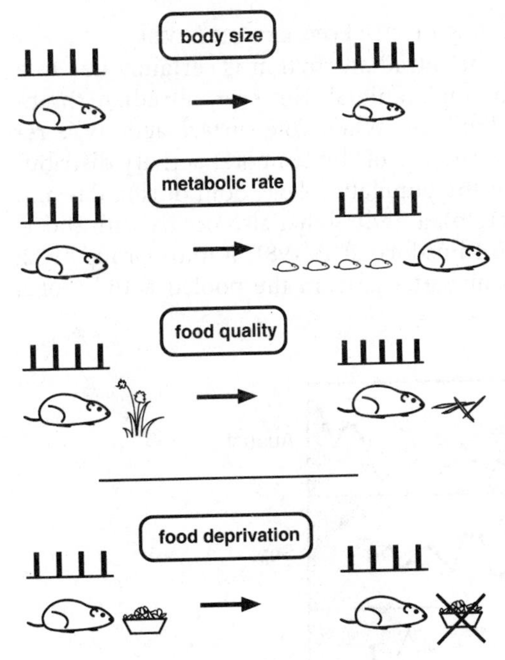

Fig. 11.1. Schematic overview of known features of microtine ultradian rhythmicity as established by experiments. The frequency of the rhythm is higher in smaller animals, with higher metabolic rates due to lactation, and with low food quality. Furthermore, the ultradian rhythm is persistent during food deprivation. (Adapted from Halle and Stenseth 1994)

enclosure recordings while non-cyclic field voles from South Sweden did not, and hybrids revealed an intermediate behaviour (Rasmuson et al. 1977).

The ultradian rhythm during both day and night is a conspicuous feature in almost every cage recording with microtine rodents. It is masked by wheel running during 'light off' and the subjective night in continuous darkness, respectively, but when the circadian rhythm has disappeared, either spontaneously or by electrolytic lesions of the suprachiasmatic nuclei (SCN), only the ultradian rhythm remains (Gerkema et al. 1990; Gerkema and van der Leest 1991). Ultradian activity bouts are always related to feeding (making up more than 50% of the time being active in bank voles, Mironov 1990), drinking and excretion (Lehmann 1976), and also heart rate and metabolic turnover fluctuate with the high frequency of the ultradian rhythm (e.g. Wiegert 1961; Ishii et al. 1993). Even coprophagy, i.e., the reingestion of hind gut faeces, is integrated with the short-term time schedule (Kenagy and Hoyt 1980; Ouellette and Heisinger 1980). Thus increasing evidence suggests that the ultimate explanation of the short-term rhythm in microtines is causally related to specific metabolic and/or digestion needs, probably as a consequence of specialisation on bulky and low-energy food resources (for a more extensive treatment of this aspect see Daan and Slopsema 1978; Daan 1981; Daan and Aschoff 1981, 1982; Halle and Stenseth 1994).

11.3.1.2 Activity Synchronisation at the Population Level

To summarise the above section, the ultradian rhythm is certainly due to a short-term clock, ticking in each single animal. However, ultradian rhythmicity does not only concern individuals. When vole surface activity is recorded in the field, the ultradian structure of the temporal activity distribution is also surprisingly distinct at the population level both during day and night and in the long term (Fig. 11.2b), a finding that also occurs with short-term trapping (Fig. 11.2a; see also Rijnsdorp et al. 1981; Raptor Group 1982; Hoogenboom et al. 1984). Clear-cut patterns from the pooled activity of a

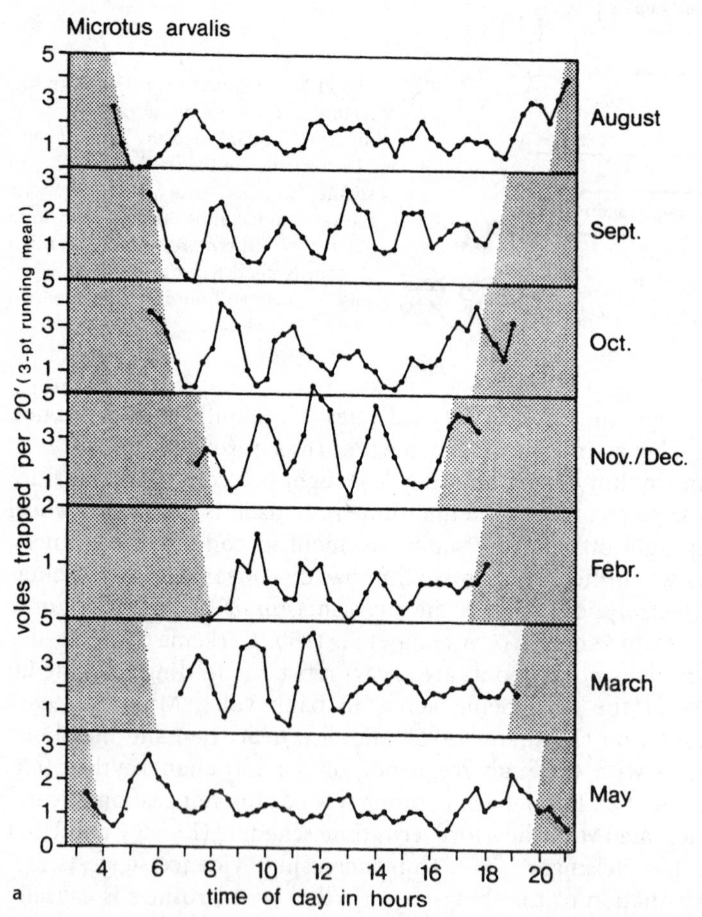

Fig. 11.2. Pattern of surface activity of microtine populations as recorded **a** in the field by short-term trapping (*Microtus arvalis*, traps checked every 20 min, patterns plotted for 1-month intervals; from Daan and Slopsema 1978), and **b** by automatic recording with passage counters (*M. oeconomus* recorded in a 50 × 100 m outdoor enclosure, percent of recorded passages per 30-min interval, patterns plotted for 2-week intervals; adapted from Halle 1995a). Data in both graphs were smoothed by a running mean for three data points. In panel **a** night is indicated by the *shaded* area, in panel **b** by the rows of sticks which stand for sunrise (*SR*) and sunset (*SS*). Panel **b** is a double plot, i.e., the same pattern is drawn twice side by side to show the complete night-time pattern in the middle

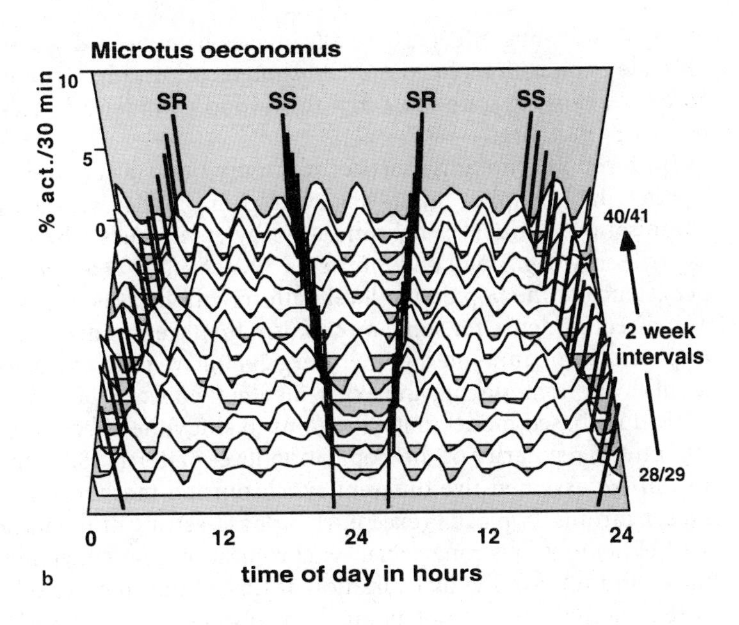

Fig. 11.2b

great number of individuals over several days will only occur with a high degree of behavioural synchrony in populations, i.e., when more or less all individuals are active on the surface and resting in their nests, respectively, at more or less the same times.

Synchronous surface activity has far-reaching consequences. First, in territorial species, as most voles in fact are, it means that space as a limited resource is used inefficiently. During activity peaks the habitat is crowded, especially when density is high, while it lies almost abandoned during the activity troughs. However, synchronous activity also increases the probability that territory keepers will meet frequently at the borderlines, which may help to maintain the spatial organisation of the population on low costs by restraining territorial conflicts (Lehmann and Sommersberg 1980). In addition, synchronised resting in nests and burrows obviously gives energetic benefits for group members due to thermoregulation (Wiegert 1961; Trojan and Wojciechowska 1968; Webster and Brooks 1981).

Second, synchronised ultradian activity severely affects predator-prey interactions. For the predator the short-term activity pattern of its prey results in an odd temporal pattern of food availability. Prey is abundant at some hours of the day, but for the rest of the time the chances for successful foraging are very low and energetic expenditures for hunting will probably exceed the prospective energy gain (cf. Chap. 4). For microtine prey, synchronous activity may reduce the relative individual risk to predation due to safety-in-number effects (Daan and Slopsema 1978; Daan and Aschoff 1981; see, however, Raptor Group 1982), due to an exchange of warning signals (Gerkema and Verhulst 1990), and possibly also due to confusing the predator by a swarming effect. These advantages have been discussed as possible ultimate explanations for the establishment of ultradian rhythmicity in microtines, referring to the fact that this group is under an exceptionally high predatory pressure.

However, interpreting short-term activity bouts as an anti-predator strategy inevitably implies that it has evolved in parallel with active social synchronisation. Hints at synchronisation by social cues where found in cage experiments (e.g. Daan and Slopsema 1978; Stubbe et al. 1986), and also the persistence of the population level pattern with increasing density was taken as indirect evidence for some mechanism that keeps individual activity bouts in phase (Lehmann and Sommersberg 1980). We have evaluated these arguments by both modelling and experimental field work. First, we developed an individual-based model that allowed us to study the effect of increasing density on the regularity of the population level pattern (Halle and Halle 1999). In this we assumed five different synchronising mechanisms, ranging from no synchronisation at all (except from phase-setting of the ultradian rhythm, see below) to very strong social synchronisation that efficiently reduces both intra- and inter-individual variation. It turned out that periodicity in the recorded activity pattern of populations always increased with density, irrespective of the assumed mechanism of synchronisation. This simply was a consequence of recording probability being highest at times when the activity phases of individuals overlap at most. This effect increased much more with the number of animals moving around than the accumulation of inter-individual variation which, in contrast, would corrupt the population level pattern.

In a second approach we tested the hypothesis that direct social contacts are crucial for synchronous activity. We studied the activity of root voles (*M. oeconomus*) in two large outdoor enclosures (50 × 100 m) with six habitat patches of identical size, but with increasing distance of open coverless matrix of short grass between patches (7.5–60 m; Halle, unpubl.). We pooled the activity recordings from two plots at a time and analysed the degree of synchronisation in the resultant patterns. In this, we found no negative relationship between the degree of synchronisation and distance between patches, which had been expected under the assumption of synchronisation by social contacts. In particular, no difference in the degree of synchronisation was found in the patterns from recordings in two plots within the same habitat patch, compared with recordings from two plots in different enclosures, a situation in which individuals never had direct contact. The latter finding in particular strongly argues for synchronisation due to entrainment by comprehensive environmental signals.

To draw a conclusion, the present picture about synchronous activity in microtine populations is the following (Fig. 11.3): the basic trigger for activity is the endogenous short-term rhythm in each single individual. Synchronisation by social cues most probably occurs on the level of family groups, dominated by adult and resident females. Synchronisation among family

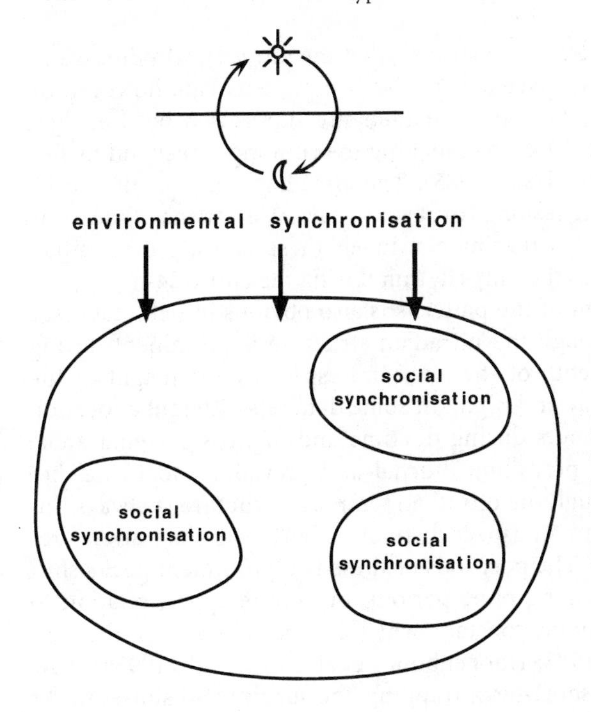

Fig. 11.3. Schematic view of the suggested levels of activity synchronisation in microtine populations. Synchronisation within family groups is mainly due to social cues, while synchronisation among family groups is achieved by entrainment through the environment, i.e., the daily light/dark cycle

groups, however, which is the crucial feature for synchronisation at the population level, is achieved by entrainment through the environment. If so, synchronisation will also occur among populations in a given area, making the task of foraging even more difficult for a predator because prey activity is low or high at specific times of the day all over the hunting grounds. Our results from up to six experimental enclosure populations studied in parallel support this conclusion, but this prediction still has to be tested under open-field conditions.

11.3.2 The Circadian Component

Although the ultradian rhythm runs more or less constantly during day and night, it has been known for a long time that the short-term rhythm is coupled with the basic circadian system (Stebbins 1975; cf. Gerkema et al. 1993), seemingly characteristic of all mammals (see Chaps. 1 and 2). In cage experiments, wheel running is performed almost exclusively during 'light off' periods (actually voles run the wheel all night through), while feeding, drinking and excretion occur in ultradian bouts during 'light on' periods. In constant conditions (darkness or light), both components are free-running with a characteristic endogenous period close to but not exactly at 24 h, denoted as τ (e.g. Lehmann 1976; Daan and Slopsema 1978). Wheel running and ultradian feeding bouts are still prevalent during the shifting subjective days and nights, respectively, demonstrating that the ultradian rhythm is under control of the circadian pacemaker. The two systems can, however, be uncoupled: wheel running can be experimentally suppressed by dim light during the night (Lehmann 1976), by blocking the running wheel and by low temperatures (Gerkema and Daan 1985). The circadian component can be eliminated by hypothalamic lesions (Gerkema et al. 1990) or may fade away spontaneously in long-term experiments. In all these cases, the ultradian component alone remains as the only rhythm during the entire 24-h day.

The circadian component of the patterns is also obvious in field data (see Figs. 11.2 and 11.6a). Although the ultradian structure is prominent during both day and night, the depth of the lows and especially the height of the peaks can differ between day and night. In some data sets different allocation of activity to short-term bouts during daytime and night is so remarkable that one can speak about prevailing diurnal and prevailing nocturnal behaviour, respectively, although the ultradian temporal structure is always underlying. As a second circadian aspect distinct activity peaks are always related to sunrise and sunset. The positions of these two prominent peaks shift on the time axis with changing photoperiods, but the temporal relation to 'light on' and 'light off' remains constant, which is especially true of the sunrise peak (Raptor Group 1982; Hoogenboom et al. 1984; Halle 1995a). Particularly in data sets from short-term trapping, the sunrise and sunset peaks are often much higher than all other activity peaks of the 24-h day, adding a marked crepuscular element to the pattern (Halle, unpubl. data). Finally, our analysis of root vole (*M. oeconomus*) activity has revealed a small but statistically significant difference in the period length of the short-term rhythm during day and night, being about 30 min longer at night (Halle 1995a). This raises the interesting question as to whether there are two different ultradian oscillators, one for the day and one for the night, or whether the period length of one and the same oscillator is altered by the environmental signals of 'light on' and 'light off', respectively.

The obvious function of the circadian component is to reset the ultradian rhythm and to phase-lock it to the daily LD cycle. For sunrise this has been known for a long time and verified by good evidence. The ultradian rhythm on the group or population level is very clear-cut after sunrise, but gradually fades away as a function of time elapsed since the last reset (Daan and Slopsema 1978; Lehmann and Sommersberg 1980; Madison 1981; Rijnsdorp et al. 1981; Hoogenboom et al. 1984; Gerkema and Daan 1985; Halle 1995a). On the basis of extensive laboratory experiments Gerkema et al. (1993) concluded "that the most likely mechanism of ultradian entrainment is that of a light-insensitive ultradian oscillator, reset every dawn by the termination of the activity phase controlled by the circadian pacemaker, which is itself entrained by the light/dark cycle" (single-circadian-cog hypothesis). A detailed analysis of the fine structure of root vole activity patterns in the field, however, clearly established that phase-setting in fact occurs twice per 24-h day, i.e., at sunrise and at sunset (Halle 1995a; cf. Lehmann and Sommersberg 1980), which is in

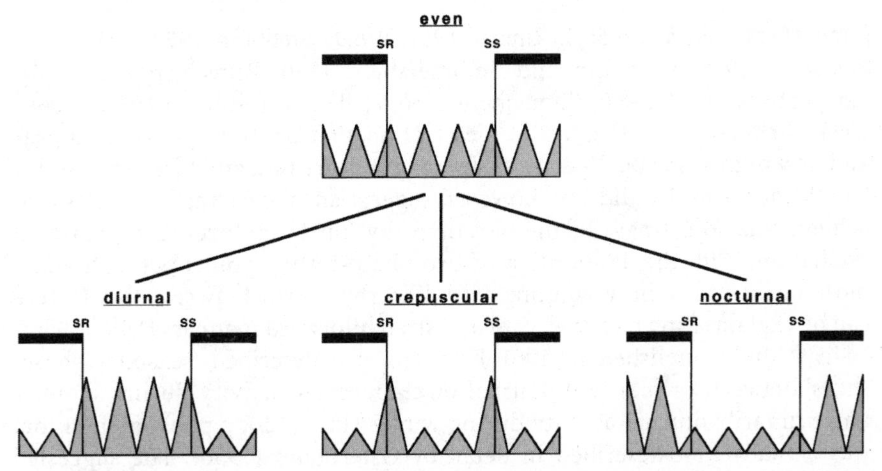

Fig. 11.4. Schematic view of the four possible pattern characteristics that can be derived from the ultradian rhythm of microtine rodents. Different activity levels at specific short-term bouts of the 24-h day will result in an either even, diurnal, crepuscular or nocturnal pattern, while the basic ultradian structure of the pattern remains unchanged. *Black bars* to the *top* of the panels indicate night-time (*SR* sunrise, *SS* sunset)

obvious accordance with the 'two-oscillator hypothesis' supposed by Pittendrigh (1960) and Daan and Berde (1978).

The summation of recordings from several days results in a clear-cut pattern on the population level, which demonstrates that phase-setting occurs every morning and evening, and entrainment with the environment happens in all individuals simultaneously. As verified by modelling, this feature seems to represent the decisive mechanism for activity synchronisation (Halle and Halle 1999). The twice repeated phase-setting per 24-h day and the coupling of the ultradian with the circadian system results in a highly flexible timing mechanism for activity, such that patterns may easily show adaptive responses to specific ecological situations (see below). In particular, while keeping the relation to the zeitgeber signals and the principal temporal structure of short-term activity and rest constant, different allocation of activity to specific activity bouts will either result in an even activity distribution, a primarily diurnal, a primarily crepuscular, or a primarily nocturnal pattern (Fig. 11.4).

11.3.3 The Seasonal Component

An evident advantage of the remarkably complex mechanism for triggering activity is that it allows the short-term rhythm to adjust to changing day length, which is dramatic in the high north. Long-term recordings of vole activity revealed that the pattern changes considerably with photoperiod, but that the short-term rhythm is always systematically related to the photic regime (Ostermann 1956; Erkinaro 1961, 1969; Stebbins 1972; Daan and Slopsema 1978; Lehmann and Sommersberg 1980; Rijnsdorp et al. 1981; Hoogenboom et al. 1984; Blumenberg 1986; Halle and Lehmann 1987; Drabek 1994). Erkinaro (1969) gave the description that daytime peaks disappear and new night-time peaks arise as day length decreases, and vice versa as day length increases. He did not, however, suggest an underlying mechanism to achieve this adjustment of the ultradian rhythm to the overruling seasonal oscillation. With the assumption of two phase-setting points per 24 h and a more-or-less constantly running ultradian rhythm in between, this feature can be explained in an easy, elegant and straightforward manner (Halle 1995a).

In a study published in 1961 Erkinaro also described 'seasonal phase-shifts' between primarily nocturnal or crepuscular activity during summer and primarily diurnal activity during winter (Fig. 11.5), a phenomenon that was actually first described in detail by Ostermann (1956). The suggested adaptive function of changing the main activity phase in autumn and spring was to avoid adverse temperature conditions. Lehmann (1976) found a seasonal variation in the sensitivity to light in common voles (*M. arvalis*) indicated by the threshold of dim light that was able to suppress wheel running at night efficiently (0.5 lx during summer but 5 lx during spring and autumn). It was speculated that differences in the light perceiving trait may hint at the mechanism that achieves the seasonal change of the prevailing activity period.

Seasonal phase-shifts were confirmed in a large number of studies, including latitude transitions experiments (e.g. Grodzinski 1963; Stebbins 1972, 1974; Herman 1977; Rowsemitt et al. 1982; Hoogenboom et al. 1984; Rowsemitt 1991; Reynolds and Gorman 1994, and many others). However, all strongly confirming evidence stems from cage experiments, while most data from the field (mainly trapping records) showed this effect only weakly, if at all (e.g. Webster and Brooks 1981; Hoogenboom et al. 1984; Ylönen 1988; Drabek 1994). Additionally, we never found any hint of a regular seasonal phase-shift in our continuous long-term recordings with passage counters in enclosed and free-ranging vole populations (Lehmann and Sommersberg 1980; Halle and Lehmann 1987, 1992; cf. Blumenberg 1986). There are, however, some remarkable exceptions: on the Orkney Islands, Reynolds and Gorman (1994) found no seasonal variation in the degree of diurnality in captive Orkney voles (*M. arvalis orcadensis*) but a marked seasonal change in the activity pattern occurred instead in free-living voles as revealed by live-trapping, with the highest degree of diurnality from late winter to spring (see also Bäumler 1975). Another piece of evidence comes from an enclosure experiment with radio-tracked *Clethrionomys gapperi*, in which Herman (1977) observed a shift from a diurnal rhythm in winter to a nocturnal and crepuscular rhythm in summer.

So undoubtedly, something real is measured with cage experiments and other techniques, since seasonal phase-shifts occur consistently in various kinds of studies on different species. But undoubtedly as well, what is meas-

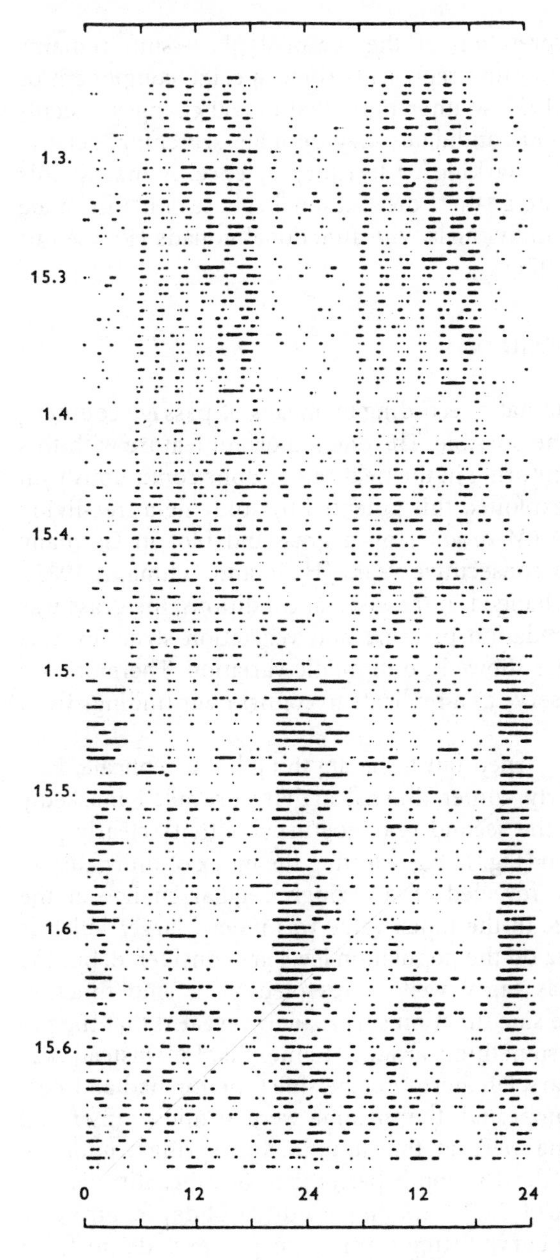

Fig. 11.5. Long-term recording (129 days) of wheel-running activity in a caged field vole (*Microtus agrestis*), showing the transition from diurnal winter activity to nocturnal summer activity; i.e., 'seasonal phase-shift'. The graph is a double plot with days 1 and 2 in the first row, days 2 and 3 in the second row, and so on. (From Erkinaro 1969)

ured is not simply activity, because the seasonal phase-shift cannot be definitely established in the field. The story of seasonal phase-shifts clearly demonstrates that results from cage studies alone must be treated extremely carefully, and that confirmation from field data is always essential before sound ecological conclusions can been drawn (see also Appendix; cf. Pearson

1962). At present, the interpretation of the seasonal phase-shift remains speculative. Probably wheel running represents some specific component of activity (De Kock and Rohn 1971; Mather 1981) that in fact changes systematically over the seasons. Good candidates may be mate searching, general exploratory activity, territory marking or territory defence. A hint at this realm is given by the finding that the seasonal phase-shift in male *M. montanus* only occurs in individuals with functional gonads (Rowsemitt 1986; cf. De Kock and Rohn 1971).

11.3.4 A Long-Term Component?

With the development of automatic recording by means of passage counters, a novel kind of survey became possible. The one important feature with this technique is that activity monitoring in the field can be done continuously on a long-term scale. We have exploited this potential by studying a free-living population of common vole (*M. arvalis*) on a grassland plot in Germany (50° 51' N, 6° 24' E) for two consecutive years (Halle and Lehmann 1987). During this period nothing changed in the general conditions, the grass was unmown the whole time, predation pressure and vegetation structure was comparable, and even weather showed no dramatic variation. Therefore, we expected the second year to serve as a replicate to confirm our findings from the first year.

However, the second year turned out to be anything but a replicate. Particularly, in autumn and early winter diurnal activity prevailed markedly during the first year, but in the second year nocturnal activity clearly prevailed during the same season (Fig. 11.6a). Plotting the index of diurnality I_D (see Appendix) against time revealed an amazingly regular change of the main activity phase. Variance in the index value was indeed large, which is not a surprise when field data on the population level are considered, but the general long-term trend was fairly well represented by a sine function (Fig. 11.6b). Curve-fitting the sine function surprisingly revealed that the period was not, as expected, something around 12 months, but instead was clearly 18 months, i.e., 1.5 years. Furthermore, the sine function had an interesting phase relation to photoperiod: the maxima were at midsummer and midwinter, while the minima were at the autumn and spring equinoxes. Hence, prominent points of the 18-month periodicity of diurnality fell together with prominent points of the 12-month periodicity of day length.

After we had checked the curve-fitting several times to exclude misfit or calculation mistakes in the analysis, we had a look at other long-term data sets from passage counter recording. One was a 13-month series on a field vole (*M. agrestis*) population living on a small and isolated meadow surrounded by woodland, some kilometres away from the above mentioned *M. arvalis* population. In this data set the I_D index also varied considerably, although the level was generally shifted to higher diurnality. A sine function with an 18-month period length and with exactly the same phase relation to

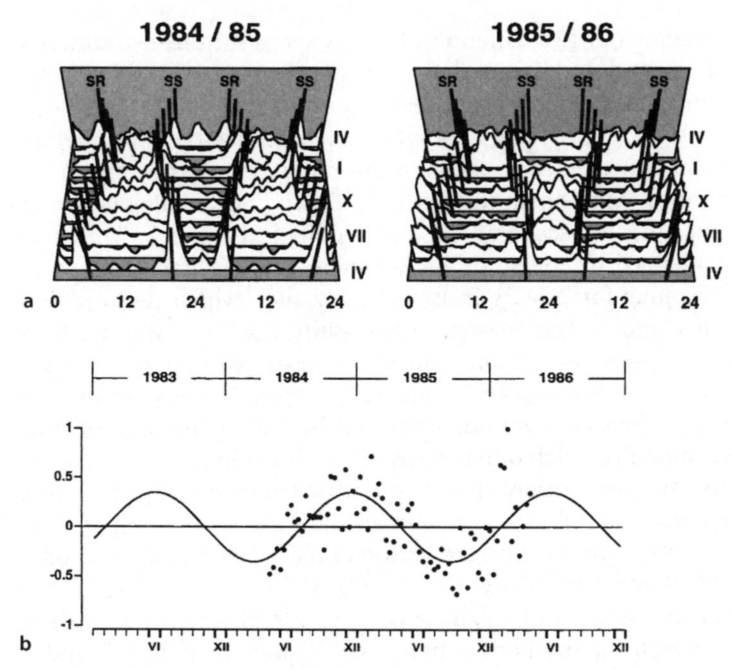

Fig. 11.6. a Activity patterns of a free-ranging common vole (*Microtus arvalis*) population as monitored for two consecutive years in the same study plot by automatic recording with passage counters (graphic representation as in Fig. 11.2b, but for monthly intervals). **b** Depiction of the diurnality index I_D, calculated for 10-day intervals for the entire 2-year study. Positive index values indicate prevailing diurnal activity (maximum +1: all activity during daytime), negative index values indicate prevailing nocturnal activity (minimum –1: all activity during night-time)

photoperiod (I_D minimum at the autumn equinox) gave a much better fit to the data than any sine function with a 12-month period length (Halle and Lehmann 1987; see, however, Reynolds and Gorman 1994).

The third data set that we analysed was again from *M. agrestis*, but from a cyclic population in North Sweden (63° 35' N, 19° 50' E), so under very different ecological conditions. Again, activity was monitored with passage counters in a meadow habitat and on a long-term scale (over 7 years), but with long interruptions in between due to the original objective of the study to compare activity behaviour among cycle phases (Halle and Lehmann 1987, 1992). However, the I_D index again showed large variations, for which a fourier analysis revealed a sine function as the first component, once again with a period length of 18 months and the same specific phase relation to photoperiod as in the other data sets. The second component was a new type of function not found before, which we interpreted as the behavioural response to changing densities in the course of the population cycle (see below).

So actually we detected identical long-term changes in the distribution of activity to day and night in all available data sets from passage counter recording. They stem from two different *Microtus* species, from two different latitudes, and from non-cyclic as well as cyclic populations. We also searched the literature for data to support or question our results, but unfortunately, there were almost no comparable data sets available, since no other method gives the information needed, i.e., long-term continuous recording in the field. So a long-term component in microtine activity behaviour may possibly be common, but undiscovered for methodological reasons. This view, however, is challenged by the study of Reynolds and Gorman (1994) with monthly live-trapping for 2 years in two Orkney sites, which revealed prevailing diurnality and a clear seasonal phase-shift, that even was synchronous between two sites. So the question of a long-term component in microtine activity must at present be considered as open, but as with the seasonal phase-shifts, apparent contradictions may be due to different methods of activity recording that catch different aspects of behaviour.

Nevertheless, the assumption of a long-term component implies several problems. First, voles are short-lived animals with life spans ranging from 3 to about 9 months, depending on the season of birth. Hence, information to generate the 18-month periodicity must be transmitted from generation to generation, but this seems to be principally possible in small mammals by entrainment through the mother's activity (see Viswanathan and Chandrashekaran 1985). Second, signals that trigger the 18-month periodicity cannot be derived directly from photoperiod information or any other environmental cue. However, in a theoretical analysis (Lehmann and Halle 1987) we have shown that it may be possible to generate the 18-month periodicity from the 12-month periodicity by a set of simple 'rules', for which the peculiar phase relation between the two periodicities is crucial. This hypothesis remains highly speculative while unverified by experiments, but 'seasonal phase-shifts' may be seen as misinterpretations of long-term changes in the amount of diurnality, resulting from the 'natural' expectation to find an annual characteristic (see discussion in Lehmann and Halle 1987).

Provided that the observed long-term component is not an artefact or just a strange coincidence, one may ask for a reasonable hypothesis as to how such a strange pattern could have evolved in microtine rodents. One idea is to interpret the change of the activity phase as a strategy to address the extremely high predation pressure. Obviously, the spectrum of predators changes depending on prevailing activity during either day or night. Together with a period different to 12 months, the regular change of the main activity phase results in temporally unpredictable appearance – or better, unpredictable availability. Microtines are a reliable food resource for diurnal predators for some time, but 9 months later they are more available for nocturnal predators. The essential point is that with an 18-month periodicity this change in availability is not related to specific seasons, so there is no pattern the predators can 'learn' and adapt to. The 'unpredictability' is indeed greatly enhanced when the change in activity behaviour is combined with the dramatic density changes in cyclic populations.

So the 'evolutionary advantage' of a long-term component may be that all predators are forced to keep an option for alternative prey for those phases when microtines are not available because they are active at times when the predator is not. This may be of particular importance for prey with a strict ultradian structure of activity. If a predator could rely on microtines as always available prey in sufficient numbers, the next step would be to adapt to the short-term rhythm by adjusting hunting efforts to the confined intervals of high profitability. Long-term changes in the main activity phase may prevent this kind of specialisation which would result in highly efficient vole predators (cf. Halle 1993).

11.4 Ecological Factors Affecting Microtine Activity Behaviour

With the combination of ultradian, circadian and seasonal rhythmic components - and perhaps a long-term change in the main activity phase in addition - the activity pattern of microtine rodents represents an exceptionally complicated case of activity timing. This may be due partly to the special feature of ultradian rhythmicity as the most prominent pace in this group of small mammals. The short-term rhythm has no counterpart in a periodic change of the environment, hence synchronisation with the seasonal course of photoperiod is a challenging task and may demand a sophisticated triggering system. On the other hand, the system of coupled rhythmic components results in high flexibility of activity timing which allows for adaptive behavioural responses to various ecological conditions.

The hypothesis that flexible activity timing helps microtines to cope with their environment is supported by a number of observations in which the activity pattern was under the influence of ecological factors. Interestingly, the factors that naively may be regarded as the most obvious to affect microtine activity, i.e., weather conditions, disturbances and moonlight, have been shown to be of relatively little importance (Halle 1995b). Only low temperatures and new moon increased the general activity level, but did not alter the temporal pattern structure, i.e., the relative distribution of activity over the 24-h day. Probably, the weak effect of these factors is due to damping by dense vegetation cover which forms the primary habitat for most microtine species.

In contrast, habitat structure and quality seem to have a profound effect on activity timing. In an extensive experimental study with passage counters we found that root voles (*M. oeconomus*) are more active at night in fragmented habitats than in continuous habitats. Furthermore, the degree of population synchrony was considerably lower in the fragmented habitats, which was, however, not a direct consequence of interrupted social contacts (Halle, unpubl. data). In the same study we also found that activity in open areas with no vegetation cover was almost entirely restricted to the night, with particularly high activity in the first hours after sunset. Another effect of the habitat structure, although more difficult to interpret, was described by Hoogenboom et al. (1984) for common voles in the Netherlands. Trapping success was higher during the daytime in a habitat with low grass vegetation on sandy soil than in a habitat dominated by reeds on clay soil. To explain this difference in activity, or – to be correct – in trappability distribution, the authors assumed the underground tunnel system to be an important factor, as tunnels were abundant in clay soil but were absent in the sandy habitat.

A second ecological factor that affects the activity pattern of microtines is interspecific competition, an aspect of diel activity behaviour which is recently under debate again (see Chaps. 4 and 9). Several studies deal with competition between microtines or between microtines and other small rodents, and the results are somewhat puzzling: Pearson (1962) reported little or no indication of an effect of competition on the activity pattern of grey-sided voles (*Clethrionomys rufocanus*) and bank voles (*C. glareolus*) although both species share the same habitat. Danielson and Swihart (1987) found spatial instead of temporal segregation between prairie voles (*M. ochrogaster*) and southern bog lemming (*Synaptomys cooperi*) in an experiment in northeastern Kansas, a species pairing in which *M. ochrogaster* was supposed to be dominant. Glass and Slade (1980), however, found just the contrary with another pairing in the same study area. Competing prairie voles and cotton rats (*Sigmodon hispidus*) shared the same habitat, but used it at different times of the day. In this case, however, *S. hispidus* was considered to be dominant over *M. ochrogaster*. Glass and Slade (1980) took their results as an example where Schoener's (1974) arguments against temporal partitioning do not hold, following a reasoning very similar to the ideas presented in Chap. 4.

Brown (1956) found wood mice (*Apodemus sylvaticus*) to be strictly nocturnal and bank voles (*C. glareolus*) to be strictly diurnal when these two highly competitive species share the same habitat. With *M. agrestis* and relaxed competition, peak activity of bank voles occurred at dusk and dawn, which seem to be favourable activity times for both species. Similar results from Greenwood (1978) and Buchalczyk (1964) with *C. glareolus* and *A. sylvaticus* and *A. flavicollis*, respectively, support the view that the vole avoids times when the antagonistic mouse is active. Wójcik and Wołk (1985) found the same at high densities of the two species, but not at low density situations. From this finding they concluded that the polyphasic activity pattern gives microtines the option to avoid severe interference competition by fine tuning activity. Taking different competitive situations as outlined above into consideration, it appears that the response of microtines to interspecific competition depends on being dominant or subordinate, with shifting of the activity phase occurring when in a subordinate position.

Intraspecific interaction is the third factor that affects vole activity patterns. We found distinct changes of the diel activity distribution among different phases of *M. agrestis* population cycles in North Sweden (Halle and

Lehmann 1987, 1992; cf. Rasmuson et al. 1977; Nygren 1978), with high diurnality during the increase and peak phase and a marked phase-shift towards prevailing nocturnality during the late decline and early low phase. Analysis of the diurnality index I_D revealed that long-term changes in the index value may be described by superimposing a basic 18-month periodicity (see above) with a second function, resembling a saturation characteristic of increasing diurnality with a sharp collapse towards night activity when it reaches its maximum. We hypothesised that this function represents the effect of changing density and/or social environment on activity behaviour, with taking the shift to nocturnality in the late decline phase as an indication for social stress (Halle and Lehmann 1992; see below). Nevertheless, whatever the underlying mechanisms are, phase-dependent changes in the preferred activity phase will considerably affect temporal predator-prey interactions in the course of the cycles (cf. Halle 1993), which has consequences for the current hypotheses on the driving force behind microtine density fluctuations.

Some findings from the field indicate that individual voles do dedicate different kinds of activity to different times of the day, so behavioural differentiation may be a fourth ecological factor. By visual observation in small indoor enclosures Hoogenboom et al. (1984) found feeding activity during both day and night bouts in common voles (*M. arvalis*) while other activities like running and digging occurred predominantly during the night. Nocturnality was enhanced in stress situations after changing the enclosure environment. From this and other dispersed pieces of evidence it appears that daily routines are performed in the surface runway system and follow a strict ultradian rhythm, while anything unusual, like exploration outside the runway system, inspection of new objects in the home range, or activity under heavily disturbed conditions or social stress, for instance, is preferably performed by night. From passage counter recording we also have first hints at a temporally differential use of different parts of the home range (i.e., some parts of the total area are only used at specific times of the day), although this needs further verification by radio-tracking in the future. Nevertheless, highly flexible activity timing as a response to external conditions has also been verified in cage experiments by Lehmann (1976), who was able to make individual common voles run the exercise wheel day or night, respectively, by changing the cage conditions.

Finally, changes in activity timing were reported as consequences of age and/or social status. In some studies there was a tendency found for juveniles to be more nocturnal than adults (Bäumler 1975; Reynolds and Gorman 1994) and to be not synchronous with the rest of the population during the first days after they left the nest (Lehmann and Sommersberg 1980; Mironov 1990). Reynolds and Gorman (1994) suggest that this feature is due to the subordinate social status and that juvenile activity timing may reduce potentially harmful interactions and feeding competition with adults. Very little is known about sex differences in activity behaviour, but most authors assume that they are small, if there are any. However, Hansen (1957) found a tendency towards higher ultradian frequencies in males, while the average amount of activity was higher for females. In addition, the sexes showed slightly different responses to the change in photoperiod (see also Pearson 1962; Rowsemitt et al. 1982; Rowsemitt 1991 and discussion in Drabek 1994). Mironov (1990) stated that lactating females had the most distinct short-term rhythm, and that their activity pattern was the pacemaker for male activity. Studying activity behaviour with respect to sex, age and social status in the field is methodologically most challenging because detailed long-term, maybe even life-time individual records are required. Further developments and miniaturisation in radiotransmitters might, however, allow such surveys in the times to come.

11.5 Conclusion

Obviously, the activity behaviour in microtine rodents is extraordinarily flexible as far as small mammals are concerned. This flexibility, which is not due an arrhythmic habit but to various distinct and interacting rhythms instead, must be seen in relation to the specific living conditions of microtines, of which the high and ever present predation risk is probably the most demanding. Other peculiarities are wide distribution ranges with consequent high diversity of ecological situations, and the dramatically changing social environment in the course of the population cycles in the north. All this might require high adaptability in general, which includes temporal strategies to maximise survival and fitness under various constraints.

The key to understanding the potential for temporal flexibility of microtines is the coupling of two endogenous rhythms, i.e., the circadian rhythm characteristic of all small mammals as the basic biological clock, and the predominant ultradian rhythm. As shown in Fig. 11.4, a variety of different activity patterns can be derived from this combination while keeping the general temporal structure of activity unchanged. The ultradian rhythm is undoubtedly the ruling pace of microtine life, but allocating different amounts of activity to specific short-term bouts at different times of the 24-h day allows fine-tuning according to the particular needs. So, microtines are an exceptionally complicated case for studying the ecological implications of activity patterns, but on the other hand, the high flexibility makes them a most suitable system for learning how ecological factors shape activity timing in the field.

References

Airoldi J-P (1979) Etude du rythme d'activité du campagnol terrestre, *Arvicola terrestris scherman* Shaw. Mammalia 43:25-52

Ashby KR (1972) Patterns of daily activity in mammals. Mammal Rev 1:171-185

Banks EM, Brooks RJ, Schnell J (1975) A radiotracking study of home range and activity of the brown lemming (*Lemmus trimucronatus*). J Mammal 56:888-901

Bäumler W (1975) Activity of some small mammals in the field. Acta Theriol 20:365-377

Blumenberg D (1986) Telemetrische und endoskopische Untersuchungen zur Soziologie, zur Aktivität und zum Massenwechsel der Feldmaus, *Microtus arvalis* (Pall.). Z Angew Zool 20:301-344

Brown LE (1956) Field experiments on the activity of the small mammals, *Apodemus*, *Clethrionomys* and *Microtus*. Proc Zool Soc Lond 126:549-564

Buchalczyk T (1964) Daily acivity rhythm in rodents under natural conditions. Acta Theriol 9:357-362

Calhoun JB (1945) Diel activity rhythms of the rodents, *Microtus ochrogaster* and *Sigmodon hispidus hispidus*. Ecology 26:251-273

Daan S (1981) Adaptive daily strategies in behavior. In: Aschoff J (ed) Biological rhythms. Handbook of behavioral neurobiology, vol 4. Plenum Press, New York, pp 275-298

Daan S, Aschoff J (1981) Short-term rhythms in activity. In: Aschoff J (ed) Biological rhythms. Handbook of behavioral neurobiology, vol 4. Plenum Press, New York, pp 491-498

Daan S, Aschoff J (1982) Circadian contributions to survival. In: Aschoff J, Daan S, Groos GA (eds) Vertebrate circadian systems. Springer, Berlin Heidelberg New York, pp 305-321

Daan S, Berde C (1978) Two coupled oscillators: simulations of the circadian pacemaker in mammalian activity rhythms. J Theor Biol 70:297-313

Daan S, Slopsema S (1978) Short-term rhythms in foraging behaviour of the common vole, *Microtus arvalis*. J Comp Physiol A 127:215-227

Danielson BJ, Swihart RK (1987) Home range dynamics and activity patterns of *Microtus ochrogaster* and *Synaptomys cooperi* in syntopy. J Mammal 68:160-165

Davis DHS (1933) Rhythmic activity in the short-tailed vole, *Microtus*. J Anim Ecol 2:232-238

De Kock LL, Rohn I (1971) Observations on the use of the exercise-wheel in relation to the social rank and hormonal conditions in the bank vole (*Clethrionomys glareolus*), and the Norway lemming (*Lemmus lemmus*). Z Tierpsychol 29:180-195

Drabek CM (1994) Summer and autumn temporal activity of the montane vole (*Microtus montanus*) in the field. Northwest Sci 68:178-184

Erkinaro E (1961) The seasonal change of the activity of *Microtus agrestis*. Oikos 12:157-163

Erkinaro E (1969) Der Phasenwechsel der lokomotorischen Aktivität bei *Microtus agrestis* (L.), *M. arvalis* (Pall.) und *M. oeconomus* (Pall.). Aquilo Ser Zool 8:1-31

Erkinaro E (1972) Seasonal changes in the phase position of the circadian activity rhythms in some voles, and their endogenous component. Aquilo Ser Zool 13:87-91

Gerkema MP, Daan S (1985) Ultradian rhythms in behavior: The case of the common vole (*Microtus arvalis*). Exp Brain Res [Suppl] 12:11-31

Gerkema MP, van der Leest F (1991) Ongoing ultradian activity rhythms in the common vole, *Microtus arvalis*, during deprivations of food, water and rest. J Comp Physiol A 168:591-597

Gerkema MP, Verhulst S (1990) Warning against an unseen predator: a functional aspect of synchronous feeding in the common vole, *Microtus arvalis*. Anim Behav 40:1169-1178

Gerkema MP, Groos GA, Daan S (1990) Differential elimination of circadian and ultradian rhythmicity by hypothalamic lesions in the common vole, *Microtus arvalis*. J Biol Rhythms 5:81-95

Gerkema MP, Daan S, Wilbrink M, Hop MW, van der Leest F (1993) Phase control of ultradian feeding rhythms in the common vole (*Microtus arvalis*): the roles of light and the circadian system. J Biol Rhythms 8:151-171

Glass GE, Slade NA (1980) The effect of *Sigmodon hispidus* on spatial and temporal activity of *Microtus ochrogaster*: evidence for competition. Ecology 61:358–370
Greenwood PJ (1978) Timing of activity in the bank vole *Clethrionomys glareolus* and the wood mouse *Apodemus sylvaticus* in a deciduous woodland. Oikos 31:123–127
Grodzinski W (1963) Seasonal changes in the circadian activity of small rodents. Ekol Polska Ser B 9:3–17
Grodzinski W, Makomaska M, Tertil R, Weiner J (1977) Bioenergetics and total impact of vole populations. Oikos 29:494–510
Halle S (1993) Diel pattern of predation risk in microtine rodents. Oikos 68:510–518
Halle S (1995a) Diel pattern of locomotor activity in populations of root voles, *Microtus oeconomus*. J Biol Rhythms 10:211–224
Halle S (1995b) Effect of extrinsic factors on activity of root voles, *Microtus oeconomus*. J Mammal 76:88–99
Halle S, Halle B (1999) Modelling activity synchronization in free-ranging microtine rodents. Ecol Modelling 115: 165–176
Halle S, Lehmann U (1987) Circadian activity patterns, photoperiodic responses and population cycles in voles. I. Long-term variations in circadian activity patterns. Oecologia 71:568–572
Halle S, Lehmann U (1992) Cycle-correlated changes in the activity behaviour of field voles, *Microtus agrestis*. Oikos 64:489–497
Halle S, Stenseth NC (1994) Microtine ultradian rhythm of activity: an evaluation of different hypotheses on the triggering mechanism. Mammal Rev 24:17–39
Hamilton WJ (1937) Activity and home range of the field mouse, *Microtus pennsylvanicus pennsylvanicus* (Ord.). Ecology 18:255–263
Hansen RM (1957) Influence of daylength on activity of the varying lemming. J Mammal 38: 218–223
Herman TB (1977) Activity patterns and movements of subarctic voles. Oikos 29:434–444
Hoogenboom I, Daan S, Dallinga JH, Schoenmakers M (1984) Seasonal change in the daily timing of behaviour of the common vole, *Microtus arvalis*. Oecologia 61:18–31
Ishii K, Kuwahara M, Tsubone H, Sugano S (1993) Diurnal fluctuations of heart rate and locomotive activity in the vole (*Microtus arvalis*). Exp Anim 42:33–39
Johst V (1973) Das Aktivitätsmuster der Schermaus *Arvicola terrestris* (L.). Zool Jb Physiol 77: 98–106
Kenagy GJ, Hoyt DF (1980) Reingestion of feces in rodents and its daily rhythmicity. Oecologia 44:403–409
Lehmann U (1976) Short-term and circadian rhythms in the behaviour of the vole, *Microtus agrestis* (L.). Oecologia 23:185–199
Lehmann U, Sommersberg CW (1980) Activity patterns of the common vole, *Microtus arvalis* – automatic recording of behaviour in an enclosure. Oecologia 47:61–75
Lehmann U, Halle S (1987) Circadian activity patterns, photoperiodic responses and population cycles in voles. II. Photperiodic responses and population cycles. Oecologia 71:573–576
Madison DM (1981) Time patterning of nest visitation by lactating meadow voles. J Mammal 62:389–391
Madison DM (1985) Activity rhythms and spacing. In: Tamarin RH (ed) Biology of New World *Microtus*. Am Soc Mammal Spec Publ 8, Shippensburg, pp 373–419
Mather JG (1981) Wheel-running activity: a new interpretation. Mammal Rev 11:41–51
Miller RS (1955) Activity rhythms in the wood mouse, *Apodemus sylvaticus* and the bank vole, *Clethrionomys glareolus*. Proc Zool Soc Lond 125:505–519
Mironov AD (1990) Spatial and temporal organization of populations of the bank vole, *Clethrionomys glareolus*. In: Tamarin RH, Ostfeld RS, Pugh SR, Bujalska G (eds) Social systems and population cycles in voles. Birkhäuser, Basel, pp 181–192
Mossing T (1975) Measuring small mammal locomotory activity with passage counters. Oikos 26:237–239
Moyat P (1957) Über den Einfluß von Licht und Aktivität auf endogene Stoffwechselrhythmen bei Kleinsäugern und Vögeln. Z Vergl Physiol 40:397–414

Nygren J (1978) Interindividual influence on diurnal rhythms of activity in cyclic and noncyclic populations of the field vole, *Microtus agrestis* L. Oecologia 35:231-239

Ostermann K (1956) Zur Aktivität heimischer Muriden und Gliriden. Zool Jb Physiol 66: 355-388

Ouellette DE, Heisinger JF (1980) Reingestion of feces by *Microtus pennsylvanicus*. J Mammal 61:366-368

Pearson AM (1962) Activity patterns, energy metabolism, and growth rate of the voles *Clethrionomys rufocanus* and *C. glareolus* in Finland. Ann Zool Soc Vanamo 24:1-58

Peterson RM, Batzli GO, Banks EM (1976) Activity and energetics of the brown lemming in its natural habitat. Arct Alp Res 8:131-138

Pittendrigh CS (1960) Circadian rhythms and the circadian organization of living systems. Cold Spring Harbor Symp Quant Biol 25:159-184

Raptor Group (1982) Timing of vole hunting in aerial predators. Mammal Rev 12:169-181

Rasmuson B, Rasmuson M, Nygren J (1977) Genetically controlled differences in behaviour between cycling and non-cycling populations of field vole (*Microtus agrestis*). Hereditas 87: 33-42

Reynolds P, Gorman ML (1994) Seasonal variation in the activity patterns of the Orkney vole, *Microtus arvalis orcadensis*. J Zool Lond 233:605-616

Rijnsdorp A, Daan S, Dijkstra C (1981) Hunting in the kestrel, *Falco tinnunculus*, and the adaptive significance of daily habits. Oecologia 50:391-406

Rowsemitt CN (1986) Seasonal variations in activity rhythms of male voles: mediation by gonadal homones. Physiol Behav 37:797-803

Rowsemitt CN (1991) Activity rhythms in female montane voles (*Microtus montanus*). Can J Zool 69:1071-1075

Rowsemitt CN, Petterborg LJ, Claypool LE, Hoppensteadt FC, Negus NC, Berger PJ (1982) Photoperiodic induction of diurnal locomotor activity in *Microtus montanus*, the montane vole. Can J Zool 60:2798-2803

Saarikko J, Hanski I (1990) Timing of rest and sleep in foraging shrews. Anim Behav 40:861-869

Salvioni M (1988) Rythmes d'activité de trois espèces de *Pitymys*: *Pitymys multiplex*, *P. savii*, *P. subterraneus* (Mammalia, Rodentia). Mammalia 52:483-496

Schoener TW (1974) Resource partitioning in ecological communities. Science 185:27-39

Stebbins LL (1972) Seasonal and latitudinal variations in circadian rhythms of red-backed vole. Arctic 25:216-224

Stebbins LL (1974) Response of circadian rhythms in *Clethrionomys* mice to a transfer from 60°N to 53°N. Oikos 25:108-113

Stebbins LL (1975) Short activity periods in relation to circadian rhythms in *Clethrionomys gapperi*. Oikos 26:32-38

Stubbe A, Blumenauer V, Schuh J, Dawaa N (1986) Aktivitätsrhythmen von *Microtus brandti* (Radde, 1861) bei unterschiedlicher Gruppengröße. Wiss Beitr Universität Halle-Wittenberg 1985/18:49-57

Szymanski JS (1920) Aktivität und Ruhe bei Tieren und Menschen. Z Allg Physiol 18:105-162

Trojan P, Wojciechowska B (1968) The influence of darkness on the oxygen consumption of the nesting European common vole *Microtus arvalis* (Pall.). Bull Acad Pol Sci (Sér Sci biol) 16:111-112

Viswanathan N, Chandrashekaran MK (1985) Cycles of presence and absence of mother mouse entrain the circadian clock of pups. Nature 317:530-531

Webster AB, Brooks RJ (1981) Daily movements and short activity periods of free-ranging meadow voles *Microtus pennsylvanicus*. Oikos 37:80-87

Wiegert RG (1961) Respiratory energy loss and activity patterns in the meadow vole, *Microtus pennsylvanicus pennsylvanicus*. Ecology 42:245-253

Wójcik JM, Wołk K (1985) The daily activity rhythm of two competitive rodents: *Clethrionomys glareolus* and *Apodemus flavicollis*. Acta Theriol 30:241-258

Ylönen H (1988) Diel activity and demography in an enclosed population of the vole *Clethrionomys glareolus* (Schreb.). Ann Zool Fenn 25:221-228

12 Djungarian Hamsters – Small Graminivores with Daily Torpor

Thomas Ruf and Gerhard Heldmaier

12.1 Djungarian Hamsters: Through Cold Winters on Hairy Feet

The species of interest in this chapter, the Djungarian dwarf hamster *(Phodopus sungorus)*, is sometimes also called the hairy-footed hamster. This name, which refers to fur covering even the soles of the feet in this species, indicates an adaptation to extremely harsh climatic conditions in its natural habitat, the subarid steppes of continental Asia. Among these steppes, the Djungarian (or [D]sungarian, as in '*sungorus*') basin in northern China represents the southernmost area of its distribution. These Djungarian plains, as well as the other regions inhabited by *P. sungorus*, such as grasslands and vermouth-steppes in northern Kasakhstan, southern Siberia, the Altai, and Mongolia, are characterised by pronounced seasonal climatic changes. While in summer ambient temperature (T_a), e.g. in the Siberian steppe, often rises above +30 °C, it regularly drops below –30 °C during winter (Köppen and Geiger 1939). In Mongolia, mean monthly surface temperatures in January are as low as –25 °C, while the snow cover is thin or absent (Weiner 1987a). Given these environmental conditions, one might expect that Djungarian hamsters, like many other small mammals of the temperate zone, restrict surface activity to the summer months and survive winters via permanent retreat and hibernation in thermally buffered burrows. Surprisingly though, Djungarian hamsters are known to be active, predominantly at night, throughout the year and forage at the surface even in winter at air temperatures of down to –40 °C (Flint 1966).

A seasonal strategy that is based on continued foraging activity in this harsh environment has severe obstacles, in particular for a 'dwarf' hamster, weighing only 20 to 40 g. In a mammal of this size, locomotor activity in the cold at high body temperature (T_b) leads to tremendous rates of heat loss and energy expenditure, due to its adverse surface-to-volume ratio. Even for a hairy-footed and generally well-insulated hamster, further improvement in insulation by lengthening of the fur is limited by its small size. Thus, maximum energy expenditure in the cold is required at a time of the year when food availability is lowest. Seeds of grass and shrubs (e.g. *Stipa capillata*, *Potentilla*, *Nitraria*) represent the sole food source during winter, while the

Ecological Studies, Vol. 141
Halle/Stenseth (eds.) Activity Patterns in Small Mammals

diet is supplemented by insects only in summer (Flint 1966; Heldmaier, unpubl. obs.). During the cold season, dwarf hamsters apparently do not forage below the snow cover, but collect seeds from plants penetrating the snow, sometimes at distances of several hundred meters from their burrows (Flint 1966). Hence, Djungarian hamsters actually experience the full range of seasonal climatic changes but obviously gain sufficient energy during foraging throughout the year.

This sustained activity and apparent unresponsiveness towards harsh winter conditions may lead one to the conclusion that Djungarian hamsters are characterised by a lack of seasonal adjustment. However, the opposite is true. The capability of this species to exploit its habitat during all seasons only results from a complex set of physiological and behavioural adaptations (for reviews see Hoffmann 1979; Heldmaier 1989; Heldmaier et al. 1989).

First, reproduction as one of the energetically most costly activities, and even the physiological capability of being sexually active, is restricted to summer. The reproductive season is limited to a few months, from April/May to September, during which females give birth to up to four litters (Flint 1966). Also, aggressive behaviour both within and between sexes, causing additional energetic costs, seems to occur in summer only. While summer-adapted hamsters are highly aggressive (except for a short period during lactation when females tolerate males inside the nest), huddling and social thermoregulation have been observed during winter, at least under laboratory conditions (Müller 1982; Ruf et al. 1991, and unpubl. obs.). Second, as winter approaches, dwarf hamsters significantly improve their ability to maintain high T_bs in the cold, in particular by increasing their capacity for non-shivering thermogenesis in the brown adipose tissue. In the fully winter-acclimatised state, *P. sungorus* may reach its cold limit at environmental temperatures as low as –80 °C (Heldmaier et al. 1982, 1989). This ability to at least temporarily withstand extreme cold loads is one of the major prerequisites for foraging activity in winter. Third, the risk of predation is lowered by a gradual change from a brown summer fur to a white pelage in winter. Since Djungarian hamsters are not always strictly nocturnal, but can be active during daylight as well, this adaptation will reduce exposure to both nocturnal and diurnal predators. Accordingly, known predators of *P. sungorus* include diurnal as well as nocturnal birds and mammals (e.g. *Bubo bubo*, *Aquila nipalensis*, *Falco tinnunculus*, *Falco cherrug*, *Vulpes corsac*; see Flint 1966). Fourth, energy requirements are lowered during fall by a substantial decrease of body weight. This – at first sight paradoxical – adjustment, effectively reduces the body mass to be heated, which (together with some improvement of fur density) more than compensates for the associated relative increase in heat loss (Heldmaier 1989). Finally, during midwinter, Djungarian hamsters save additional energy by exhibiting short daily torpor. Bouts of torpor in this species occur during the day only, and are characterised by a state of complete inactivity with significantly reduced metabolic rates (MRs) and low T_bs that lasts several hours. Unlike deep hibernators, however, Djungarian hamsters always rewarm and become active around dusk each day (Ruf et al. 1991, 1993).

Obviously, all of these specialised adaptations primarily serve to adjust the animal's physiological capabilities and energy requirements to seasonal changes of both food availability in the environment and metabolic costs of locomotor (and specifically foraging) activity. The present chapter therefore centers around locomotor activity as probably the most important variable in energy budgets of Djungarian hamsters. The energetic significance of locomotor activity is not predominantly due to the costs of mass transportation, but to costs for thermoregulation during foraging outside the burrow. Since these metabolic costs of surface activity may become a limiting factor for survival in the cold, we herein particularly investigate its role as an energetic factor in hamsters living in winter conditions. Also, because in winter-adapted hamsters periods of activity typically alternate with bouts of hypothermia, special emphasis is placed on the interaction between nocturnal locomotion and daily torpor in this species.

Data on locomotor activity, food intake, T_b and MR presented here were not gathered in Djungaria or the Siberian steppe, but recorded in laboratory environments or in hamsters exposed to more natural climatic fluctuations in outdoor cages. This is simply because up to now, many of the physiological variables of interest such as, for example, the time course of energy expenditure in individuals, cannot be precisely obtained from free-living animals. However, laboratory measurements can certainly yield insights into basic mechanisms, limitations and interrelations of physiological and behavioural adjustments. These insights may provide a useful framework for understanding the principal forces and constraints acting on free-ranging animals in the wild. While appreciating the inherent limitations, we therefore not only report laboratory findings, but also attempt to infer on their significance for hamsters living in their natural habitat.

12.2 Measurements of Locomotion, Temperature and Metabolic Rate

Djungarian hamsters were bred and raised under a natural photoperiod (50° 49' N, 09° E) and kept either individually or as pairs in plastic cages (size 42 × 26 cm or 26 × 20 cm). Animals were exposed to both constant climatic conditions in environmental chambers or to fluctuating photoperiod and temperature in cages maintained in outdoor shelters. If not stated otherwise, hamsters were provided with food (rodent chow, slices of apple) and water ad libitum. Normally, hamsters were kept on wood-shavings as bedding material and given wooden boxes and nesting material. Exceptions to these housing conditions are stated in the text.

Locomotor activity was recorded continuously by the use of infrared movement-detectors that generate an electrical signal whenever the animal moves more than approximately 1 cm (for details see Ruf et al. 1991). Body temperature was measured at 6-min intervals via temperature-sensitive transmitters (Minimitter Model XM, accuracy ± 0.1 °C) that were implanted in the visceral cavity (see Ruf et al. 1991). Metabolic rates were determined in an open-flow, automated 6-channel respirometric system with an accuracy of 0.001 vol% O_2 and CO_2 (see Ruf and Heldmaier 1992).

Locomotor activity is presented as number of impulses generated in 6-min bins. Total daily activity duration was estimated from the sum of all 6-min intervals with at least one event occurring. Differences in the variance of mean activity duration and number of events among hamsters exposed to different environmental conditions were tested with Levene's test. Since variances differed significantly (see Table 12.1), the Kruskal-Wallis ANOVA was used to test for differences in activity measures between groups. Groups under identical photoperiods but at different T_as were further compared by the use of the Mann-Whitney U-test. Unless stated otherwise, results are given as means ± one SD.

An animal was considered to enter daily torpor once T_b decreased below 32 °C. Accordingly, torpor duration was calculated from the time spent at T_bs below 32 °C. Resting metabolic rates (RMRs) represent the median of the three lowest successively recorded MRs in inactive hamsters. The average metabolic rate, including both active and inactive periods over a day or certain parts of a day, is called ADMR and AMR, respectively.

12.3 Model Calculations

To investigate the interrelation between energy gain and energy expenditure during rest and foraging, and their consequences for daily activity duration, we employed several model calculations. The basic assumptions underlying these calculations are virtually identical to the general model outlined by Weiner (Chap. 3), only that in the present chapter we largely focus on the impact of daily torpor on total energy expenditure and its consequences for foraging. Calculations are based on measurements of MRs in hamsters at various behavioural states and environmental conditions that were determined previously or are presented below. Resting MRs of hamsters without nesting material at different T_as were calculated from the equation

$$\mathrm{RMR}\ [\mathrm{ml\ O_2\ g^{-1}\ h^{-1}}] = 5.27 - 0.14 \cdot T_a. \qquad \text{(Heldmaier et al. 1982)}$$

Metabolic rates of hamsters provided with bedding material at different T_as in both the normothermic state (i.e., $T_b > 35$ °C) and during daily torpor were calculated from

$$\mathrm{MR}\ [\mathrm{ml\ O_2\ g^{-1}\ h^{-1}}] = 0.01 + 0.108 \cdot (T_b - T_a). \qquad \text{(Ruf and Heldmaier 1992)}$$

Model calculations are based on the assumption that T_b during daily torpor is lowered to 15 °C (typical range: 13–22 °C; Ruf et al. 1993). Total MRs were calculated assuming an average body weight of winter-adapted hamsters of 25 g (Heldmaier and Steinlechner 1981a).

12.4 Adjustment of Locomotor Activity Patterns to Environmental Changes

Djungarian hamsters are principally nocturnal animals, as illustrated in Fig. 12.1. In hamsters at moderate (> 18 °C) temperatures with free access to food, 80 to 90% of locomotion occurs during the night. This daily pattern of activity is based on a circadian rhythm with an average period length just below 24 h (Puchalski and Lynch 1988), that is entrained by the LD cycle. Thus, the daily time course of locomotion is altered by seasonal changes of photoperiod (Fig. 12.1). Total amount and net duration of activity stays, however, fairly stable throughout the year (Table 12.1).

In *P. sungorus*, locomotion occurs in 'bursts' that last from several minutes to about 2 h and alternate with episodes of rest and sleep (Heldmaier et al. 1989; Deboer et al. 1994; see also Fig. 12.1). Similar ultradian rhythms of locomotion and rest also characterise patterns of the emergence from, and retreat into artificial burrows (Heldmaier, unpubl. data). Hence, these patterns most likely correspond to a periodically occurring surface activity in free-living hamsters. Also, laboratory measurements that always show a certain

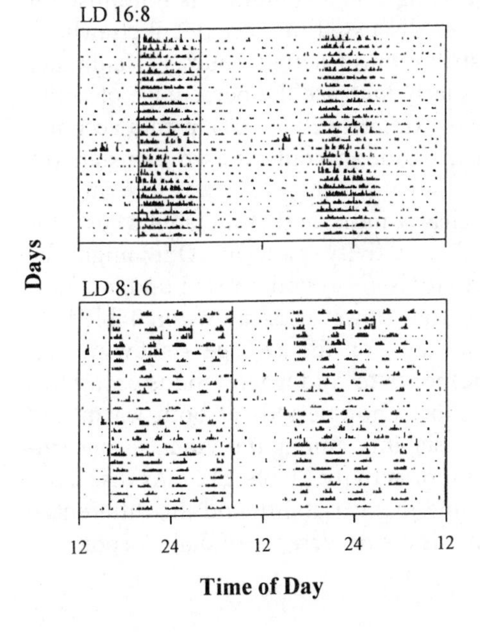

Fig. 12.1. Double plots of locomotor activity in a Djungarian hamster exposed to long photoperiods (LD 16 : 8, *upper panel*) and short photoperiods (LD 8 : 16, *lower panel*). Locomotion was recorded continuously by the use of infrared movement-detectors. The graphs show the sum of events (movements) within 6-min intervals over 26 days each. *Vertical lines* in the *left* half of each graph indicate beginning and end of the dark period

percentage of activity occurring during the daylight hours (Fig. 12.1) seem to reflect natural behaviour. For example, Nekipelow (1960; cited after Flint 1966) found in field studies that surface activity during the day is not unusual in free-living hamsters.

Previous studies demonstrated that in *P. sungorus*, the daily and ultradian time course of general locomotion largely reflects feeding activity (Ruf and Heldmaier 1993). The only obvious disproportion between feeding and general motor activity can be found in summer-adapted hamsters under long photoperiods when food uptake, but not general locomotion, is most intense in the last part of the night. This observation may indicate that, as in other small mammals during summer, locomotion related to sexual activity (namely in males) occurs predominantly in the early part of the night, while foraging and food uptake is delayed (Daan and Aschoff 1982). In winter, when hamsters are sexually completely inactive, aggression and territorial behaviour appear to be largely absent (Müller 1982), and foraging is probably the sole purpose of locomotion.

This conclusion is supported by both field and laboratory studies. As mentioned before, direct observations and analyses of stomach contents have shown that the natural diet of hamsters is basically comprised of seeds from grass and other plants (Flint 1966). Hence, under natural conditions in the Siberian steppe, food comes in small 'bits' with relatively low energy content. Consequently, food uptake will require extensive foraging activity if daily energy demands are to be met. Further, laboratory recordings have shown that locomotion in hamsters can be strongly affected by changes in either metabolic costs of foraging or food availability in the environment. For instance, acclimatisation to low T_as in summer-adapted hamsters leads to an overall decrease of activity (Table 12.1), suggesting that locomotion is minimised if metabolic costs for foraging increase. As discussed in more detail elsewhere, cold T_as also affect the temporal distribution of activity bursts (Heldmaier et al. 1989). Cold temperatures particularly increase the percentage of activity occurring during daylight from approximately 20 to 30% in cold acclimatised animals, and to more than 50% in suddenly cold exposed hamsters (cf. Weiner in Chap. 3).

These effects of T_a on locomotion clearly indicate that metabolic costs for heat production can significantly alter activity patterns. One might be tempted to conclude that foraging activity is *always* minimised under conditions of high cold load outside the burrow. However, as shown in Table 12.1, short-day adapted hamsters during winter – on the average – show no significant change of motor activity in response to cold when food is unlimited. Instead, locomotion patterns under short photoperiods at both warm and cold T_as are characterised by a large variability among individuals, concerning both the duration and total amount of activity (Table 12.1). As discussed in detail below, the key to understanding this variability and the differential effects of low T_as in summer and winter is the occurrence of daily torpor.

Table 12.1. Duration of locomotor activity (h) and number of events (movements) recorded per day in Djungarian hamsters exposed to long (*LD 16 : 8*) and short photoperiods (*LD 8 : 16*) at warm and cold temperatures (n = 8–16 per group). Cold exposure leads to a reduction of mean activity duration ($p < 0.003$) and intensity ($p < 0.05$) in hamsters under a long photoperiod, but not under a short photoperiod (ns). In long days, cold temperatures also tended to lower the variance among individuals (duration: $p = 0.10$; events: $p < 0.01$), while in hamsters under short days the variability between individuals was unaffected by T_a ($p > 0.30$; see SD and range)

		LD 16 : 8		LD 8 : 16	
		23 °C	5 °C	23 °C	5 °C
Duration	Mean	6.2	4.8	5.3	4.9
	SD	0.94	0.47	1.26	1.06
	Range	3.3	1.4	3.7	4.1
		(4.4–7.7)	(4.4–5.8)	(3.2–6.9)	(2.8–6.9)
Events	Mean	410	244	529	582
	SD	199	51	342	400
	Range	633	180	838	1 244
		(191–824)	(158–338)	(70–908)	(182–1 426)

Torpor as an energy saving mechanism is absent in long-day adapted hamsters and therefore provides no ‘escape’ mechanism for energetic challenges in summer. This blockade of voluntary hypothermia in long photoperiods is probably due to an incompatibility of low T_bs with maintaining sexual activity and functional gonads (Barnes et al. 1986). Not surprisingly then, Djungarian hamsters defend high T_bs in summer even under conditions of severe shortage of food (Steinlechner et al. 1986). Erratic bouts of hypothermia may eventually occur in summer after prolonged food restriction, but only following a severe decrease in body weight. Before that, a shortage of food (similar to low T_as) causes significant extensions of locomotor activity into the daylight. This change of daily rhythms may even lead to a complete inversion of the day/night cycle, with activity and elevated T_bs occurring predominantly during the day (Steinlechner et al. 1986; Ruf and Heldmaier 1987; K. Gwinner and G. Heldmaier, unpubl.).

Interestingly, this degree of flexibility in activity patterns that can be modulated by environmental conditions seems to be largely species specific. For instance, similar food restriction-induced extensions of locomotion into the day were also found in house mice, but not in another Cricetid species, the deer mouse (Blank and Desjardins 1985). The fact that a more flexible response is present in Djungarian hamsters suggests that in this species, energetic demands can override the impact of predation risks on daily partitioning of locomotion. Thus, in *P. sungorus*, energy availability and particularly metabolic costs of foraging activity, represent most important variables in the regulation of duration and daily patterns of locomotion. Further understanding of these regulating mechanisms, therefore, requires information on the actual energetic costs of motor activity in this species.

12.5 Energy Requirements for Activity – The Impact of Cold Load

Metabolic costs for the transport of mass are directly related to body weight (e.g. Taylor et al. 1970). Since Djungarian hamsters reduce body weight towards winter, energy requirements for locomotion decrease, on the average by 23% (Heldmaier and Steinlechner 1981a). However, as pointed out by several authors, by far the most important factor for metabolic costs of activity is the time an animal spends outside its thermally buffered burrow (Wunder 1978; Bronson 1989). These costs dramatically increase at low T_as during winter. Figure 12.2 shows MRs of 12 hamsters exposed to T_as between +23 °C and −50 °C with maximum values reflecting intensive locomotion. As indicated by the upper curve through mean maximum values of the 12 individuals at each T_a, activity-induced metabolism first increases as T_a is lowered and then declines at T_as below −30 °C. At even lower T_as (< −30 °C) hamsters begin to encounter an important constraint to activity in the cold, namely their peak MR. In this case of cold acclimatised hamsters in summer, peak MRs varied around 15 ml O_2 g^{-1} h^{-1} (but see Heldmaier 1989).

The initial increase in the difference between activity and resting metabolism is probably due to an impairment of thermal insulation in hamsters that move and run, instead of sitting in a curled position (cf. Wunder 1975). However, MRs during intense activity at T_as between +23 °C and −30 °C stayed fairly close to a factor of 2 above RMRs (1.89 ± 0.97). This factor is very similar to estimates of metabolic costs of activity derived from continuous long-term recordings of oxygen consumption. In *P. sungorus* living under moderate cold load (T_a 0–15 °C) mean ADMRs are typically 25 to 30% higher than RMRs (Wiesinger 1989; Ruf and Heldmaier 1992). Since mean activity

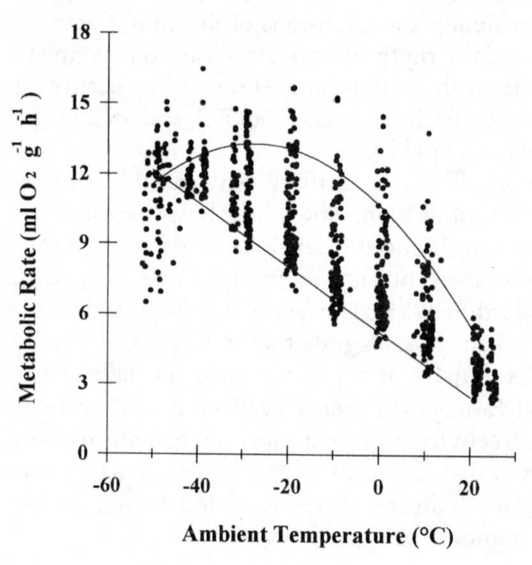

Fig. 12.2. Metabolic rates (MR) of 12 Djungarian hamsters exposed to ambient temperatures between +23 °C and −50 °C. The metabolic rate was measured at 6-min intervals. The *linear regression line* was fitted to resting MRs in euthermic hamsters. Sharply decreasing MRs at T_a −50 °C originate from hamsters that became hypothermic. The *upper curve* was fitted to means of individual maxima at each temperature step

duration ranges around 5 to 6 h per day (Table 12.1) this total elevation means that during activity, average MRs were 2–2.5 times above RMRs.

Hence, both of the above approaches yield a factor of about two times RMR for metabolic costs of activity, which probably represents a conservative measure, because MRs were measured in dry environments without wind. This magnitude of locomotion-induced increase of MR is well in the range of observations in other small mammals and, according to the model proposed by Wunder (1975), should be expected in a hamster running at an average speed of 3 km/h. To illustrate the impact of these metabolic costs one might, for example, compare the energy expenditure of a hamster in its nest at T_a 0 °C and foraging on the surface at T_a –20 °C. In the first situation, metabolism consumes about 75 J g^{-1} h^{-1}, while during locomotion outside the burrow energy requirements are more than four times higher, i.e., 320 J g^{-1} h^{-1}. Therefore, if environmental temperatures are low, any mechanism suited to reduce the need for foraging should be highly adaptive. Not surprisingly then, as T_a in the natural environment decreases towards winter, Djungarian hamsters do in fact develop such a mechanism, namely daily torpor.

12.6 Locomotion and Daily Torpor: Interactions

Daily torpor in *Phodopus* - although it ultimately represents a response to cold load and shortage of food - is induced by another, proximate signal, the shortening of the photoperiod in fall (Heldmaier and Steinlechner 1981b). This mechanism is, however, responsible for the majority of seasonal adaptations in Djungarian hamsters (for reviews see Hoffmann 1979; Heldmaier 1989). In naturally changing photoperiods, torpor first occurs in October, and then increases in depth, duration, and frequency until midwinter. Spontaneous torpor is almost completely restricted to the day (the photophase), and locomotor activity is resumed each night. An example for the daily alternation between torpor and nocturnal normothermia and activity is illustrated in Fig. 12.3.

Torpor episodes vary considerably, in particular between individuals. Torpor depth may range from 12 to 30 °C, and torpor duration from only 15 min to more than 12 h. Most importantly, hamsters significantly differ in their tendency to exhibit torpor. Certain individuals may be found torpid every day (e.g. Fig. 12.3), while others exhibit torpor only occasionally, at least if food is abundant (Heldmaier and Steinlechner 1981b; Ruf et al. 1991). Similar to other species, Djungarian hamsters immediately increase their frequency of torpor episodes in response to food restriction (Tucker 1966; Hudson 1978; Tannenbaum and Pivorun 1987; Ruf et al. 1993). Thus, daily torpor, during which MRs are lowered down to 30% of normothermic RMRs, clearly helps to counteract short-term energetic jeopardy.

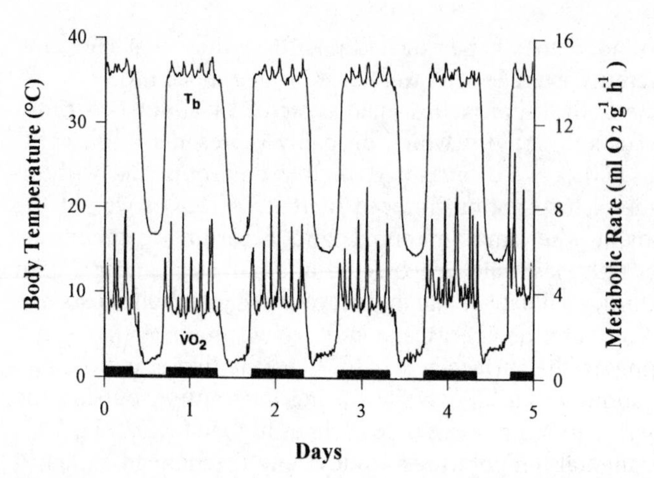

Fig. 12.3. Records of body temperature and metabolic rate in a winter-adapted hamster that displays daily torpor at T_a 5–15 °C. During torpor, body temperature and metabolic rate are significantly lowered for several hours. At night, activity is resumed as indicated by nocturnal peaks in metabolic rate. *Black bars* on the abscissa indicate hours of darkness

Energy savings via daily torpor linearly increase with its daily duration, and (at T_as of 0–15 °C) prolonged torpor episodes with a duration of 9 h lower total energy requirements by 30% (Ruf and Heldmaier 1992). However, this estimate is based on comparisons of days with and without torpor within the same individuals and ignores important associated behavioural differences between hamsters, in particular concerning the intensity of locomotion. If those interactions between torpor and activity are considered, the total benefits of combined thermoregulatory and behavioural adjustments turn out to be much more significant.

At first sight it may seem that there is little relation between torpor and activity, since torpor occurs during the daylight resting phase of hamsters. If anything, one might expect that the loss of about 10 to 20% daylight activity, which is replaced by torpor, might be compensated for by increased locomotion in the subsequent night. However, as illustrated in Fig. 12.4, the opposite is true. This comparison among individuals of long-term (4 weeks) mean daily times spent in torpor and at locomotor activity reveals that the daily activity period significantly decreases as torpor increases. For example, in this group hamsters that constantly remained normothermic were active for about 6.2 h per day, whereas hamsters that spent an average of 6 h per day in torpor (that is, including both days with and without torpor) also reduced their daily activity period by 37% to 3.7 h only. Earlier studies have shown that it is in fact mainly *nocturnal* activity that is reduced in both intensity and duration in hamsters that frequently exhibit torpor (Ruf et al. 1991; Ruf and Heldmaier 1993). Expectedly, as shown in Fig. 12.4 (upper line), torpor also replaces a certain amount of resting periods, but, in view of the above

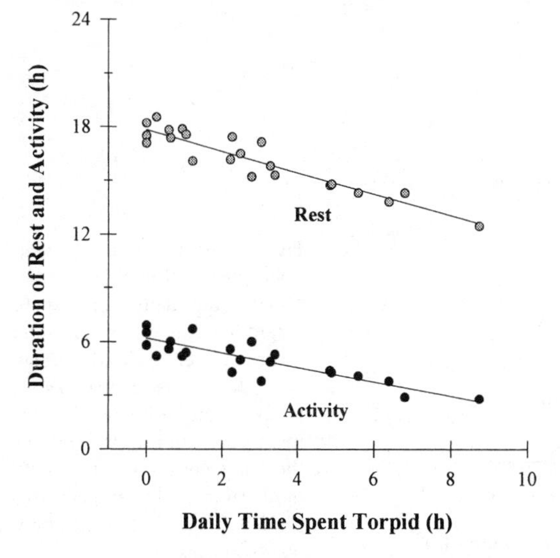

Fig. 12.4. Duration of daily periods of rest (*grey dots*) and activity (*black dots*) as a function of the daily time spent torpid. Data points represent means from continuous 3–4-week recordings in 22 hamsters

data on the impact of activity on energy expenditure, the reduction of locomotion periods will be energetically much more significant.

An estimate of the exact magnitude of additional energy savings by a torpor-linked decrease in activity can be obtained from measurements of ADMRs in hamsters with different long-term tendencies for torpor. Since individuals also differ slightly in fur insulation and body weight, and consequently in total thermal conductance (cf. Ruf et al. 1993), comparisons of these energetic effects are best based on calculations of the purely activity-induced increase of metabolism, i.e., the difference between average and resting MRs. Figure 12.5 shows 3-week means of this measure in hamsters with a different proneness for torpor. If locomotor activity was unaffected by the exhibition of torpor, activity-induced metabolism should stay constant and equal among individuals during the normothermic part of the day. However, since torpor is accompanied by a reduction of locomotion, energy expenditure during normothermia (mainly during night) also significantly decreases as the mean time spent hypothermic increases (Fig. 12.5, upper line). From this separate calculation of total daily and nocturnal (activity-induced) metabolism it becomes clear that the magnitude of energy savings via reduced activity is about equal to the immediate effect of torpor via reduction of T_b and heat loss during the day. In other words, even in hamsters under moderate cold load (at T_as above 0 °C) each Joule of energy saved by torpor during daytime saves at least another Joule at night, because it allows for a shortening of foraging activity. This relation will further change in favor of nocturnal, activity-related energy savings when T_a is lower than in the experiment presented here.

It should be emphasised that Figs. 12.4 and 12.5 are based on long-term comparisons between individuals that showed distinct ‘preferences’ for either

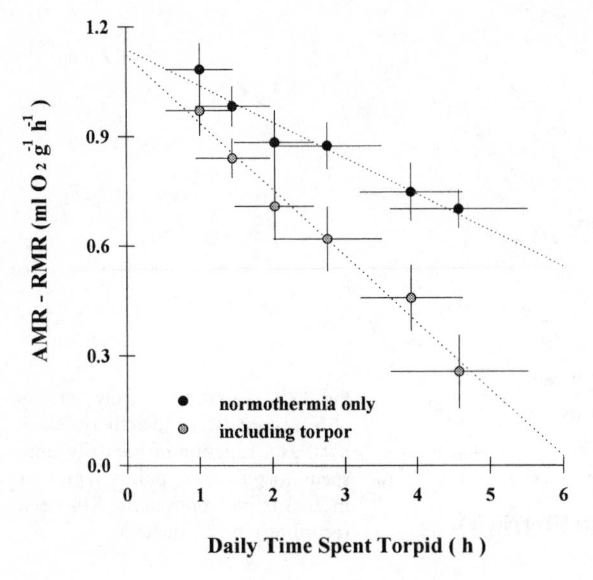

Fig. 12.5. Activity-induced metabolic rate (*AMR-RMR*) as a function of the daily time spent torpid (means of 6 hamsters recorded over 20 days). Energy expenditure was calculated separately for the entire day (including torpor, *grey dots*) and for the daily period spent at normothermic body temperatures (> 35 °C, *black dots*), i.e., mainly night-time hours

high levels of activity and low torpor frequency or vice versa. These differences also explain the tremendous variability of locomotion under short days pointed out above (see Table 12.1). These individual characteristics lead to long-term energy requirements and rates of food uptake that can differ by more than 70% between highly active, normothermic, and fairly inactive, torpor-prone hamsters (Ruf et al. 1991). Hence, at least under conditions of unlimited food supply and temperatures above freezing, Djungarian hamsters employ highly variable individual energetic strategies to cope with identical environmental conditions. As discussed below, this large range of possible 'solutions' for the problem of maintaining stable energy budgets may, however, shrink under natural conditions in the Djungarian or Siberian winter.

12.7 Budgeting of Time and Energy Under Natural Conditions

The above information on energetic impacts, relations, and feedback interactions of locomotion, food uptake, rest, and torpor, may be integrated in some model calculations that also attempt to include the principal environmental conditions experienced by hamsters in the wild. Prior to a discussion of these calculations, several important constraints and limitations to budgeting of time and energy have to be pointed out.

First, since peak MR in *Phodopus* never exceeds about 18 ml O_2 g^{-1} h^{-1} (Heldmaier et al. 1989), extensive locomotion becomes virtually impossible at very low ambient temperatures (at –50 °C or below). Second, an increase of activity duration is limited because hamsters normally spent more than 12 h per day asleep (Deboer et al. 1994). Any significant reduction of this period will be prevented by the mechanisms responsible for homeostatic sleep-regulation (Borbély 1982; Daan et al. 1984; Deboer and Tobler 1994). Third, either the size and capacity of the digestive tract or other constraints in energy turnover (see Hammond and Diamond 1997) limit maximum daily energy assimilation, in the case of winter-adapted hamsters to approximately 80 kJ per animal per day (calculated from Weiner 1987b). If energy balance, namely stable fat stores and body weights, are to be maintained energy expenditure must not exceed this limit (cf. Weiner in Chap. 3).

Based on these constraints and measurements of MRs during activity, rest, and torpor, the required minimum duration of foraging activity and corresponding daily energy expenditure can be estimated (Fig. 12.6). These calculations were carried out for two environmental temperatures outside the burrow, 0 °C and –30 °C, while a constant T_a of 0 °C inside the burrow was assumed. The other important environmental variable, accessibility of food or, more precisely, energy gain per hour of foraging, was varied continuously between 0 and 30 kJ h^{-1}. Energy gains around 30 kJ h^{-1} seem to be typical for a terrestrial seed eater with a body weight of 25 g (Weiner, Chap. 3). Thus, the range of foraging rates investigated here reflects very low up to average food abundance, as can be expected to occur in winter environments. The assumptions underlying these calculations may of course not accurately represent conditions and metabolic costs for free-living hamsters. This, however, will not affect the principal relations between variables and conclusions discussed here.

In general, the upper panel of Fig. 12.6 illustrates that the daily minimum time period required for foraging increases exponentially as energy availability decreases. This is because each minute spend outside the burrow not only increases total energy intake, but also has additional metabolic costs that require further foraging. Secondly, it becomes clear that temperatures outside the nest dramatically affect the range of possible energy gains under which hamsters can still keep daily MRs below maximum energy assimilation, and thus maintain stable energy budgets (Fig. 12.6, lower panel). For example, while in this model a hamster that always stays normothermic can survive at T_a 0 °C with energy gains above 8 kJ h^{-1}, at T_a –30 °C outside the nest energy balance turns negative if energy gain falls below 18 kJ h^{-1}.

Figure 12.6 also shows that, although it might be advantageous in terms of energy expenditure to minimise locomotion at any given T_a, a significant *increase* of foraging activity in response to scarcity of food is possible only under conditions of fairly moderate temperatures. At very low T_as, energy expenditure reaches the maximum sustained metabolic rate within short periods of foraging activity, and any further time spent outside the burrow would only facilitate depletion of fat and other internal energy stores.

Finally, Fig. 12.6 illustrates the principal effect of daily torpor, namely to extend the range of environmental conditions that can be tolerated with a

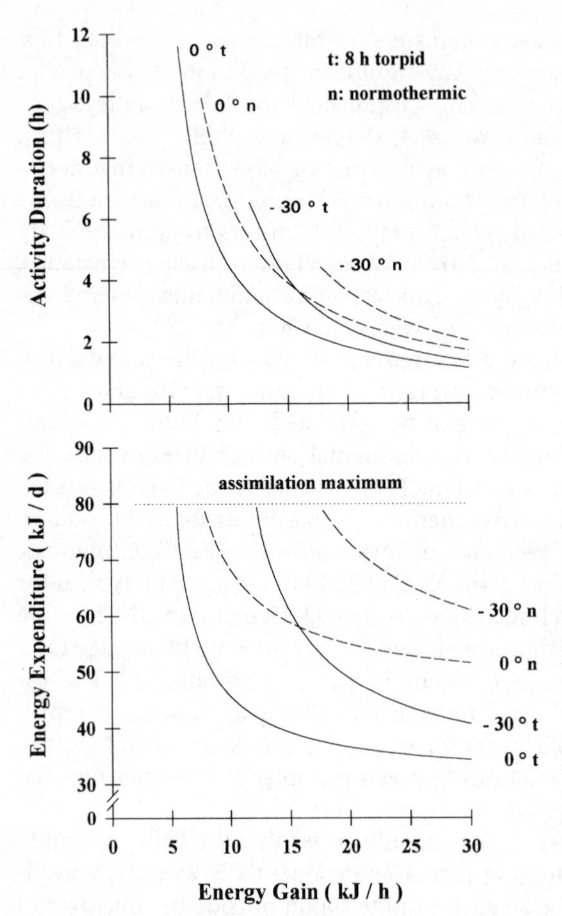

Fig. 12.6. A model for the effect of daily torpor on the relations between energy gain during foraging and minimum required daily activity duration (*upper graph*), as well as daily energy expenditure (*lower graph*). Calculations were carried out for hamsters staying either normothermic (*n*) or exhibiting 8-h episodes of daily torpor (*t*), and encountering temperatures outside their burrow of 0 or –30 °C

positive energy balance towards lower temperatures and decreased food accessibility. At any given T_a and energy gain, torpor significantly lowers both the minimum required foraging time and resulting daily metabolic costs. As indicated by the two crossing lines in the lower panel of Fig. 12.6, these energy savings may compensate for a 30 °C drop in environmental temperature. Most importantly, however, as food sources decrease, only the use of daily torpor will in turn allow for an *increase* of foraging activity and thus guarantee survival in extremely harsh winters (see also Weiner in Chap. 3). Daily torpor thus helps not only to cope with high metabolic costs and scarcity of food but also enables hamsters to flexibly respond to varying and unpredictable energy resources in the environment.

Taken together, these results underline that activity, food intake, and energy expenditure during locomotion, rest, and torpor, represent linked variables of energy balance, that continuously feed back on each other. This feedback is, of course, not unidirectional. For instance, while torpor affects the duration of required nocturnal activity, foraging success and energy uptake during this period will in turn also influence the subsequent exhibition of daily torpor.

12.8 Constraints, Unknowns, and Alternative Strategies

The above discussion of seasonal strategies in Djungarian hamsters leaves several open questions that require the incorporation of additional factors and physiological constraints. For instance, with respect to torpor, why, if food is abundant, not all hamsters exhibit episodes of hypothermia and reduce energy turnover? The following argument may explain this phenomenon: deep abdominal temperatures during hypothermia probably reduce nutrient assimilation in the intestine and thereby lower total daily energy assimilation rates (Carey 1989). If food resources are high, torpor will therefore counteract maximum energy intake and particularly hinder the formation of fat and other energy stores in the body. There is evidence from other species that large energy stores and increased body weights, in particular in males, can provide a better 'starting position' for sexual activity and competition when environmental conditions improve in spring (e.g. Kenagy 1986). Hence, hamsters employing a torpor-avoidance, high-energy-turnover strategy of winter survival may well represent a 'high-risk' phenotype, that has selective advantages after survival of mild winters with abundant food.

Also, while this chapter has centered around short-term behavioural adjustments, the regulation of energy budgets in a small mammal obviously involves an entire set of additional factors. In particular, a variety of more static seasonal adaptations, such as changes of body weight and fur insulation, certainly play an important role for the energetic costs of locomotor activity. Further, both internal and external energy stores, i.e., body fat and food caches, certainly can have a strong impact on budgeting of time and energy, and deserve further attention. As far as external energy stores are concerned we have, however, evidence that hoarding behaviour also largely differs between individual hamsters, and that intense food caching is again closely correlated with an individual's disposition for intense locomotor activity and low torpor tendency (Ruf and Heldmaier 1993). Thus, while hoarding of food certainly can provide an important energetic buffer and therefore represents an alternative to energy savings via daily torpor, it does not seem to principally alter the relations between locomotion, energy uptake and metabolic costs outlined above. Basically, hoarding is merely another behavioural adaptation designed to avoid or minimise foraging under unfavourable conditions. It is interesting, though, that even in the presence of large food caches the frequency of daily torpor keeps closely correlated to total food uptake (that is, to both digested and hoarded food; Ruf and Heldmaier 1993; see also Tannenbaum 1993). This observation indicates that the 'decision making' processes involved in balancing of energy budgets as described above, can rely on surprisingly precise mechanisms for the assessment of not only time, actual environmental conditions, and the animal's present energetic state, but is also able to keep track of, and incorporate, past variations of foraging success.

In view of these complex regulatory mechanisms required for a seasonal strategy of sustained foraging and its inherent risks, one might be tempted to ask why natural selection in Djungarian hamsters has not resulted in the seemingly simpler solution of prolonged, deep hibernation during winter? It could be argued that possibly, for a small graminivore, the subarid plains of Djungaria, Siberia and Mongolia - if the entire year is considered - are simply too harsh for a hibernation strategy that relies on gathering of substantial energy stores prior to winter (cf. Weiner 1987a). It is conceivable that both their body size, which excludes large fat depots, and the distribution and accessibility of small seeds in their natural habitat prevent Djungarian hamsters from storing quantities of energy that are sufficient to survive long winters, even at hibernation-typical metabolic rates. Even more so since *P. sungorus* lives in shallow burrows at depths of 25 to 30 cm only (Flint 1966). Therefore, cold nest temperatures (Weiner 1987a) would require relatively high rates of heat production during hibernation to avoid freezing of the body, which again increases the need for large energy stores. On the other hand, these factors may just as well be completely irrelevant, since there are several examples of mammals of similar size, feeding habit, and apparent general ecology that may hibernate or stay normothermic throughout winter in the same habitat. Our lack of understanding of these differences, or of the specific evolutionary processes leading to intermediate patterns such as the rapid shifts between hypothermia and foraging observed in Djungarian hamsters, indicates that we are still largely ignorant as to the actual selective forces and constraints that have shaped different strategies of winter survival in small mammals.

Acknowledgements. This study was supported by the DFG - Special Research Center 305, 'Ecophysiology'.

References

Barnes BM, Kretzmann M, Licht P, Zucker I (1986) The influence of hibernation on testis growth and spermatogenesis in the golden mantled ground squirrel, *Spermophilus lateralis*. Biol Reprod 35:1289–1297

Borbély AA (1982) A two-process model of sleep regulation. Human Neurobiol 1:195–204

Blank JL, Desjardins C (1985) Differential effects of food restriction on pituitary-testicular function in mice. Am J Physiol 248:R181–R189

Bronson FH (1989) Mammalian reproductive biology. University of Chicago Press, Chicago

Carey HV (1989) Seasonal variation in intestinal transport in ground squirrels. In: Malan A, Canguilhem B (eds) Living in the cold. Libbey, London, pp 225–233

Daan S, Aschoff J (1982) Circadian contributions to survival. In: Aschoff J, Daan S, Groos GA (eds) Vertebrate circadian systems. Springer, Berlin Heidelberg New York, pp 305–321

Daan S, Beersma DGM, Borbély AA (1984) Timing of human sleep: recovery process gated by a circadian pacemaker. Am J Physiol 246:R161–R178

Deboer T, Tobler I (1994) Sleep EEG after daily torpor in the Djungarian hamster: similarity to the effects of sleep deprivation. Neurosci Lett 166:35–38

Deboer T, Franken P, Tobler I (1994) Sleep and cortical temperature in the Djungarian hamster under baseline conditions and after sleep deprivation. J Comp Physiol A 174:145–155

Flint WE (1966) Die Zwerghamster der paläarktischen Fauna. Ziemsen, Wittenberg

Hammond KA, Diamond J (1997) Maximal sustained energy budgets in humans and animals. Nature 386:457–462

Heldmaier G (1989) Seasonal acclimation of energy requirements in mammals: functional significance of body weight control, hypothermia, torpor and hibernation. In: Wieser W, Gnaiger E (eds) Energy transformations in cells and organisms. Thieme, Stuttgart, pp 129–139

Heldmaier G, Steinlechner S (1981a) Seasonal control of energy requirements for thermoregulation in the Djungarian hamster (*Phodopus sungorus*), living in natural photoperiod. J Comp Physiol 142:429–437

Heldmaier G, Steinlechner S (1981b) Seasonal pattern and energetics of short daily torpor in the Djungarian Hamster, *Phodopus sungorus*. Oecologia 48:265–270

Heldmaier G, Steinlechner S, Rafael J (1982) Nonshivering thermogenesis and cold resistance during seasonal acclimatization in the Djungarian hamster. J Comp Physiol B 149:1–9

Heldmaier G, Steinlechner S, Ruf T, Wiesinger H, Klingenspor M (1989) Photoperiod and thermoregulation in vertebrates: body temperature rhythms and thermogenic acclimation. J Biol Rhythms 4:251–265

Hoffmann K (1979) Photoperiod, pineal, melatonin and reproduction in hamsters. In: Kappers JA, Pévet P (eds) The pineal gland of vertebrates including man. Elsevier, New York, pp 397–415

Hudson JW (1978) Shallow, daily torpor: a thermoregulatory adaptation. In: Wang LCH, Hudson JW (eds) Strategies in cold: natural torpidity and thermogenesis. Academic Press, New York, pp 67–108

Kenagy GJ (1986) Strategies and mechanisms for timing of reproduction and hibernation in ground squirrels. In: Heller HC, Musacchia XJ, Wang LHC (eds) Living in the cold. Elsevier, London, pp 383–402

Köppen W, Geiger R (1939) Klimakunde von Rußland in Europa und Asien. Bornträger, Berlin

Müller D (1982) Populationsökologie und soziale Organisation des Dsungarischen Zwerghamsters *Phodopus sungorus*. Diploma thesis, Tech University Braunschweig

Puchalski W, Lynch GR (1988) Characterization of circadian function in Djungarian hamsters insensitive to short photoperiod. J Comp Physiol 162:309–316

Ruf T, Heldmaier G (1987) Computerized body temperature telemetry in small animals: use of simple equipment and advanced noise supression. Comput Biol Med 17:331–340

Ruf T, Heldmaier G (1992) The impact of daily torpor on energy requirements in the Djungarian Hamster, *Phodopus sungorus*. Physiol Zool 65:994–1010

Ruf T, Heldmaier G (1993) Individual energetic strategies in winter-adapted Djungarian hamsters: the relation between daily torpor, locomotion, and food consumption. In: Carey C (ed) Life in the cold. Ecological, physiological, and molecular mechanisms. Westview Press, Boulder, pp 99–107

Ruf T, Klingenspor M, Preis H, Heldmaier G (1991) Daily torpor in the Djungarian hamster (*Phodopus sungorus*): interactions with food intake, activity, and social behaviour. J Comp Physiol B 160:609–615

Ruf T, Stieglitz A, Steinlechner S, Blank JL, Heldmaier G (1993) Cold exposure and food restriction facilitate physiological responses to short photoperiod in Djungarian hamsters (*Phodopus sungorus*). J Exp Zool 267:104–112

Steinlechner S, Heldmaier G, Weber C, Ruf T (1986) Role of photoperiod/pineal gland interaction in torpor control. In: Heller HC, Musacchia XJ, Wang LHC (eds) Living in the cold. Elsevier, London, pp 301–308

Tannenbaum MG (1993) Spontaneous daily torpor in *Peromyscus*: contributing factors. Proceedings of the symposium Living in the cold III, Crested Butte, Colorado, p 36
Tannenbaum MG, Pivorun EB (1987) Differential effects of food restriction on the induction of daily torpor in *Peromyscus maniculatus* and *Peromyscus leucopus*. J Therm Biol 12:159–162
Taylor CR, Schmidt-Nielsen K, Raab JL (1970) Scaling energetic cost of running to body size in mammals. Am J Physiol 219:1104–1107
Tucker VA (1966) Diurnal torpidity in the California pocket mouse. Science 136:380–381
Weiner J (1987a) Limits to energy budgets and tactics in energy investments during reproduction in the Djungarian hamster (*Phodopus sungorus sungorus* Pallas 1770). In: Loudon ASI, Racey PA (eds) Reproductive energetics in mammals. Symp Zool Soc Lond 57:167–187
Weiner J (1987b) Maximum energy assimilation rates in the Djungarian hamster (*Phodopus sungorus*). Oecologia 72:297–302
Wiesinger H (1989) Kälteakklimatisation beim Dsungarischen Zwerghamster, *Phodopus sungorus*. Doctorate thesis, Philipps University Marburg
Wunder BA (1975) A model for estimating metabolic rate of active or resting mammals. J Theor Biol 49:345–354
Wunder BA (1978) Implications of a conceptual model for the allocation of energy resources by small mammals. In: Snyder D (ed) Populations of small mammals under natural conditions. Spec Publ Carn Mus Nat Hist 5:165–172

13 Shrews – Small Insectivores with Polyphasic Patterns

Joseph F. Merritt and Stephen H. Vessey

13.1 Introduction

Daily activity patterns of shrews are controlled by metabolic requirements commensurate with their diminutive body mass and resultant high surface-to-mass ratios: they must forage often to avoid exhaustion of their energy stores. To remain homeothermic, shrews must partition a 24-h period into multiple bouts of foraging, rest, and sleep.

Shrews typically display polyphasic activity patterns shorter than 24 h (ultradian rhythms, Aschoff 1981). A variety of field and laboratory studies during the past 50 years elucidated the polyphasic rhythm of shrews (Table 13.1). Comparing experiments is compromised by differences in techniques and biological attributes of individuals and species. However, we present a diagram of a 'typical' activity pattern in order to discuss variations within the family Soricidae.

The classic laboratory study of Crowcroft (1954), working with *Sorex araneus*, *S. minutus*, and *Neomys fodiens*, is commonly cited as a reference point in studies of diel activity. His diagram of the daily activity of *S. araneus* is representative for our discussion (Fig. 13.1). The activity pattern of one *S. araneus* consisted of about ten active periods lasting from 1–2 h (Fig. 13.1a). Two periods of maximum activity occurred with the major peak between 20:00 and 4:00 and a lesser peak between 7:00 and 11:00. (Fig. 13.1b). Peaks are reported to represent the periods of sunrise and sunset (Crowcroft 1954). Shrews demonstrated periods of rest and sleep, with most active foraging in darkness and with an afternoon lull in activity. This depiction of the 24-h activity pattern of *S. araneus* permits resolution of the number and duration of foraging bouts plus calculation of the percent of activity occurring per day. Most studies of soricid activity presented in Table 13.1 have employed the technique of Crowcroft (1954) or a comparable one. Our discussion will focus initially on a survey of the different methods employed to study activity periods and then address the factors influencing variation in the polyphasic activity patterns of shrews.

Ecological Studies, Vol. 141
Halle/Stenseth (eds.) Activity Patterns in Small Mammals

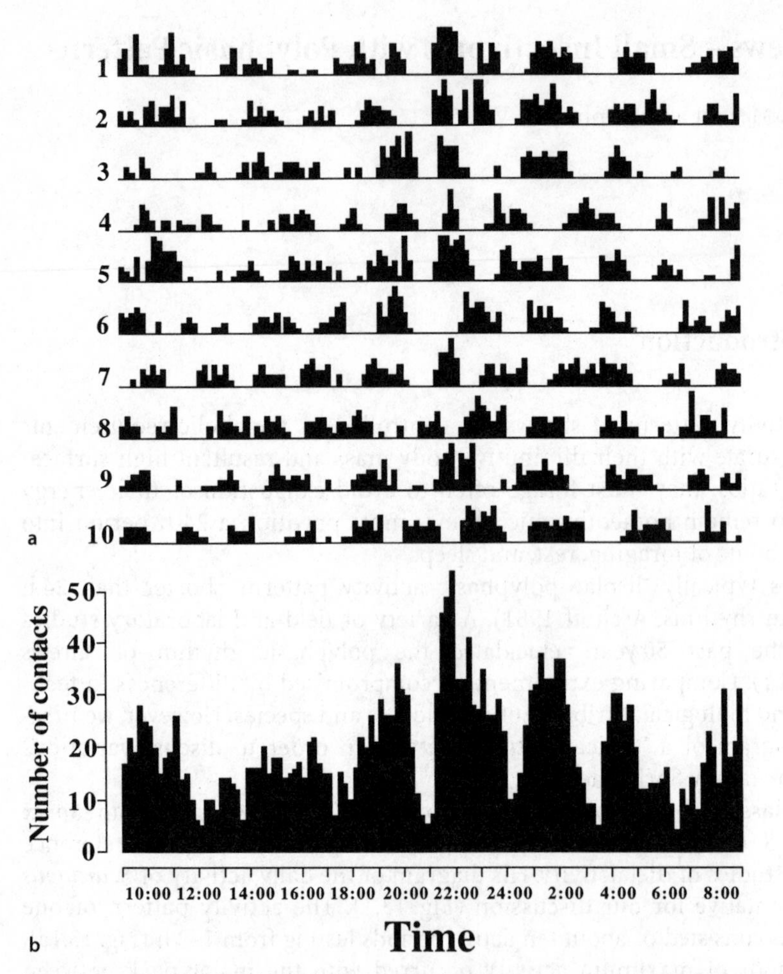

Fig. 13.1. a The activity rhythm of one common shrew (*Sorex araneus*) during 10 consecutive days. Each daily record shows ca. ten 'blocks' of activity. b Summed activity of the ten records, illustrating a 24-h rhythm. (Adapted from Crowcroft 1954)

13.2 Methods

Because of their diminutive size and high caloric requirements shrews must allocate large amounts of energy to foraging activities. Many different techniques have been employed to quantify activity patterns; representative studies and methods are presented in Table 13.1. Investigators conducting field studies of regional faunas often have detailed activity patterns as secondary to the main objective of their study. Trapping was performed by snap- and live-trapping techniques (Clothier 1955; Getz 1961; Shillito 1963; Buckner 1964; Randolph 1973; Pernetta 1977). By monitoring traps at regular intervals, time of activity could be approximated (Mystkowska and Sidorowicz 1961; Cawthorn 1994). However, an accurate picture of activity is compromised by restraining animals in traps, avoidance or habitual visitation of traps by shrews for bait, or the inability of shrews to activate capture mechanisms of live traps due to their small body mass (Crowcroft 1957; Michielsen 1966; Pernetta 1977; Churchfield 1980; Cawthorn 1994). Sand-tracking provides a more sensitive measure of changes in activity than does live-trapping but alters the microhabitat, and delineation of individual time budgets is not possible (Bider 1968; Doucet and Bider 1974; Vickery and Bider 1978).

Field studies were refined by many imaginative techniques in the laboratory. Initially, laboratory studies detailed periods of activity by direct observation (Hamilton 1944; Davis and Joeris 1945; Rood 1958; Crowcroft 1959; Goulden and Meester 1978; Yoshino and Abe 1984). However, this procedure required long hours of constant monitoring by researchers and thus techniques were developed to track activity mechanically. The classic studies of Crowcroft (1954, 1957) employed electrical recording devices to more precisely track activity patterns in the laboratory. Techniques include tilting activity cages and cage vibration (Mann and Stinson 1957; Loxton et al. 1975), activity chambers with infrared photocells (Richardson 1973), activity boxes with microswitches (Rust 1978), wheel running and feeding experiments (Antipas et al. 1990), nest temperature monitoring and video taping (Genoud and Vogel 1981), radar (Godfrey 1978), and computer-facilitated tracking of feeding activity (Saarikko and Hanski 1990). Physiologists have employed oxygen consumption techniques contributing greatly to our understanding of the amount of energy expended on daily activity (Table 13.1; Vogel 1976; Genoud 1988). Further, Kenagy and Vleck (1982) studied temporal organisation of metabolism of shrews compared with 17 species of small mammals.

Although laboratory studies are able to quantify specific activity patterns, laboratory-acclimation may occur and consequently activity rhythms may not accurately reflect behaviour in the natural environment. Therefore, investigators have designed experiments that expose animals to simulated natural environments, measuring activity patterns of shrews housed in cages outside, exposed to natural photoperiods and temperatures (Ingles 1960; Buchalczyk 1972; Pankakoski 1979; Churchfield 1982; Martin 1983). Recording devices ranged from electromagnetic switches and polygraphs to time-lapse photography. Information from studies of soricids residing in large enclosures has augmented our understanding of activity patterns in the natural environment. Sorenson (1962) detailed activity of *Sorex palustris* in a large enclosure in the laboratory by direct observation and timed captures while Platt (1976) employed radioactive tracking of *Blarina brevicauda* in a large outdoor enclosure. Continuous monitoring of core body temperature in *B. brevicauda* residing in outdoor enclosures (Merritt and Bozinovic 1994) involved implanted transmitters and computer facilitation. Implanted radiotransmitters, however, are criticised because small mammals may show weight loss (Webster and Brooks 1980) or develop infection associated with

Table 13.1. Study conditions, length, duration, and number of foraging bouts and techniques employed in studying activity patterns of 23 species of shrews

	Study			Activity descriptors			
Author (year)	n	Condition	Technique	Rhythm	Total length (% of day)	Duration of foraging bouts (h)	Number of foraging bouts per day
Blarina brevicauda							
Antipas et al. (1990)	13	L	Terrarium (microswitch), Wheel running	N	–	–	–
Buckner (1964)	9	L	Observation chamber	N	7	4.5	11
Cawthorn (1994)	32	F (Sp, S, A)	Pitfall live-traps	D	–	–	–
Chandler (1989)	13	F (Sp, S, A)	Radiotelemetry	A.H.	35	1–2	–
Mann and Stinson (1957)	9	L	Tilting activity cage	A.H.	> 10	0.75	–
Martin (1983)	4	F (W, Sp, S, A)	Outdoor cages, Microswitches	N	7–31 (W-S)	–	–
Martinsen (1969)	8	L	Terrarium	A.H.	16	–	8–9
Merritt and Adamerovich (1991)	6	F	Radiotelemetry	A.H.	–	–	–
Merritt and Bozinovic (1994)	4	F	Radiotelemetry	A.H.	–	–	–
Osterberg (1962)	–	F (Sp, A)	Photographic recording	N (B)	–	–	–
Pearson (1947)	3	L	Respirometer	N (B)	–	–	–
Platt (1976)	27	F	Live-trapping, Radioisotopes	D	28–38	–	–
Richardson (1973)	10	L	Activity chamber (Photo-electric cell)	N	–	0.3	9.5
Crocidura flavescens							
Goulden and Meester (1978)	36	L	Terraria	C	–	≤ 1	–
Crocidura mariquensis							
Goulden and Meester (1978)	3	L	Terraria	N	–	–	–
Crocidura russula							
Genoud and Vogel (1981)	13	F, L	Radioactive tracking (F) Nest temp. + video (L)	A.H.	33	0.60	12
Godfrey (1978)	2	L	Radar	A.H.	39	0.1–0.5	27–49

Table 13.1 *(continued)*

Crocidura suavelens							
Godfrey (1978)	2	L	Radar	N	20	0.1–0.3	13–30
Cryptotis parva							
Hamilton (1944)	3	L	Wash tubes	A.H.	–	–	–
Myosorex varius							
Goulden and Meester (1978)	31	L	Terraria	N	–	1–2 (N)	–
						2–3 (D)	
Neomys fodiens							
Buchalczyk (1972)	10	F (W, Sp)	Outdoor cages (Polygraph)	D	–	4.3 (W)	–
Crowcroft (1954)	1	L	Electric recording	C	–	3	–
Sorex araneus							
Buchalczyk (1972)	48	F + L	Laboratory + outdoor cages	N; A.H.	–	–	–
		(S, A, W, Sp)	(Polygraph)				
Churchfield (1982)	15 (S)	L	Metabolic cage	N	28	0.1	12
	20 (W)		(Time-lapse photography)		19	0.05	–
Crowcroft (1954)	9	L	Electric recording	N	42	0.2–3	12
Gębczyński (1965)		L	Metabolic chamber		–	–	–
	9	S (old ad.)	(O_2 consumption)	N (U)	–	3.2	–
	19	S (young ad.)		N (U)	–	2.0	–
	12	A (young ad.)		N (B)	–	2.0	–
	14	W (young ad.)		N (B)	–	1.8	–
	16	Sp (old ad.)		N (B)	–	2.6	–
Loxton et al. (1975)	7 (S)	L	Vibrating cage	N	–	0.4–0.6	–
	7 (W)					0.3–0.4	
Pankakoski (1979)	3	F	Terraria, running wheel,	N (B)	–	–	–
		(A)	Electric circuit				
Pernetta (1977)	–	F (W, S)	Live-trapping	–	–	–	–
Saarikko and Hanski (1990)	17	L	Arena with feeders	A.H.	42	0.8	–
Shillito (1963)	–	F	Live-trapping	A.H.	–	–	–

Table 13.1 *(continued)*

Author (year)	Study			Activity descriptors			
	n	Condition	Technique	Rhythm	Total length (% of day)	Duration of foraging bouts (h)	Number of foraging bouts per day
Sorex arcticus							
Buckner (1964)	33	L	Observation chamber	N	8	3.3	14
Sorex caecutiens							
Yoshino and Abe (1984)	4	L	Cages	N	34	–	–
Saarikko and Hanski (1990)	2	L	Arena with feeders	A.H.	–	1.0	–
Sorex cinereus							
Buckner (1964)	15	L	Tilting runways	A.H.	15	2.1	19
Cawthorn (1994)	86	F (Sp, S, A)	Live-trapping	D	–	–	–
Douchet and Bider (1974)	–	F (Sp, S, A)	Sand transects	N (U)	–	–	–
Morrison and Pearson (1946)	1	L	Respirometer	A.H.	–	1.4	–
Pearson (1947)	1	L	Respirometer	N	–	1.5	–
Sorex coronatus							
Genoud (1984)	8	F	Radioactive tracking	A.H.	54	1.3	11
Sorex fumeus							
Hamilton (1940)	–	F (Sp, S, A)	Live-trapping	A.H.	–	–	–
Cawthorn (1994)	26	F (Sp, S, A)	Live-trapping	D	–	–	–
Sorex hoyi							
Buckner (1964)	1	L	Tilting runways	A.H.	17	2.9	–
Sorex isodon							
Saarikko and Hanski (1990)	2	L	Arena with feeders	A.H.	–	1.7	–
Sorex minutus							
Buchalczyk (1972)	11	F (W, Sp)	Outdoor cages (Polygraph)	D	–	–	–

Table 13.1 *(continued)*

Crowcroft (1954)	1	L	Electric recording	A.H.	–	2	12
Pernetta (1977)	–	F (W, S)	Live-trapping	D	–	–	–
Sorex ornatus							
Newman and Rudd (1978)	55	L	Metabolic chamber	N (Sp, S) D (W)	–	≤ 10	–
Pearson (1959)	–	F	Photographic recording	N	–	–	–
Rust (1978)	9	L	Activity box	N	24	1.0	–
Sorex palustris							
Sorenson (1962)	13	L	Enclosure (O_2 consumption)	N	> 15	0.5	16
Sorex trowbridgii							
Rust (1978)	7	L	Activity box	N	39	1.0	–
Sorex unguiculatus							
Yoshino and Abe (1984)	12	L	Cages	N	23	0.5–1.5 (N) 2.0 (D)	–
Sorex vagrans							
Clothier (1955)	–	F (S, Sp, A, W)	Snap-trapping	A.H.	–	–	–
Ingles (1960)	3	F (S)	Outdoor cages (Electromagnetic recording)	N (B)	–	0.7 (D) 1.0 (N)	9–15
Suncus murinus							
Richardson (1973)	10	L	Activity chamber (Photo-electric cell)	N	22	0.8	9

Study: *F* = field, *L* = laboratory. **Seasons:** *W* = winter, *Sp* = spring, *S* = summer, *A* = autumn. **Rhythm** (primarily): *D* = diurnal, *N* = nocturnal, *C* = crepuscular, *A.H.* = all hours, *U* = unimodal, *B* = bimodal.

surgery (Reynolds 1992), or the mass of the transmitter may impair normal locomotor ability of shrews. In addition, the cost of transmitters precludes large sample sizes.

Although shrews in outdoor enclosures are subjected to natural events, they are still confined and thus their movements are restricted. Few studies have monitored truly free-ranging shrews. The early studies of Pearson (1959) and Osterberg (1962) employed electronic flash photography to document activity periods of shrews using surface runways. Photographic monitoring has great potential but is compromised by the inability to recognise individuals and the subsurface foraging zones of many species. More recently, investigators employed radioisotopes to analyse time budgets and ecological energetics of free-ranging *Crocidura russula* and *Sorex coronatus* (Genoud and Vogel 1981; Genoud 1984, 1985). Use of isotopes mitigate some of the shortcomings of implanted radiotransmitters and present an accurate depiction of activity patterns of shrews occurring in the wild. However, acute doses of radiation may adversely affect small mammals (Mihok et al. 1985). Many ingenious methods have been employed to assess activity patterns in shrews and each possesses inherent biases.

13.3 Factors Influencing Activity Rhythms

13.3.1 Food

The high metabolic demands of shrews necessitate an activity profile typified by numerous intense and sustained foraging periods spanning both day and night. The polyphasic activity pattern of shrews is dictated by their diet of invertebrate animals, and they typically cannot survive more than a few hours without feeding (Churchfield 1990). Shrews are active from 7 to 54% of the day with foraging bouts ranging from 0.1 to 3 h in length (Table 13.1). Periods of activity alternate with periods of rest during which shrews may exhibit depressed metabolic rates (Platt 1974).

The short-term rhythm of shrews is clearly linked to nutritional requirements. Studies of oxygen consumption have quantified caloric requirements of shrews and clarified their daily activity patterns. Temporal organisation of metabolism corresponds well to food requirements in nature (Morrison et al. 1957; Kenagy and Vleck 1982). Vlasák (1980) employed a large outdoor enclosure to study surface activity patterns of *S. araneus* in southern Czechoslovakia and found that temperature influenced them by affecting availability of invertebrate prey. Starvation times ranged from 3 to 12 h, commensurate with body mass. Activity levels are adjusted in response to food availability with foraging bouts interrupted by brief rest periods averaging about 7 min in length with its duration of rest related to fullness of the stomach or intestine (Hanski 1985; Saarikko and Hanski 1990). Periods of rest may confer significant energy savings and Martinsen (1969) concluded that *B. brevicauda* spent 84% of the day at low metabolic rates. Foraging bouts are followed by longer periods of sleep lasting from 12 min to 2 h (Sorenson 1962; Saarikko and Hanski 1990). Length and frequency of waking/sleep states were elucidated for *Suncus murinus*, *B. brevicauda* and *Crytotis parva* behaviourally and by use of polygraphic techniques (Allison et al. 1977).

A further tactic of energy conservation of some species of shrews is to undergo periods of shallow daily torpor. This short-term depression in body temperature is found in the Crocidurinae, namely, *Crocidura russula*, *C. suaveolens*, *Diplomesodon pulchellum*, and *Suncus etruscus* (Vogel 1976; Nagel 1977; Genoud 1985, 1988; Taylor 1998). However, members of the Soricinae, typified by higher metabolic rates, rarely undergo torpidity. When mildly food-stressed, *Notiosorex crawfordi* is reported to restrict activity to the nest site and to decrease metabolic rate and body temperature (Lindstedt 1980). Further, during metabolic experiments *Sorex ornatus* showed torpor-like behaviour as indicated by inactivity, slowed breathing, and a decline in metabolic rate (Newman and Rudd 1978). Both species reside in geographic areas of the US characterised by warm climates and physiological adjustments may represent mechanisms enhancing water conservation, at least in the case of *N. crawfordi*.

Hoarding behaviour is documented for 30 families of mammals (Vander Wall 1990). Within the Soricidae seven species are reported to cache food: *B. brevicauda*, *S. araneus*, *S. minutissimus*, *S. minutus*, *S. palustris*, *S. pacificus* and *Neomys fodiens* (Crowcroft 1957; Sorenson 1962; Buchalczyk and Pucek 1963; Maser and Hooven 1974; Wołk 1976; Skaren 1978; Martin 1984). Although the ability to cache food confers the adaptive advantage of decreasing energy required for foraging, most shrews do not hoard food and the behaviour has been demonstrated primarily under laboratory conditions with food provided ad libitum. The northern short-tailed shrew (*B. brevicauda*) is reported to cache a variety of food items ranging from earthworms and plant material to small mammals (Platt 1974; Tomasi 1978; Martin 1981; Robinson and Brodie 1982; Martin 1984). These shrews possess a venom, produced by the submaxillary gland, which is delivered by biting and immobilises prey in a comatose state (Tomasi 1978; Martin 1981; Merritt 1986). Cached prey are marked by urination and defecation, available to shrews as a fresh source of food for a period of time after capture, and may serve as a predictable source of energy, readily available as a buffer against food scarcity. Because food caches are located below ground in thermally-stable microclimatic zones, shrews may in this way reduce foraging activity plus minimise energy used for thermoregulation during winter months (Merritt and Adamerovich 1991). The water shrew (*S. palustris*) is reported to hoard specific types of food (Sorenson 1962), economising the length of activity periods and restricting foraging to zones possessing established caches.

Coprophagy, or refection, i.e., the habit of feeding upon faeces, is well known in the orders Lagomorpha and Rodentia and has also been documented for about nine species of shrews (Crowcroft 1957; Loxton et al. 1975; Goulden and Meester 1978; Skaren 1978). This behaviour may reduce the time spent in foraging. For *S. araneus*, coprophagy was reported to occur during daytime when food consumption and activity were low (Loxton et al. 1975). While in lagomorphs and rodents coprophagy is related to refection of bulky food by endosymbionts, the adaptive significance of coprophagy for shrews in unknown. It may represent a technique of reducing daily food intake and extracting certain essential nutrients from available food.

13.3.2 Light, Weather, and Seasonal Factors

The daily activity patterns of shrews are affected by changes in climatological factors. Investigators have worked both in the field and laboratory to assess the importance of these factors, but results are compromised in that many climatological events covary and thus are not mutually exclusive. Daily and seasonal changes in the activity pattern of many shrews are primarily influenced by photoperiod, typified by two nocturnal peaks in foraging (Crowcroft 1957; Ingles 1960; Gębczyński 1965; Buchalczyk 1972).

Students of soricid biology commonly acknowledge precipitation as a driving force in dictating pulses of activity. Rainfall was reported to strongly increase the activity of *S. araneus*, *S. caecutiens* and *S. minutus* in Poland (Mystkowska and Sidorowicz 1961). Many meteorological factors enhanced captures. Getz (1961) noted an increased trappability of *B. brevicauda* on cloudy rather than sunny or rainy days, but temperature had no influence on activity patterns. The studies of Doucet and Bider (1974) and Vickery and Bider (1978) employed sand transects to monitor the activity of *S. cinereus*. Initiation of rainfall marked the start of foraging and explained 27% of the variation in activity. Nocturnal hourly activity was influenced by rainfall and activity increased on cloudy nights. In contrast, the radiotelemetry studies of Chandler (1989) indicate that activity of *B. brevicauda* was negatively correlated with rainfall. High relative air humidity correlated positively with the activity of *S. araneus* housed in outdoor cages (Pankakoski 1979); the influence of rainfall could not be evaluated because animals were maintained in a roofed space. Relative humidity did not seem to be important in regulating activity patterns of *S. araneus* studied by Vlasak (1980).

Daily and seasonal temperature changes are known to influence activity patterns of shrews. Since temperature changes influence the activity of invertebrates within the soil, foraging by the soricid predator may also be influenced. The role of temperature on activity and food consumption of *S. araneus* was evaluated with time-lapse photography (Churchfield 1982). Shrews were active throughout the day and night with activity peaks every 1–2 h. Activity rhythms predominated during darkness. Summer activity outside the nest was 28% but dropped to 19% during winter, associated with a decrease in food consumption. The reduction in caloric intake and foraging activity may represent an overwinter survival tactic of *S. araneus*. Buchalczyk

(1972) studied the activity of four species of shrews from Poland housed under laboratory and semi-laboratory conditions: activity was highest in summer and lowest in winter. Vlasák (1980) assessed the influence of average day and night temperatures on the surface activity of *S. araneus* residing in a large outdoor enclosure in southern Czechoslovakia. In contrast to the above studies, he found that patterns of activity were not primarily dependent on temperature, and surface activity was highest in winter and lowest in summer. Activity of shrews was closely associated with an interaction of temperature and photoperiod on the availability of invertebrate food. In support of this view, trends in seasonal maintenance metabolism for *B. brevicauda* show a metabolic increase from a low in summer to a maximum in autumn (Randolph 1980; Merritt 1986) and winter (Randolph 1973; Platt 1974; Merritt 1986). Seasonal changes in activity are difficult to ascribe to a single environmental factor. Activity is linked to a complex interaction of temperature, humidity, barometric pressure, and precipitation (Osterberg 1962).

The climatological regime of the subterranean and subnivean foraging zone also regulates activity patterns of shrews. The ecology and year-round activity of shrews has been documented for northern regions. In such areas, snow cover may persist for up to 7.5 months of the year reaching depths of 200 cm. Temperatures above the snow may reach –40 °C, while subnivean temperatures remain stable at ca. 0 °C (Pruitt 1957), and shrews are active within the subnivean space (Merritt and Merritt 1978). Coulianos and Johnels (1962) emphasised the importance of the subnivean air space for locomotor activity and feeding dynamics of voles, mice and shrews. Studies of the activity of *B. brevicauda* in snowy regions delimit the following adaptive mechanisms enhancing overwinter survivorship: (1) foraging restricted to stable microclimatic zones below an insulatory layer of snow, (2) utilisation of nest chambers with temperatures close to thermoneutrality, (3) activity patterns characterised by lengthy resting periods at low metabolic rates and constant body temperatures, (4) utilisation of underground food caches, and (5) ability to employ rapid thermogenic responses for foraging bouts due to non-shivering thermogenesis (Martinsen 1969; Platt 1974; Randolph 1980; Martin 1983; Merritt 1986; Hyvärinen 1994; Merritt and Bozinovic 1994). Variations in the daily activity patterns of shrews residing in northern regions may in part be attributable to some or all of the above survival tactics.

Two major 'metabolic levels' characterise Soricidae (Vogel 1976; Genoud 1988). These levels show many different attributes (see Table 6.5 in Churchfield 1990) that influence activity patterns of shrews. The Soricinae are characterised by high metabolic rates and body temperatures; they are solitary, and do not torpor or share nests. In contrast, the Crocidurinae possess moderate and variable metabolic rates and body temperature, are social, undergo torpor, and share nests. The Soricinae are distributed in more northerly regions whereas the Crocidurinae occur in warmer, more southern latitudes. Genoud (1984) compared the activity patterns of sympatric *C. russula* (Crocidurinae) and *S. coronatus* (Soricinae) using radioactive tracking techniques. *S. coronatus* exhibited long activity periods (54%), interspersed with short resting periods – an intense diurnal activity rhythm was noted. In contrast, total daily activity of *C. russula* amounted to just 33%. By undergoing short-term torpor, crocidurine shrews may reduce periods of activity. The higher energy levels of the Soricinae may have evolved in response to cold stress associated with northern environments (Taylor 1998). Selective advantages include increased thermogenesis and homeostasis in cold and optimisation of foraging ability (Genoud 1988; Churchfield 1990).

As noted above, most shrews do not exhibit hibernation, although some show slight seasonal depression in body temperature and metabolic rate. In contrast, many species of bats, of similar body mass as shrews, readily reduce body temperature and metabolism and undergo long seasonal bouts of hibernation. For hibernating bats, aerial insects form a staple in their diet, and abrupt seasonal decreases of this high energy food in winter, necessitates inactivity in the form of hibernation. However, shrews subsist on ground-dwelling invertebrates, abundant year-round, and thus, unlike in bats, availability of prey does not seem to represent a sufficient selective pressure to have caused seasonal inactivity to have evolved in this group.

13.3.3 Sex and Reproduction

Few investigators have depicted the daily activity pattern of shrews according to age, sex, and reproductive status. Gębczyński (1965) examined year-round trends in oxygen consumption of *S. araneus* in Poland. Bimodal activity patterns occurred in autumn, winter, and spring, but old adults showed a unimodal pattern in summer. The length of the foraging bouts ranged from 1.5 to 2.5 h; young adults displayed the shortest foraging durations in winter and old adults exhibited longer bouts in summer (Table 13.1). Older shrews exhibited a longer duration in foraging than younger animals, but young animals were more active as indicated by metabolic rate. Year-round locomotor and feeding activity were studied in *S. araneus*, *S. minutus*, *Neomys anomalus*, and *N. fodiens* in outdoor cages in Poland (Buchalczyk 1972). During summer, young adults were 83% more active than old adults. Activity decreased in autumn and winter but shrews showed an abrupt increase in activity in spring associated with reproductive activity. Rust (1978) used activity chambers to study patterns in *S. ornatus* and *S. trowbridgii* from the western US. Breeding adults were twice as active as non-breeding young and old adults. Age and sex of shrews did not influence level of activity. Breeding shrews were primarily nocturnal. Chandler (1989) used radiotelemetry to monitor activity patterns of free-ranging *B. brevicauda* and reported that males were more active than females. This difference may have been attributable to males' searching for mates.

13.3.4 Optimal Foraging and Competitors

A number of laboratory studies have explored the roles that competitors have in foraging decisions made by shrews, in terms of both diet selection and patch choice. For instance, *S. araneus* become less selective in the type of prey they consume in the presence of a competitor (Barnard and Brown 1981). As predicted by theory, shrews become risk-prone (choosing the more variable rather than the more constant reward) when not allowed to meet their energy requirement (Barnard and Brown 1985a). In the presence of a competitor, however, they become risk-indifferent (Barnard and Brown 1985b). As regards choice of food patches and patch residence times, shrews failing to meet their food requirements choose residence times that maximise variance in intake rate, thus increasing the chances of a high reward (Barnard and Brown 1987). They spend little time in poor patches, moving quickly to good ones. It is not clear how these variables might translate into the various activity patterns described in Table 13.1. Presumably, some of the differences in lengths of foraging bouts are caused by the types of food items available, their distribution in space and time, and the presence of competitors, but field data are lacking.

One laboratory study specifically addressed the effect of conspecific competitors on foraging activity. Because shrews tend to be dependent on a short-term food supply, they might be expected to be particularly sensitive to the presence of competitors, showing an increase in foraging and caching behaviour. Barnard et al. (1983) tested this hypothesis in the laboratory with *S. araneus*. When a conspecific was visible, the resident foraged more, captured more, and ate more fly pupae than when the conspecific was confined to one end of the apparatus and out of sight. Frequencies (but not durations) of other behaviour patterns such as alert rest, defecating, and grooming also increased significantly, while sleep and rest decreased when the competitor was visible. The number of pupae cached was significantly higher when the competitor was visible as well. Estimates of caloric expenditure confirmed that shrews used more energy when competitors were present.

Interspecific competition among soricids has received some attention, and sympatric species of shrews seem to segregate by body size, food size, and foraging mode (Churchfield and Sheftel 1994); however, there are no experimental field studies on the role of competition in structuring shrew communities (Kirkland 1991). Temporal (i.e., nocturnal versus diurnal) segregation among sympatric species of shrews is unlikely due to their small body mass and high metabolic rate. Sympatric species do differ in mass and metabolic rate, and small species have shorter periodicities in activity bouts than do larger species (Yoshino and Abe 1984); however, there is presently no evidence showing that activity patterns are influenced by interspecific competition. A review of interspecific competition in the Soricidae is provided by Rychlik (1998).

13.4 Conclusions

Numerous laboratory and field studies confirm that shrews have ultradian activity cycles. The usual pattern seems to be bouts of as little as 0.1–3 h in length, with 10–15 bouts per day (Table 13.1). For intensively-studied species, such as *B. brevicauda* and *S. araneus*, there is little agreement as to whether rhythms are predominantly diurnal or nocturnal. Among the factors influencing these activity rhythms, food seems to be the most crucial, considering shrews' high metabolic rate and diet of invertebrates. Some studies show an effect of photoperiod on the timing of activity, but others do not. Precipitation seems to be positively associated with onset of activity, although there are exceptions. Temperature is both positively and negatively associated with activity within the same species. Shrews have a number of adaptations for overwintering in the subnivean space that make them less dependent on the supranivean climate than other small mammals. The use of runways and tunnels might enable shrews to remain active in the daytime without suffering heavy predation, as is the case with many species of voles. Members of the subfamily Soricinae have high metabolic rates and longer activity bouts compared with members of the Crocidurinae; the latter group is also generally found at lower latitudes. Several studies report that breeding adult shrews are more active than non-breeding animals, and there is one report that breeding males are more active than females. Shrews adjust activity patterns in response to the abundance and distribution of food resources as do other species of vertebrates, and they increase foraging activity in response to conspecific competitors. Conventional wisdom argues that shrews must engage in short, frequent bouts of activity, given their high metabolic rates.

References

Allison T, Gerber SD, Breedlove SM, Dryden GL (1977) A behavioral and polygraphic study of sleep in the shrews *Suncus murinus*, *Blarina brevicauda*, and *Cryptotis parva*. Behav Biol 20: 354–366

Antipas AJ, Madison DM, Ferraro JS (1990) Circadian rhythms in the short-tailed shrew, *Blarina brevicauda*. Physiol Behav 48:255–260

Aschoff J (1981) A survey on biological rhythms. In: Aschoff J (ed) Biological rhythms. Handbook of behavioural neurobiology, vol 4. Plenum Press, New York, pp 3–10

Barnard CJ, Brown CAJ (1981) Prey size selection and competition in the common shrew (*Sorex araneus* L.). Behav Ecol Sociobiol 8:239–243

Barnard CJ, Brown CAJ (1985a) Risk-sensitive foraging in common shrews (*Sorex araneus* L.). Behav Ecol Sociobiol 16:161–164

Barnard CJ, Brown CAJ (1985b) Competition affects risk-sensitivity in foraging shrews. Behav Ecol Sociobiol 16:379–382

Barnard CJ, Brown CAJ (1987) Risk-sensitive foraging and patch residence time in common shrews, *Sorex araneus* L. Anim Behav 35:1255–1257

Barnard CJ, Brown CAJ, Gray-Wallis J (1983) Time and energy budgets and competition in the common shrew (*Sorex araneus* L.). Behav Ecol Sociobiol 13:13–18

Bider JR (1968) Animal activity in uncontrolled terrestrial communities as determined by a sand transect technique. Ecol Mongr 38:269–308

Buchalczyk A (1972) Seasonal variation in the activity of shrews. Acta Theriol 17:221–243

Buchalczyk A, Pucek Z (1963) Food storage of the European water shrew, *Neomys fodiens* (Pennant 1771). Acta Theriol 7:376–379

Buckner CH (1964) Metabolism, food capacity, and feeding behavior in four species of shrews. Can J Zool 42:259–279

Cawthorn JM (1994) A live-trapping study of two syntopic species of *Sorex*, *S. cinereus* and *S. fumeus*, in southwestern Pennsylvania. In: Merritt J, Kirkland GL Jr, Rose RK (eds) Advances in the biology of shrews. Spec Publ Carnegie Mus Nat Hist 18, Pittsburgh, pp 39–43

Chandler MC (1989) The population ecology and temporal and spatial activity of three species of shrews (*Sorex cinereus*, *S. fumeus*, and *Blarina brevicauda*) in southwestern Pennsylvania. PhD thesis, Bowling Green State University

Churchfield S (1980) Population dynamics and the seasonal fluctuations in numbers of the common shrew in Britian. Acta Theriol 25:415–424

Churchfield S (1982) The influence of temperature on the activity and food consumption of the common shrew. Acta Theriol 27:295–304

Churchfield S (1990) The natural history of shrews. Helm, London

Churchfield S, Sheftel BI (1994) Food niche overlap and ecological separation in a multispecies community of shrews in the Siberian taiga. J Zool Lond 234:105–124

Clothier RR (1955) Contribution to the life history of *Sorex vagrans* in Montana. J Mammal 36:214–221

Coulianos CC, Johnels AG (1962) Note on the subnivean environment of small mammals. Arkiv Zool 15:363–370

Crowcroft P (1954) The daily cycle of activity in British shrews. Proc Zool Soc Lond 123: 715–729

Crowcroft P (1957) The life of the shrew. Reinhardt, London

Crowcroft P (1959) A simple technique for studying activity rhythms of small mammals by direct observation. Acta Theriol 3:105–111

Davis WB, Joeris L (1945) Notes on the life-history of the little short-tailed shrew. J Mammal 26:136–138

Doucet GJ, Bider JR (1974) The effects of weather on the activity of the masked shrew. J Mammal 55:348–363

Gębczyński M (1965) Seasonal and age changes in the metabolism and activity of *Sorex araneus* Linnaeus 1758. Acta Theriol 10:303–331

Genoud M (1984) Activity of *Sorex coronatus* (Insectivora, Soricidae) in the field. Z Säugetierkd 49:74–78

Genoud M (1985) Ecological energetics of two European shrews: *Crocidura russula* and *Sorex coronatus* (Soricidae: Mammalia). J Zool Lond 207:63–85

Genoud M (1988) Energetic strategies of shrews: ecological constraints and evolutionary implications. Mamm Rev 18:173–193

Genoud M, Vogel P (1981) The activity of *Crocidura russula* (Insectivora, Soricidae) in the field and in captivity. Z Säugetierkd 46:222–232

Getz LL (1961) Responses of small mammals to live traps and weather conditions. Am Midl Nat 66:160–170

Godfrey GK (1978) The activity pattern in white-toothed shrews studied with radar. Acta Theriol 23:381–390

Goulden EA, Meester J (1978) Notes on the behaviour of *Crocidura* and *Myosorex* (Mammalia: Soricidae) in captivity. Mammalia 42:197–207

Hamilton WJ Jr (1940) The biology of the smoky shrew (*Sorex fumeus fumeus* Miller). Zoologica 25:473–492

Hamilton WJ Jr (1944) The biology of the little short-tailed shrew, *Cryptotis parva*. J Mammal 25:1–7

Hanski I (1985) What does a shrew do in an energy crisis? In: Sibly RM, Smith RH (eds) Behavioral ecology. Blackwell, Oxford, pp 247–252

Hyvärinen H (1994) Brown fat and the wintering of shrews. In: Merritt J, Kirkland GL Jr, Rose RK (eds) Advances in the biology of shrews. Spec Publ Carnegie Mus Nat Hist 18, Pittsburgh, pp 259–266

Ingles LG (1960) A quantitative study on the activity of the dusky shrew (*Sorex vagrans obscurus*). Ecology 41:656–660

Kenagy GJ, Vleck D (1982) Daily temporal organization of metabolism in small mammals: adaptation and diversity. In: Aschoff J, Daan S, Groos GA (eds) Vertebrate circadian systems. Springer, Berlin Heidelberg New York, pp 322–338

Kirkland GL Jr (1991) Competition and coexistence in shrews (Insectivora: Soricidae). In: Findley JS, Yates TL (eds) The biology of the Soricidae. Mus Southw Biol, University of New Mexico, Albuquerque, pp 15–22

Lindstedt SL (1980) Energetics and water economy of the smallest desert mammal. Physiol Zool 53:82–97

Loxton RG, Raffaelli D, Begon M (1975) Coprophagy and the diurnal cycle of the common shrew. J Zool Lond 177:449–453

Mann PM, Stinson RH (1957) Activity of the short-tailed shrew. Can J Zool 35:171–177

Martin IG (1981) Venom of the short-tailed shrew (*Blarina brevicauda)* as an insect immobilizing agent. J Mammal 62:189–192

Martin IG (1983) Daily activity of short-tailed shrews (*Blarina brevicauda*) in simulated natural conditions. Am Midl Nat 109:136–144

Martin IG (1984) Factors affecting food hoarding in the short-tailed shrew *Blarina brevicauda*. Mammalia 48:65–71

Martinsen DL (1969) Energetics and activity patterns of short-tailed shrews (*Blarina)* on restricted diets. Ecology 50:505–510

Maser C, Hooven EF (1974) Notes on the behavior and food habits of captive Pacific shrews, *Sorex pacificus pacificus*. Northw Sci 48:81–95

Merritt JF (1986) Winter survival adaptations of the short-tailed shrew (*Blarina brevicauda*) in an Appalachian montane forest. J Mammal 67:450–464

Merritt JF, Adamerovich A (1991) Winter thermoregulatory mechanisms of *Blarina brevicauda* as revealed by radiotelemetry. In: Findley JS, Yates TL (eds) The biology of the Soricidae. Mus Southw Biol, University of New Mexico, Albuquerque, pp 47–64

Merritt JF, Bozinovic F (1994) Thermal biology of free-ranging shrews as revealed by computer-facilitated radiotelemetry: energetic implications. In: Merritt JF, Kirkland GL Jr, Rose RK (eds) Advances in the biology of shrews. Spec Publ Carnegie Mus Nat Hist 18, Pittsburgh, pp 163–169

Merritt JF, Merritt JM (1978) Population ecology and energy relationships of *Clethrionomys gapperi* in a Colorado subalpine forest. J Mammal 59:576–598

Michielsen NC (1966) Intraspecific and interspecific competition in the shrews *Sorex araneus* L. and *Sorex minutus* L. Arch Neerland Zool 17:73–174

Mihok S, Schwartz B, Iverson SL (1985) Ecology of red-backed voles (*Clethrionomys gapperi*) in a gradient of gamma radiation. Ann Zool Fennici 22:257–271

Morrison PR, Pearson OP (1946) The metabolism of a very small mammal. Science 104:287–289

Morrison PL, Pierce M, Ryser FA (1957) Food consumption and body weight in the masked and short-tail shrews. Am Midl Nat 57:493–501

Mystkowska ET, Sidorowicz J (1961) Influence of the weather on captures of Micromammalia. II. Insectivora. Acta Theriol 5:263–273

Nagel A (1977) Torpor in the European white-toothed shrews. Experientia 33:1455–1456

Newman JR, Rudd RL (1978) Minimum and maximum metabolic rates of *Sorex sinuosus*. Acta Theriol 23:371–380

Osterberg DM (1962) Activity of small mammals as recorded by a photographic device. J Mammal 43:219–229
Pankakoski E (1979) The influence of weather on the activity of the common shrew. Acta Theriol 24:522–526
Pearson O (1947) The rate of metabolism of some small mammals. Ecology 28:127–145
Pearson O (1959) A traffic survey of *Microtus-Reithrodontomys* runways. J Mammal 40:169–180
Pernetta JC (1977) Population ecology of British shrews in grassland. Acta Theriol 22:279–296
Platt WJ (1974) Metabolic rates of short-tailed shrews. Physiol Zool 47:75–90
Platt WJ (1976) The social organization and territoriality of short-tailed shrew (*Blarina brevicauda*) populations in old-field habitats. Anim Behav 24:305–318
Pruitt WO Jr (1957) Observations on the bioclimate of some taiga mammals. Arctic 10:130–138
Randolph JC (1973) Ecological energetics of a homeothermic predator, the short-tailed shrew. Ecology 54:1166–1187
Randolph JC (1980) Daily metabolic patterns of short-tailed shrews (*Blarina*) in three natural seasonal temperature regimes. J Mammal 61:628–638
Reynolds PS (1992) White blood cell profiles as a means of evaluating transmitter-implant surgery in small mammals. J Mammal 73:178–185
Richardson JH (1973) Locomotory and feeding activity of the shrews, *Blarina brevicauda* and *Suncus murinus*. Am Midl Nat 90:224–227
Robinson DE, Brodie ED Jr (1982) Food hoarding behavior in the short-tailed shrew *Blarina brevicauda*. Am Midl Nat 108:369–375
Rood JP (1958) Habits of the short-tailed shrew in captivity. J Mammal 39:499–507
Rust AK (1978) Activity rhythms in the shrews, *Sorex sinuosus* Grinnell and *Sorex trowbridgii* Baird. Am Midl Nat 99:369–388
Rychlik L (1998) Evolution of social systems in shrews.. In: Wójcik JM, Wolsan M (eds) Evolution of shrews. Mammal Res Inst, Polish Acad Sci, Białowieża, Poland, pp 347–406
Saarikko J, Hanski I (1990) Timing of rest and sleep in foraging shrews. Anim Behav 40:861–869
Shillito JF (1963) Field observations on the growth, reproduction and activity of a woodland population of the common shrew *Sorex araneus* L. Proc Zool Soc Lond 140:99–114
Skarén U (1978) Feeding behavior, coprophagy and passage of foodstuffs in a captive least shrew. Acta Theriol 23:131–140
Sorenson MW (1962) Some aspects of water shrew behavior. Am Midl Nat 68:445–462
Taylor JRE (1998) Evolution of energetic strategies in shrews. In: Wójcik JM, Wolsan M (eds) Evolution of shrews. Mammal Res Inst, Polish Acad Sci, Białowieża, Poland, pp 309–346
Tomasi TE (1978) Function of venom in the short-tailed shrew, *Blarina brevicauda*. J Mammal 59:852–854
Vander Wall SB (1990) Food hoarding in animals. University of Chicago Press, Chicago
Vickery WL, Bider JR (1978) The effect of weather on *Sorex cinereus* activity. Can J Zool 56: 291–297
Vlasák P (1980) Seasonal changes in the surface activity of the common shrew, *Sorex araneus* (Insectivora). Vest Spolec Zool 44:306–319
Vogel P (1976) Energy consumption of European and African shrews. Acta Theriol 21:195–206
Webster AB, Brooks RJ (1980) Effects of radiotransmitters on the meadow vole, *Microtus pennsylvanicus*. Can J Zool 58:997–1001
Wołk K (1976) The winter food of the European water-shrew. Acta Theriol 21:117–129
Yoshino H, Abe H (1984) Comparative study on the foraging habits of two species of soricine shrews. Acta Theriol 29:35–43

14 Bats – Flying Nocturnal Mammals

Hans G. Erkert

14.1 Introduction

Bats constitute the second largest mammalian order, the Chiroptera. It comprises a total of some 950 extant species, all of which are nocturnal. The evolution of true, i.e., active flight more than 50 million years ago, as well as the development of echolocation in the Microchiroptera and in the megachiropteran genus *Rousettus*, and of the ability of Megachiroptera to see well in dim light, allowed the bats to adapt to a variety of niches largely unoccupied at night. Nocturnality for the bats entailed the benefit of minimising the predator pressure by diurnal birds of prey such as falcons, hawks, etc. during the activity time. It also provided access to important food resources otherwise scarcely exploited because only very few bird species were able to specialise for nocturnal foraging. Therefore these flying nocturnal mammals could evolve a multitude of food habits. Most bat species (approximately 70%) are insectivorous and catch their prey either on the wing, by foliage gleaning or by terrestrial acquisition (Hill and Smith 1988). Carnivory is found in at least six species. Four of them belong to the large New World family of leaf-nosed bats (Phyllostomatidae), and the other two to the 'false vampires' (Megadermatidae), distributed in India, Southeast Asia, and Australia. Piscivory is reported for the two species *Noctilio leporinus* (Noctilionidae) and *Pizonyx vivesi* (Vespertilionidae), both of which inhabit the tropical and subtropical regions of the Americas. The most extraordinary specialisation of zoovory is the sanguivory developed in the true vampires *Desmodus rotundus*, *Diaemus youngi*, and *Diphylla ecaudata*, which constitute the phyllostomatid subfamily Desmodontinae.

Vegetarian food habits in the form of frugivory and nectarivory are widely distributed in species inhabiting the tropical and subtropical regions of the world. All 175 species of the Old World family Pteropodidae (suborder Megachiroptera) consume almost exclusively fruits or nectar, parts of flowers, and pollen. In the New World bats frugivory and nectarivory evolved independently. They are found in the majority of the 147 species of the Phyllostomatidae (suborder Microchiroptera). Many of them occasionally also

Ecological Studies, Vol. 141
Halle/Stenseth (eds.) Activity Patterns in Small Mammals

consume insects, a habit presumably derived from insectivorous ancestors. No leaf-eating or granivorous species are known.

Successful exploitation of a great variety of food resources, which not only provide quite different forms and amounts of energy per item, but also vary in space and time, requires quite different foraging strategies. These, in combination with the species' and/or individuals' energetic demands and the requirements of digestion, determine the time course of the bats' nocturnal activities to a large extent. Therefore in this large group of flying small mammals ranging in body mass from 2 g in *Craseonycteris thonglongyai* to about 1 kg in the Indian fruit bat (*Pteropus giganteus*) extremely diverse nocturnal activity patterns are to be expected.

The availability of suitable roosts where the bats can spend the daytime safely and under favourable climatic conditions is another limiting factor for the Chiroptera. The marked differences in the roosting habits found in the various bat species may also affect their activity patterns. A great number of the Microchiroptera and some members of the megachiropteran genus *Rousettus* spend their daily resting time in sheltered roosts such as caves, rock crevices, tree cavities or man-made structures. Many other Microchiroptera and most of the Megachiroptera use external day roosts such as dense foliage, the underside of large leaves, branches or trunks of trees, etc. In contrast to these external roosts, which are subject to the climatic conditions of the environment, the sheltered roosts are of special advantage because they offer good protection against predation, unfavourable weather conditions and bright sunlight. Furthermore, they provide a stable microclimate and are relatively permanent (Kunz 1982). However, whereas external roosts are usually ubiquitous and abundant within the bats' foraging areas, sheltered roosts may be scattered over large areas. While external roosts are changed frequently, internal roosts such as caves, pits or buildings usually serve as traditional day resting sites of the respective bat population for very long times, often over centuries or even millenia. Such favourable day roosts in large caves or abandoned pits are at best situated within the foraging area, or several hundred meters or a few kilometers distant from it. Other day roosts may be located far away from the foraging areas and are often used, either permanently or only during the summer, by very large numbers of bats ranging up to several hundred thousands or even millions of individuals (e.g. *Tadarida plicata* in India and Southeast Asia, *Hipposideros caffer* in West Africa, *Tadarida brasiliensis* in Arizona, Texas and New Mexico; Kunz 1982; Hill and Smith 1988; Miller et al. 1988). Hence these bats have to cover long distances (up to 30 km or more) before reaching their foraging sites. Flight is an extremely energy-consuming activity, requiring up to 34 times the amount of energy necessary to maintain the bats' basal metabolic rate (Thomas and Suthers 1972; Thomas 1975, 1980). With such long flight routes from the day roost to the foraging sites and back, more time must be spent foraging to acquire enough food to compensate for the increased daily energy loss.

Other factors influencing the bats' activity patterns are, for example, whether they use separate night roosts in addition to their day roosts, whether they enter torpor in order to minimise their energy loss, whether they occupy and defend foraging territories, and whether they have to nurse offspring. Biotic and physical environmental factors reported to cause modifications of the timing and time course of bat activity are the abundance of food, local predation risk, night length, ambient temperature, precipitation, wind, and moonlight (Erkert 1982). Furthermore, it has been shown that the activity pattern obtained for a given bat species also depends largely on the method of activity recording. In this chapter the effects of such factors and relations on the time course of the flight and foraging activities of bats are considered in an attempt to summarise what is currently known about activity patterns and their variability in the Chiroptera. The emphasis here is on the expression of the activity pattern in the natural environment and its relation to ecological factors.

14.2 Flight Activity Patterns under Natural Conditions

14.2.1 Recording Methods

Thus far activity patterns in the natural habitat have only been described in about 5% of living bat species. Almost all of them belong to the Microchiroptera. The recorded activity has mostly been flight and foraging. Detailed information about social activities, which mainly occur at the roosting sites, is scarce. In most cases activity data were obtained by capturing bats with mist nets or special traps when departing from and returning to their day or night roost, while foraging on the wing, or when visiting bodies of fresh water in order to drink (Constantine 1958; Gaisler 1973; Tuttle 1974). This frequently used method is advantageous because the flight activity patterns of several species can be monitored simultaneously. It also provides demographic data on the sex ratio and age structure of the respective population as well as information about the animals' reproductive and nutritiitional state. A great disadvantage, however, is that usually only those bats can be captured that are flying no higher than 4–6 m above ground. Furthermore, a large number of insectivorous species which possess a highly efficient echolocation system may detect and avoid the nets. Once bats have had the stressful experience of being caught and handled, they may avoid the capture site and use other flight routes for the rest of the night.

Another method frequently used is direct observation of the animals departing from and returning to their roost, either with the naked eye or by using special night vision devices. In order to establish activity patterns, foraging nectarivorous and frugivorous bats have often been observed and counted, when visiting flowering or fruiting plants (Heithaus et al. 1974;

Lemke 1984). Some insectivorous species have been observed when hunting insects in the light of street-lamps (Belwood and Fullard 1984; Stutz and Haffner 1985/1986) or over the surface of ponds (Cockrum and Cross 1964; O'Farrell et al. 1967). Others were observed when hanging in wait for prey on perches and scanning their surroundings either actively by ultrasound or passively by listening for flying or crawling insects (Vaughan 1976). Except in the latter case, where individual foraging activity patterns are obtained, the observation method again usually provides only information about the flight activity of the whole population or of parts of it, unless single individuals have been caught and tagged with luminescent marks (Buchler 1976).

Population patterns are also obtained by recording the ultrasound activity at sites frequently used by foraging insectivorous or piscivorous species. During the last two decades radio-tracking has proved to be the best method to record the flight activity of individual bats. It has provided the most detailed information on the activity rhythm and about the various foraging strategies, the utilisation of different night roosts, the use of feeding perches by sit-and-wait predators, and other aspects. In some species such as the neotropical phyllostomatid *Carollia perspicillata*, or the North American vespertilionid *Eptesicus fuscus*, radio-tracking studies resulted in activity patterns quite similar to those inferred from mist-net captures or observations (Bonaccorso 1979; Fleming 1988). However, in other species, such as *Euderma maculatum* and *Lasiurus borealis*, the activity patterns of radio-tracked individuals differed substantially from those based on data from captures and observations (Fenton et al. 1993a). From radio-tracking studies it also became evident that in some species the flight activity (foraging) pattern may vary considerably from individual to individual and/or from day to day (Fenton et al. 1993a, b). Hence, most of the activity patterns described thus far cannot be considered as a genetically fixed behavioural characteristic of the respective species. They merely seem to represent a rough approximation to something like a basic pattern. Because of their strong dependence on the recording method as well as on several environmental and internal factors, bat activity patterns can and should at best be classified as tending to be 'more unimodal', 'more bimodal', 'more trimodal' or 'more multimodal' under the specific circumstances of the study.

14.2.2 Timing of Flight Activity

Most bats customarily depart from their day roost some time after sunset in the late dusk. Depending on species, characteristics of the day roost and colony size, the evening emergence phase of bat colonies may last from a few minutes up to several hours. Its timing is subject to species-specific differences. The larger and/or speedier bats seem to depart earlier than the slower smaller ones (Erkert 1978). Later emergence means flight at lower light intensities. Hence the species differences in the timing of emergence from the day roost might reflect an ecological adaptation evolved to avoid higher risk of predation by visually oriented diurnal raptors such as bat falcons (*Falco rufigularis*) which are still active during dusk and well able to catch flying bats. Throughout the annual cycle the evening emergence phase approximately parallels the time of sunset. The time of homing to the day roost may vary considerably between species and individuals as well as in the individual bats. It depends on the respective feeding habits, the foraging strategy and foraging success as well as on the actual energy demand, and on weather conditions (see below). As a rule, the times of homing of the last fliers of bat colonies to their day roost usually proceed in parallel with the time of sunrise (cf. Fig. 14.1; Erkert 1982). Due to this fairly fixed phase relationship of the time of emergence from the day roost to sunset and of the time of return of the last fliers to sunrise the annual variations in the onset and end, and hence in the duration, of the nocturnal flight activity become more pronounced with increasing latitude. However, there are some indications that in subarctic regions the daily activity time cannot be compressed to the very short midsummer nights, and that it also cannot expand to the long nights in spring and autumn (Laufens 1972; Swift 1980).

A rule established by Aschoff and Wever (1962) for birds, which states that the earlier the species or individuals start their daily activity phase the later they retire to rest, has been shown to be valid in several bat species as well (Laufens 1972; Erkert 1978). Correlation analyses have shown that there may be a closer relationship between the times of the bats' departure and homing and the time of a certain illumination intensity during dusk or dawn than between the times of departure and sunset, and between the return of the last flier and sunrise, respectively (Erkert 1978, 1982). Hence species-specific differences in the departure and homing times correspond to differences in illumination at the times of onset and end of the daily flight activity. The return to the day roost in the late night or early morning usually takes place at lower light intensities than the departure in the evening. This might be interpreted as an adaptation evolved to avoid predation by diurnal birds of prey (e.g. *Falco rufigularis*) which are hungry after a long nights' rest and become active in early dawn.

Day flights not caused by disturbance have been reported only in a few species, including *Nyctalus noctula* (Gaisler et al. 1979; Perrin 1987; Kronwitter 1988) and *Myotis mystacinus* (Nyholm 1965). They seem to represent exceptions occurring during migration periods and/or under certain environmental conditions related to low ambient temperatures at night and/or to a much higher abundance of airborne insects during the day than at night (cf. Fig. 14.1).

Despite the close coupling of the departure time to a certain range of illumination, the bats' activity rhythm is not regulated directly by the light intensity, but by way of an endogenous timing system. Its main pacemaker seems to be located in the hypothalamic suprachiasmatic nuclei (SCN). Since this circadian timing system produces a spontaneous period which usually deviates systematically from 24 h, it must be entrained by certain environ-

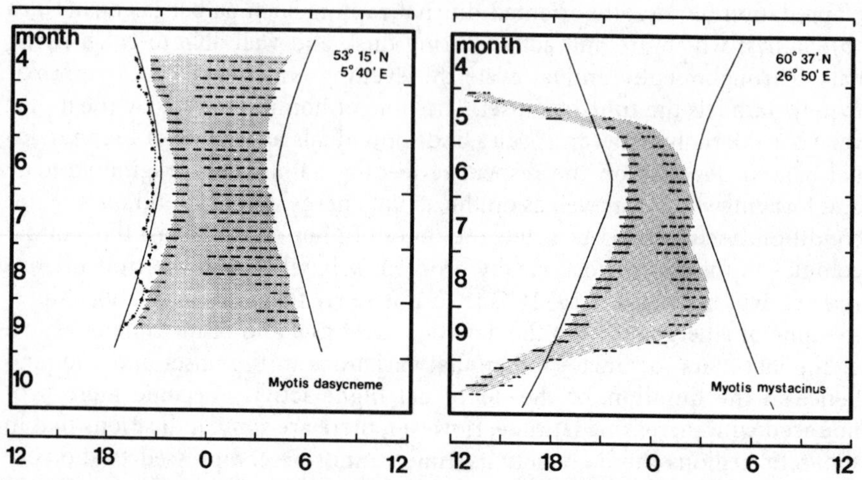

Fig. 14.1. Annual variation in the timing of flight activity of individual colonies of two palaearctic vespertilionid bat species at higher latitudes. Times of sunset and sunrise are indicated by the *vertical curves* on the *left* and *right*, respectively. *Solid horizontal lines* denote the duration of the emergence phase of the colony, and *dashed lines* denote the phase of return. *Shaded areas* represent the duration of the activity time of the colony as a whole, extending from the departure of the first flyer to the return of the last flyer. The *dots* connected by a *light dotted line* in the *left* diagram indicate the onset of the light-sampling phase of *Myotis dasycneme*. (Left diagram adapted from Voûte et al. 1974; right diagram adapted from Nyholm 1965)

mental time cues. As in other mammals, in bats the LD cycle acts as the most efficient zeitgeber, setting the phase of the animals' circadian rhythm adequately to the environmental 24-h periodicity (Erkert 1982). However, in various bat species the mechanism of circadian entrainment does not seem to function precisely enough to guarantee a correct timing of the onset of their evening flight activity. Therefore, many cave dwelling species, or bats which roost in dark rooms of buildings from which they cannot directly observe the outdoor illumination, show light-sampling behaviour in order to adjust the onset of their foraging activity to the most favourable light intensity. That is, individuals awakened too early in the evening by their circadian timing system fly or crawl to the entrance area of their roost. There or by short flights a few meters beyond it they then check repeatedly the light situation outside and do not finally depart before the illumination has fallen below a certain threshold. In a pond bat colony studied by Voûte et al. (1974) the duration of the evening light-sampling period averaged 54 ± 30 min (cf. Fig. 14.1).

14.2.3 Activity Patterns

Almost all activity patterns described represent the flight or foraging patterns of populations. Activity patterns of individual bats recorded by radio-tracking are known for only a few species.

In the insectivorous bats studied to date, bimodal (population) patterns of foraging activity have been described most frequently. However, unimodal or only weak bimodal patterns, trimodal and multimodal activity patterns do occur as well (Fig. 14.2). The maximum of flight activity is usually reached within the first 2 h after activity onset. In species showing a distinct bimodal activity pattern the second peak is normally much lower than the main peak and occurs shortly before the end of activity (Fig. 14.2d, e, g). However, as many bats seem to be quite flexible and opportunistic in foraging, local factors such as higher insect abundance at given places and/or at given times of the night might be responsible for differences in the recorded flight and foraging activity patterns. This is indicated by the differing activity patterns of *P. pipistrellus* obtained by Häussler et al. (1991) when recording the ultrasound activity over two adjacent ponds and a nearby creek in a nature reserve in southern Germany. While bats over the ponds showed a bimodal pattern with the main peak occurring at the end of the activity time, at the same time a more unimodal pattern peaking about 2 h after activity onset was recorded over the creek (Fig. 14.2f). The use of special night roosts does not seem to be a decisive factor determining the basic flight activity pattern because it occurs in insectivorous species with a more unimodal activity pattern, e.g. *Eptesicus fuscus* (Kunz 1973), as well as in species with more bimodal flight patterns, e.g. *Myotis velifer* (Fig. 14.2e; Kunz 1974; Antony et al. 1981; Barclay 1982).

Extensive radio-tracking of noctules (*Nyctalus noctula*) carried out in southern Germany (Kronwitter 1988) revealed a trimodal basic activity pattern. The individuals usually made three foraging flights per night and returned in between to their day roost. The first foraging flight was the longest, lasting on average about 180 min from mid-August to mid-September, and about 140 min before and after this time span. The last flight, which the animals began on average 95 min before sunrise, was the shortest and averaged only about 35 min. Whether the noctules made only one, two or all three foraging flights per night seemed to depend on the ambient temperature. In radio-tagged individuals of the carnivorous/insectivorous African bat *Nycteris grandis*, which sometimes forages by continuous flight and other times from perches, three or four bouts of foraging flights per night have been described. Their total duration varied between 46 and 115 min per night (Fenton et al. 1993b). Either one, two or three bouts of foraging flights totalling between 47 and 229 min per night have been recorded in radio-tagged *Noctilio albiventris*, which forages over water (Fenton et al. 1993a). Extremely short net flight times per night have been described in the territorial sit-and-wait predator *Cardioderma cor* studied in South Kenya (Vaughan 1976). The observed in-

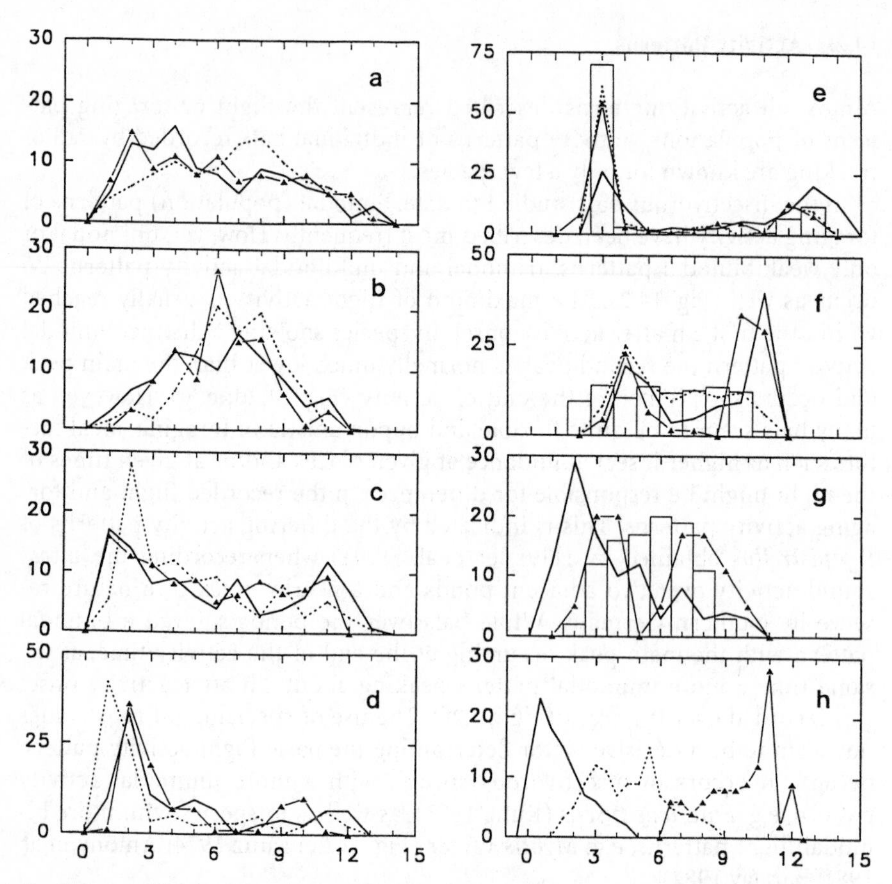

Fig. 14.2. a–h Flight and foraging activity patterns of several frugivorous, nectarivorous, and insectivorous bat species. The abscissae denote the time after sunset in hours, ordinates the relative amount of activity per h (a–g) or per 30 min (h), expressed in percent of the total amount of captures or counts per night. **a** Unimodal activity patterns of the frugivorous *Artibeus lituratus* (*solid line*; n = 86), *A. jamaicensis* (*broken line*; n = 829) and *A. phaeotis* (*triangles*; n = 132) netted at Barro Colorado Island, Panama (adapted from Bonaccorso 1979). **b** Unimodal activity patterns of the nectarivorous *Glossophaga soricina* obtained by recording the bats visiting the flowers of *Crescentia cujete* (*solid line*), *Cereus atroviridis* (*broken line*), and *Thunbergia grandiflora* (*triangles*; adapted from Lemke 1984). **c** Flight activity patterns of the phyllostomid bat *Carollia perspicillata* as obtained by mist netting (*solid line*) and radio-tracking (*triangles*) at Barro Colorado Island in Panama (adapted from Bonaccorso 1979), and by mist netting in Southeast Brasil (*broken line*; n = 75; adapted from Marinho-Filho and Sazima 1989). **d** Weakly bimodal activity patterns of three neotropical bat species with different food habits. Mist netting data from the sanguivorous *Desmodus rotundus* (*solid line*, n = 52; adapted from Marinho-Filho and Sazima 1989), the omnivorous *Phyllostomus discolor* (*broken line*, n = 58), and the insectivorous *Pteronotus parnellii* (*triangles*, n = 194; adapted from Bonaccorso 1979). **e** Bimodal flight patterns of two nearctic insectivorous vespertilionid bat species. *Myotis velifer* (*solid line*): distribution pattern of 600 adult individuals trapped at the entrance of a cave (adapted from Kunz 1974). *Myotis lucifugus*: comparison of actograms determined at a pond in New Hampshire by mist netting (*histogram*; n = 65) or by counting bat passes by means of an ultrasound detector positioned at the netting site (*broken line*; n = 935) and at another place (*triangles*; n = 1270; adapted from Kunz and Brock 1975). **f** Flight patterns of *Pipistrellus pipistrellus* over

two adjacent ponds (*solid line*; *triangles*) and a creek (*broken line*) of a small nature reserve in southern Germany as obtained by ultrasound recordings. The *curves* represent data from May 1989; the *histogram* shows the flight pattern obtained in September 1989 at the same place where a pronounced bimodal pattern was obtained in spring (*triangles*; adapted from Häussler et al. 1991). **g** Foraging patterns of the nearctic insectivorous bat species *Lasionycteris noctivagans* (*triangles*; n = 52) and *Eptesicus fuscus* (*solid line*; n = 243) obtained by mist netting in Central Iowa (USA). The *histogram* denotes the frequency distribution of visits to night roosts by *E. fuscus* (n = 620; adapted from Kunz 1973). **h** Trimodal flight activity pattern of the noctule (*Nyctalus noctula*) determined by radio-tracking in southern Germany. The curves represent the average duration of the three flights usually made per night. (Adapted from Kronwitter 1988)

dividual spent about 95% of the night on its perch waiting for large terrestrial arthropods. It left the perch for 136 short flights which covered a total of 3 200 m and lasted in total not longer than 10 min 38 sec. Nevertheless, distinct bimodal patterns could be discerned in the number of flights per hour as well as in the flight time per hour.

The sanguivorous neotropical vampire bat, *Desmodus rotundus*, has been reported by Turner (1975) to forage only a few hours per night and to return immediately thereafter to its day roost. However, other authors described a more bimodal activity pattern in this species (Brown 1968; Wimsatt 1969; Schmidt 1978; Marinho-Filho and Sazima 1989). Because the flight activity of the vampires seems to be inhibited substantially by moonlight (Fig. 14.3), such differences might be the result of moonlight effects that were not taken into consideration when mist-netting the animals and/or when evaluating the data.

In the frugivorous and nectarivorous species studied, the more unimodal patterns of foraging activity seem to dominate over the more bimodal ones (Fig. 14.2a–c). Unlike the bimodal activity patterns of the insectivorous species, which are characterised by a very low level of activity between the two peaks (Fig. 14.2d, e), the bimodal patterns described in frugivorous species, e.g. the short-tailed fruit bat (*Carollia perspicillata*), show a relatively high level of flight and foraging activity throughout the whole night (Fig. 14.2c; Bonaccorso 1979; Fleming 1982).

14.2.4 Effect of Environmental Factors on the Timing and Pattern of Flight Activity

14.2.4.1 Light

Light is the most important environmental factor involved in the regulation of bat activity rhythms. It has a twofold influence on the bats' flight and foraging activity. As the main zeitgeber, the LD cycle sets the phase of their endogenous circadian rhythm to synchronize it adequately with the external 24-h periodicity. In addition to this zeitgeber effect, the illumination also acts directly on the level of activity preprogrammed by the circadian timing system. According to the species' susceptibility and depending on the illumination intensity, this 'direct effect' or 'masking effect' of light on the activity rhythm results in a more or less pronounced inhibition of activity (Erkert 1974). In the laboratory these two different light effects on the bats' activity rhythm can be separated experimentally (Häussler and Erkert 1978). Under natural environmental conditions in the habitat the zeitgeber effect of light is indicated by the stable entrainment of the animals' circadian rhythm to the external 24-h day. The inhibitory direct effects of light become apparent in variations in the timing of the nocturnal flight activity which might (also) involve light-induced variations in the duration of the pre-departure light-sampling behaviour and/or in lunar-periodic modulations of the activity pattern (Erkert 1974, 1982).

On days with dense overcast some species depart about 5–10 min earlier from their day roost than on clear days. Such advanced emergence times, depending on the degree of cloudiness, have been reported in *Tadarida*, *Pipistrellus*, *Myotis*, *Eptesicus*, *Plecotus*, and *Molossus* (Erkert 1982). In other species, e.g. *Nyctalus noctula* and *Nyctalus lasiopterus*, the cloudiness does not seem to influence the onset of flight activity (Maeda 1974; Kronwitter 1988).

In some species moonlight can inhibit the flight activity to different degrees. The illumination intensity of the clear night sky when the moon is full amounts to about 0.1–0.5 lx but is only about 0.0001–0.0005 lx with new moon. Lunar periodic variations in the nocturnal flight and foraging activity have been described in the vampire bat (*Desmodus rotundus*, Fig. 14.3; Tamsitt and Valdivieso 1961; Wimsatt 1969; Crespo et al. 1972; Turner 1975) and other neotropical phyllostomatid bats, e.g. *Artibeus jamaicensis*, *Artibeus lituratus*, *Carollia perspicillata*, *Glossophaga soricina*, *Phyllostomus discolor* and *Vampyrops caraccioli* (Crespo et al. 1972; Heithaus et al. 1974; Heithaus and Fleming 1978; Morrison 1978; Fleming 1988) as well as in the African

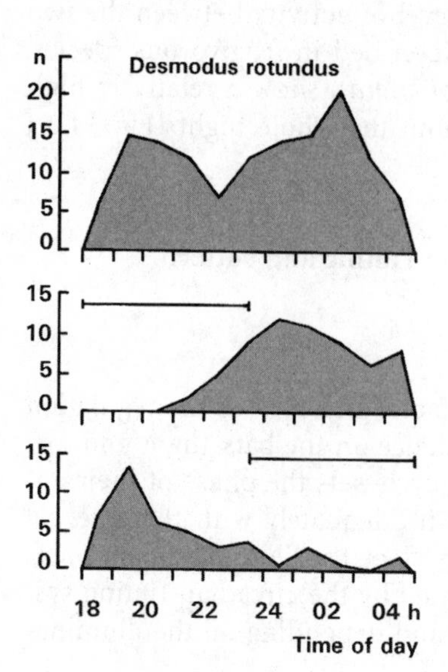

Fig. 14.3. Lunar periodic variation of the flight activity pattern of vampire bats (*Desmodus rotundus*) determined on the basis of mist nettings at new moon, (*top*; data from 12 netting nights), at waxing moon (*middle*; six netting nights) and at waning moon (*bottom*; five netting nights). Ordinate values represent the total numbers of bats caught per hour. *Horizontal bars* denote the moonlit night-times. (Adapted from Turner 1975)

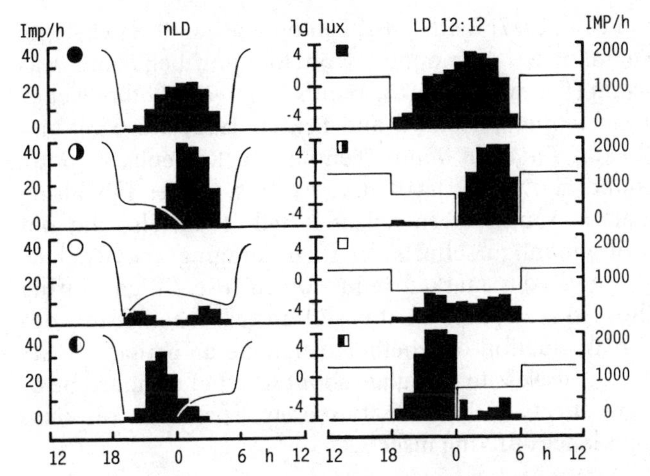

Fig. 14.4. Lunar periodic variation of the activity pattern of the neotropical leaf-nosed bat *Artibeus lituratus* when kept under natural lighting conditions (nLD, *left panels*) near the equator (lat. 4° 48' N, long. 74° 27' W). The *histograms* represent the average activity patterns recorded over a span of 15 months in the four lunar phases: new moon, waxing moon, full moon and waning moon. The *curves* show the nocturnal lighting profile in the four lunar phases at the recording site (from Erkert 1974). *Right panels*: light-induced modulation of the activity pattern of *A. lituratus* recorded under controlled conditions in the laboratory (LD 12 : 12; 25 ± 1 °C) with the nocturnal lighting conditions in the four phases of the moon simulated by square-wave LD cycles (n = 6 individuals, n' = 10 days each; from Häussler and Erkert 1978)

fruit bat (*Rousettus aegyptiacus leachi*; Jacobsen and DuPlessis 1976). In insectivorous bats a slight suppression of activity by bright moonlight and lunar-periodic variations of the foraging pattern have been reported in the African species *Scotophilus viridis*, *Eptesicus capensis*, and *Nycticeius schlieffeni* (Fenton et al. 1977).

Marked lunar-periodic variations in the activity pattern due to inhibitory effects of moonlight were also recorded in *Artibeus lituratus* (Fig. 14.4, left panels) kept in captivity under natural lighting and climatic conditions in their habitat (Erkert 1974). Similar modulations of the activity pattern in *A. lituratus* were obtained when the nightly illumination conditions at the four moon phases were simulated in the laboratory by corresponding square-wave LD cycles (Fig. 14.4, right panels; Häussler and Erkert 1978). Field observations of Usman et al. (1980) during an eclipse of the moon are in keeping with these results: there was a sharp increase in the flight activity of various bat species while the moon was in shadow and a subsequent fall to the original low full-moon level.

Insectivorous species which show no or only slight lunar-periodic variations in the amount and pattern of their nightly flight activity may displace their foraging flights during the moonlit periods from more open areas into the protecting shade of vegetation or fly shorter distances. Such moonlight avoidance behaviour has been shown in several North American and African vespertilionids (Fenton et al. 1977; Reith 1982; Fenton and Rautenbach 1986; Clark et al. 1993). No influence of moonlight on foraging behaviour was found in *Nyctalus noctula* (Kronwitter 1988), *Myotis lucifugus* (Anthony et al. 1981), and *Euderma maculatum* (Leonard and Fenton 1983; Wai-Ping and Fenton 1989), as well as in *Tadarida midas* (Fenton and Rautenbach 1986). Analysing the ultrasound activity of 8–10 insectivorous species inhabiting different desert habitats in southeastern Arizona, Bell (1980) also did not find any indications of moonlight effects on their foraging activity. Like Fenton et al. (1977), he noticed a marked reduction in insect flight during bright full-moon nights. However, because it is still unclear whether this observation reflects the real situation or whether it may be an artifact of the light-trapping method, it is useless to speculate about whether or not some of the described moonlight effects on bat activity merely represent a response to variations in the abundance of flying insects.

14.2.4.2 Ambient Temperature

Ambient temperature is well known as a crucial exogenous factor controlling the seasonal timing of flight activity and the course of hibernation in heterothermic insectivorous bats of temperate zones. During the annual activity period it may also influence the daily timing and pattern of flight and foraging activity. Many heterothermic species hunt for shorter times in cold weather. When the temperature falls below a certain threshold (especially in spring and fall), they often cease flight activity entirely. In *Antrozous pallidus* this air-temperature threshold is at about 10.5 °C (O'Shea and Vaughan 1977), in *P. pipistrellus* it is below 5 °C, and in *Pipistrellus abramus* it varies between 12 and 15 °C, depending on season and location. For *P. hesperus* an inhibition of flight activity was found below 14 °C and 19 °C, respectively (Cross 1965; Jones 1965), while O'Farrell and Bradley (1970) observed this species and *Myotis californicus* still foraging even at –8 °C. In both species these authors also noted temperature-induced changes in the temporal flight activity pattern. At average ambient temperatures below 15 °C the bats had a unimodal activity pattern shortened to about 4–5 h whereas at temperatures above 15 °C they were active throughout the whole night and displayed the bimodal flight pattern characteristic of most insectivorous species. Temperature-induced variations in the flight activity pattern were also observed in a radio-tracking study of *Nyctalus noctula* (Kronwitter 1988). At average air temperatures of 8.3 °C the noctules did not leave their day roost, at 11.2 °C they departed for only one foraging flight, and at 14.2 °C for two flights per night. Not until the temperature reached 15.9 °C did the bats depart, as usual, for three foraging flights per night (Fig. 14.2h).

Certain species, e.g. *Myotis dasycneme*, not only curtail their flight activity when it is very cold in springtime and fall but also shift into the warmer parts of the day (Nyholm 1965). At present it is not clear whether and to what extent such a transition from the normal nocturnal to a diurnal foraging habit represents a direct effect of temperature, or whether it is caused indirectly by temperature-induced changes of the flight activity time of insect prey. However, the latter explanation is assumed to be the most likely. Likewise, the occurrence of flights during the day, frequently observed in the noctule in autumn, has also been explained by unsatisfied food requirements after nights with low foraging success and/or by higher insect abundance during the day (Kronwitter 1988).

14.2.4.3 Precipitation and Wind

Precipitation and wind seem to have quite different effects on the flight activity of the various bat species. In general, light precipitation and wind have no or at most minor effects on the timing and pattern of flight activity. Heavy rainfall, however, and very strong, gusty wind may cause a delayed departure from the day roost and/or advanced homing to it, or can entirely prevent the nightly foraging activity. No effects of these environmental factors were found in some phyllostomids (Tamsitt and Valdivieso 1961) and in *Myotis velifer* (Kunz 1974). In *P. pipistrellus*, *Myotis californicus* and *Plecotus auritus* heavy rainfall and strong wind have been reported to cause a distinct reduction in flight activity (O'Farrell et al. 1967; Stebbings 1968; Frylestam 1970). Curtailment of foraging activity by heavy rain, in the form of later emergence, briefer hunting flights, or no flights at all, has been observed in several other vespertilionids, in molossid bats and in *Desmodus rotundus* (reviewed by Erkert 1982). More precise data on the effect of weather conditions on flight activity were obtained by radio-tracking of *Nyctalus noctula* (Kronwitter 1988). Whereas light precipitation only caused a slight delay of the first departure of the noctules by about 12 min, heavy rain resulted in an average delay of 85 min and a shortening of the first foraging flight from 108 ± 65 to 51 ± 23 min on average, or induced its omission (27.5%). Interestingly, the second and third foraging flight of the noctules seemed not to be affected by these unfavourable weather conditions.

14.2.4.4 Food Abundance

The effect of actual food abundance on the flight and foraging pattern of bats has not yet been studied in detail. However, as bats are small mammals with an adverse mass-to-surface relationship, a normally low energy storage capacity in their brown fatty tissue, and very high energy expenditure during flight, they can tolerate a negative energy balance only for a short while. Therefore one should assume that the abundance of food, its energy content, and the size of insect or vertebrate prey, fruits, or nectar resources do influence their activity pattern as well. In accordance with this hypothesis, pipistrelle bats cease foraging when the insect abundance falls below a certain threshold (Racey and Swift 1985). The prolongation of the nightly torpor phase on cool nights with low insect abundance described in several insectivorous bats of temperate zones (O'Shea and Vaughan 1977; Anthony et al. 1981) may also, at least partly, be a response to reduced food availability. The same might be applicable to the starvation-induced torpor observed in several small phyllostomatid bats, e.g. *Glossophaga soricina*, when unfavourable weather conditions prevent foraging (Rasweiler 1973). As already mentioned, the day flights reported to occur at high latitudes in spring and autumn in *Myotis dasycneme* and in autumn in *Nyctalus noctula* are also ascribed to reduced insect flight activity at night and a much higher abundance of airborne insect prey during the day (Nyholm 1965; Kronwitter 1988).

Even the predominantly bimodal activity pattern found in most insectivorous bats can be interpreted as an evolutionarily acquired response to times of high food availability. Various studies have shown that the activity and abundance of flying insects which may serve as prey organisms for bats also show a bimodal time course with maxima during dusk and dawn (Lewis and Taylor 1964; Funakoshi and Uchida 1975; Swift 1980; Racey and Swift 1985; Taylor and O'Neill 1988). The insectivorous Chiroptera may have adapted their foraging behaviour to this situation by evolving an endogenously programmed bimodal flight-activity pattern matching the flight pattern of their prey. This consequence considerably increases the probability that the energetically very expensive foraging on the wing will give the greatest possible energy yield with the smallest possible expenditure of energy.

14.2.5 Effect of Endogeous Factors on the Timing and Pattern of Flight Activity

Indications of age-related differences in the flight activity pattern were found in several North American vespertilionids (Kunz 1974; Belwood and Fenton 1976; Buchler 1976). Radiotelemetric studies carried out in the African vespertilionid *Scotophilus leucogaster* revealed that adult individuals foraged closer to their day roost and for a shorter time than subadults (Barclay 1985).

Minor sex-specific differences in the timing, duration, and time course of foraging activity have been reported in *Eptesicus fuscus*. According to the results of mist-nettings, the females of this species seem to depart somewhat earlier and to have a roughly 2 h longer first foraging bout than the males (Kunz 1980). In the emballonurid *Saccopteryx bilineata* the males were observed to return first to their unsheltered day roost on the trunk or large branches of trees in order to defend it against other males and/or to attract their own and other females by intense vocalisations (Bradbury 1977).

In the females of some insectivorous species the timing and pattern of the flight activity as well as the frequency and duration of foraging flights also depend on the reproductive state. Marked alterations of the flight activity pattern in the course of the annual reproductive period have been observed by Swift (1980) in a colony of female *P. pipistrellus*. When pregnant, female *Myotis lucifugus* show a distinct bimodal activity pattern and have an extended unimodal night resting phase spent outside their day roost. While lactating, however, the bats return to the day roost after each foraging flight and use night roosts only sporadically and for a short time on each visit (Anthony et al. 1981). Extensive radio-tracking in *Lasiurus cinereus* revealed that during lactation the females depart progressively earlier and hunt for increasingly longer times. After weaning the bats reduced their foraging time again. Mothers which had to suckle twins foraged significantly longer than mothers with only one young (Barclay 1989). These findings indicate that these bats are 'time minimisers' which only forage as long as is necessary to balance their actual energy demand.

14.3 Activity Patterns in the Laboratory

Under controlled environmental conditions in the laboratory the patterns of locomotor activity have been recorded in only a few species. When kept in artificial light/dark cycles (LD 12 : 12; L: 25–120 lx, D: < 10^{-3} lx) at constant ambient temperatures (20–25 ± 1 °C), and relative humidity (60 ± 5%) with water and food available ad libitum, the frugivorous Old World pteropodid bats *Rousettus aegyptiacus* and *Eidolon helvum* (Erkert 1970), as well as the frugivorous/nectarivorous neotropical phyllostomatids *Artibeus lituratus*, *Phyllostomus discolor*, *Sturnira lilium* and *Glossophaga soricina* developed distinct unimodal activity patterns (cf. Fig. 14.4, right panels; Häussler and Erkert 1978). In the insectivorous species *M. myotis* (Bay 1978), *M. molossus*, and *Molossus ater*, typical bimodal activity patterns resulted when the activity was recorded (or plotted) in 30- or 60-min intervals (Häussler 1987). However, extensive studies carried out in *M. molossus* (Häussler 1987) revealed a distinct ultradian activity rhythm underlying the bimodal pattern. At 25 ± 1 °C the daily activity phase of these small molossids was characterised by individually different numbers of 5–16 ($\bar{x}$ = 10.7 ± 2.6) ultradian cycles of 67.4 ± 19.3 min duration. They consisted on average of activity bouts lasting 23.5 ± 5.9 min and interposed resting pauses of 43.8 ± 15.9 min. Interestingly, the ultradian periods showed a significant positive correlation with the bats' body mass.

In constant darkness (DD), when the circadian activity rhythm of the molossids free-ran with spontaneous periods of only 22.3 ± 0.8 h and the activity time averaged 11.5 ± 0.9 h, the parameters of the short-term rhythm remained constant. However, a light-induced lengthening of the circadian period to 24.9 ± 0.5 h in constant dim light (LL) of 0.1 lx was parallelled by a considerable increase in the period of the short term rhythm, to 91.8 ± 25.6 min on average. Concurrently, the average number of ultradian periods per circadian cycle was reduced. The time spent really active during the circadian activity phase amounted only to about 4 h. This value is in the range of data obtained from radio-tracking field studies in some other insectivorous species. As already shown above, the results of several field studies on insectivorous species also point to the occurrence of ultradian rhythmicity during the bats' nightly foraging period (Vaughan 1976; Fenton and Rautenbach 1986; Fenton et al. 1993b).

In contrast to circadian and circannual rhythms, there is no periodicity in the natural environment to which ultradian rhythms may correspond. The origin and ultimate biological function of such short-term rhythms are still a matter of discussion. Besides the fact that in some rodent species the ultradian rhythm keeps on running even under food deprivation, a negative correlation found in successive ultradian activity periods has also been considered as evidence for an endogenous oscillator underlying ultradian periodicity (Erkinaro 1973; Gerkema and Daan 1985, cf. Chaps. 13 and 11). However, such a negative correlation could not be established in *M. molossus*. Hence, this deduction does not seem to be conclusive for these insectivorous bats. From the close relationship of ultradian periodicity with some characteristics of energy metabolism, food intake, and body mass found in several small mammals including *M. molossus* (Pearson 1947; Daan and Slopsema 1978) one might infer that in these cases metabolic and thermoregulatory processes underlie ultradian rhythmicity. In view of the animals' need to achieve a positive energy balance, the ultradian activity and foraging rhythms found in *M. molossus* and other bats may be regarded as an optimal strategy of species and individuals to adjust the temporal organisation of their behaviour to the energetic and nutritional situation in the environment.

14.4 Conclusions

Bats are unique amongst small mammals because they have developed active flight by means of highly specialised extremities, because almost all species are strictly nocturnal, and because they show quite different food habits and foraging strategies. Their unfavourable mass-to-surface area relationship as well as their energy-consuming flight make high demands on the energy balance. In many species, especially in those of temperate zones, the need to save energy not only led to the evolution of particular physiological adaptations such as torpor and hibernation, but also influences the time course of their various behavioural activities, including flight and foraging. Such constraints might have caused the Chiroptera to develop diverse and flexible activity patterns.

Whereas in the frugivorous species the more unimodal activity patterns seem to dominate over the more bimodal ones, in the insectivorous and other zoovorous bats the more bimodal patterns seem to be more frequent than unimodal, trimodal, or multimodal flight or foraging patterns. The bimodal pattern in insectivorous Chiroptera might be an evolutionary adaptation to the nightly time course of abundance of insect prey, characterized by peaks during dawn and dusk. Due to a general tendency to keep the time spent on the wing as short as possible, the daily activity time and activity pattern depend largely on foraging success. Other factors that may influence and modulate the bats' activity patterns to different degrees are the nightly illumination conditions (depending on lunar-phase, cloudiness, and vegetation), low ambient temperatures, and the bat's reproductive state. Precipitation and wind seem to play a minor role; only heavy rain or very strong and gusty wind may delay or even prevent flight and foraging activity. All these factors increase the variability of the time course of bat activity under natural environmental conditions. The recording method also crucially affects findings of species' activity patterns. Important factors are whether it measures the activity of individuals or of whole populations, and whether it includes a number of samples scattered over a long time or is based on continuous recordings over a given time span. Hence it is often quite difficult to establish the existence of a basic activity pattern which then might be related to given ecological constraints. Theoretical considerations on this subject, therefore, seem useless as long as there are insufficient reliable activity data from long-term radio-tracking studies on a large number of individual bats of several species belonging to different ecotypes.

References

Anthony ELP, Stack MH, Kunz TH (1981) Night roosting and the nocturnal time budget of the little brown bat, *Myotis lucifugus*: effects of reproductive status, prey density, and environmental conditions. Oecologia 51:151–156

Aschoff J, Wever R (1962) Beginn und Ende der täglichen Aktivität freilebender Vögel. J Ornithol 103:2–27

Barclay RMR (1982) Night roosting behavior of the little brown bat *Myotis lucifugus*. J Mammal 63:464–474

Barclay RMR (1985) Foraging behavior of the african insectivorous bat, *Scotophilus leucogaster*. Biotropica 17:65–70

Barclay RMR (1989) The effect of reproductive condition on the foraging behavior of female hoary bats, *Lasiurus cinereus*. Behav Ecol Sociobiol 24:31–37

Bay FA (1978) Light control of the circadian activity rhythm in mouse-eared bats (*Myotis myotis* Borkh. 1797). J Interdiscipl Cycle Res 9:195–209

Bell GP (1980) Habitat use and response to patches of prey by desert insectivorous bats. Can J Zool 58:1876–1883

Belwood JJ, Fenton MB (1976) Variation on the diet of *Myotis lucifugus* (Chiroptera: Vespertilionidae). Can J Zool 54:1674–1678

Belwood JJ, Fullard JH (1984) Echolocation and foraging behavior in the Hawaiian hoary bat, *Lasiurus cinereus semotus*. Can J Zool 62:2113–2120

Bonaccorso FJ (1979) Foraging and reproductive ecology in a Panamanian bat community. Bull Florida State Mus Biol Sci 24:359–408

Bradbury JW (1977) Social organization and communication. In: Wimsatt WA (ed) Biology of bats, vol 3. Academic Press, New York, pp 1–72

Brown JH (1968) Activity patterns of some neotropical bats. J Mammal 49:754–757

Buchler ER (1976) A chemiluminescent tag for tracking bats and other small nocturnal animals. J Mammal 57:173–176

Clark BS, Leslie DM, Carter TS (1993) Foraging activity of adult female ozark big-eared bats (*Plecotus townsendii* INGENS) in summer. J Mammal 74:422–427

Cockrum EL, Cross SP (1964) Time of bat activity over water holes. J Mammal 45:635–636

Constantine DG (1958) An automatic bat-collecting device. J Wildl Manage 22:17–22
Crespo JA, Linhart SB, Mitchell GC (1972) Foraging behavior of the common vampire bat related to moonlight. J Mammal 53:366–368
Cross SP (1965) Roosting habits of *Pipistrellus hesperus*. J Mammal 46:270–279
Daan S, Slopsema S (1978) Short-term rhythms in foraging behavior of the common vole, *Microtus arvalis*. J Comp Physiol A 127:215–227
Erkert HG (1974) Der Einfluss des Mondlichtes auf die Aktivitätsperiodik nachtaktiver Säugetiere. Oecologia 14:269–287
Erkert HG (1978) Sunset-related timing of flight activity in neotropical bats. Oecologia 37:59–67
Erkert HG (1982) Ecological aspects of bat activity rhythms. In: Kunz TH (ed) Ecology of bats. Plenum Press, New York, pp 201–242
Erkert S (1970) Der Einfluss des Lichtes auf die Aktivität von Flughunden (Megachiroptera). Z Vergl Physiol 67:243–272
Erkinaro E (1973) Short-term rhythm of locomotor activity within the 24 hr period in the Norwegian lemming, *Lemmus lemmus* and water vole, *Arvicola terrestris*. Aquilo Ser Zool 14: 59–67
Fenton MB, Rautenbach IL (1986) A comparison of the roosting and foraging behaviour of three species of African insectivorous bats (Rhinolophidae, Vespertilionidae, and Molossidae). Can J Zool 64:2860–2867
Fenton MB, Boyle NGH, Harrison TM, Oxley DJ (1977) Activity patterns, habitat use, and prey selection by some African insectivorous bats. Biotropica 9:73–85
Fenton MB, Audet D, Dunning DC, Long J, Merriman CB, Pearl D, Syme DM, Adkins B, Pedersen S, Wohlgenent T (1993a) Activity patterns and roost selection by *Noctilio albiventris* (Chiroptera: Noctilionidae) in Costa Rica. J Mammal 74:607–613
Fenton MB, Rautenbach IL, Chipese D, Cumming MB, Musgrave MK, Taylor JS, Volpers T (1993b) Variation in foraging behaviour, habitat use, and diet of large slit-faced bats (*Nycteris grandis*). Z Säugetierkd 59:65–74
Fleming TH (1982) Foraging strategies of plant visiting bats. In: Kunz TH (ed) Ecology of bats. Plenum Press, New York, pp 287–325
Fleming TH (1988) The short-tailed fruit bat. A study in plant-animal interactions. University of Chicago Press, Chicago
Frylestam B (1970) Studier över langörade fladdermusen (*Plecotus auritus* L.). Fauna Och Flora 65:72–84
Funakoshi K, Uchida TA (1975) Studies on the physiological and ecological adaptation of temperate insectivorous bats. I. Feeding activities in the Japanese long-fingered bats (*Miniopterus schreibersi fuliginosus*). Jpn J Ecol 25:217–234
Gaisler J (1973) Netting as a possible approach to study bat activity. Period Biol 75:129–134
Gaisler J; Hanák V, Dungel J (1979) A contribution to the population ecology of *Nyctalus noctula* (Mammalia: Chiroptera). Acta Sc Nat Brno 13:1–38
Gerkema MP, Daan S (1985) Ultradian rhythms in behavior: the case of the common vole (*Microtus arvalis*). In: Schulz H, Lavie P (eds) Ultradian rhythms in physiology and behavior. Springer, Berlin Heidelberg New York, pp 11–31
Häussler U (1987) Zeitliche Organisation von Aktivität und Verhalten der Bulldoggfledermaus *Molossus molossus* Pallas 1766 (Chiroptera, Molossidae). Doctorate thesis, University of Tübingen
Häussler U, Erkert HG (1978) Different direct effects of light intensity on the entrained activity rhythm in neotropical bats (Chiroptera, Phyllostomidae). Behav Proc 3:223–239
Häussler U, Kalko E, Bay FA (1991) Untersuchungen der Fledermaus-Fauna. In: Bay FA, Rodi D (eds) Wirksamkeitsuntersuchungen von Ausgleichs- und Ersatzmaßnahmen im Straßenbau – dargestellt am Beispiel B 29, Lorcher Baggerseen. Forschung, Straßenbau und Straßenverkehrstechnik H605, BM Verkehr, Bonn-Bad Godesberg, pp 71–84
Heithaus ER, Fleming TH (1978) Foraging movements of a frugivorous bat, *Carollia perspicillata* (Phyllostomatidae). Ecol Monogr 48:127–143

Heithaus ER, Fleming TH, Opler PA (1974) Patterns of foraging and resource utilization in seven species of bats in a seasonal tropical forest. Ecology 56:841–854
Hill JE, Smith JD (1988) Bats – a natural history. Nat Museum, London
Jacobsen NHG, DuPlessis E (1976) Observations on the ecology and biology of the cape fruit bat *Rousettus aegyptiacus leachi* in the Eastern Transvaal. S Afr J Sci 72:270–273
Jones C (1965) Ecological distribution and activity periods of bats of the Mogollon mountains area of New Mexico and adjacent Arizona. Tulane Stud Zool 12:93–100
Kronwitter F (1988) Population structure, habitat use and activity patterns of the noctule bat, *Nyctalus noctula* Schreb., 1774 (Chiroptera: Vespertilionidae) revealed by radio-tracking. Myotis 26:23–85
Kunz TH (1973) Resource untilization: temporal and spatial components of bat activity in central Iowa. J Mammal 54:14–32
Kunz TH (1974) Feeding ecology of a temperate insectivorous bat (*Myotis velifer*). Ecology 55: 693–771
Kunz TH (1980) Daily energy budgets of free-living bats. In: Wilson DE, Gardner AL (eds) Proceedings of the 5th international bat research conference. Texas Tech Press, Lubbock, pp 369–392
Kunz TH (1982) Roosting ecology of bats. In: Kunz TH (ed) Ecology of bats. Plenum Press, New York, pp 1–55
Kunz TH, Brock CE (1975) A comparison of mist nets and ultrasonic detectors for monitoring flight activity in bats. J Mammal 56:907–911
Laufens G (1972) Freilanduntersuchungen zur Aktivitätsperiodik dunkelaktiver Säuger. Doctorate thesis, University of Cologne
Lemke TO (1984) Foraging ecology of the long-nosed bat, *Glossophaga soricina*, with respect to resource availability. Ecology 65:538–548
Leonard ML, Fenton MB (1983) Habitat use by spotted bats (*Euderma maculatum*, Chiroptera: Vespertilionidae): roosting and foraging behavior. Can J Zool 61:1487–1491
Lewis T, Taylor LR (1964) Diurnal periodicity of flight by insects. Trans R Ent Soc Lond 116: 393–435
Maeda K (1974) Eco-éthologie de la grande noctule, *Nyctalus lasiopterus*, a Sapporo, Japan. Mammalia 38:461–487
Marinho-Filho JS, Sazima I (1989) Activity patterns of six phyllostomid bat species in southeastern Brazil. Rev Brasil Biol 49:777–782
Miller L, Mohl B, Brockelman WY, Andersen BB, Christensen-Dalsgaard J, Jorgensen MB, Surlykke A (1988) Fly-out count of the bat, *Tadarida plicata*, using a video recording. Nat Hist Bul Siam Soc 36:135–141
Morrison DW (1978) Lunar phobia in a neotropical fruit bat, *Artibeus jamaicensis* (Chiroptera: Phyllostomidae). Anim Behav 26:852–855
Nyholm ES (1965) Zur Ökologie von *Myotis mystacinus* (Leisl.) und *M. daubentoni* (Leisl.) (Chiroptera). Ann Zool Fenn 2:77–123
O'Farrell MJ, Bradley WG (1970) Activity patterns of bats over a desert spring. J Mammal 51:18–26
O'Farrell MJ, Bradley WG, Jones GW (1967) Fall and winter bat activity at a desert spring in southern Nevada. Southwest Nat 12:163–171
O'Shea TJ, Vaughan TA (1977) Nocturnal and seasonal activities of the pallid bat, *Antrozous pallidus*. J Mammal 58:269–284
Pearson OP (1947) The rate of metabolism in some small mammals. Ecology 28:127–145
Perrin LPA (1987) Zum Morgenflug von *Nyctalus noctula* Schreber, 1774 (Mammalia, Chiroptera). Z Säugetierkd 52:50–51
Racey PA, Swift SM (1985) Feeding ecology of *Pipistrellus pipistrellus* (Chiroptera: Vespertilionidae) during pregnancy and lactation. I. Foraging behaviour. J Anim Ecol 54:205–215
Rasweiler JJ (1973) Care and management of the long-tongued bat, *Glossophaga soricina* (Chiroptera: Phyllostomatidae), in the laboratory with observations on estivation induced by food deprivation. J Mammal 54:391–404
Reith CC (1982) Insectivorous bats fly in shadows to avoid moonlight. J Mammal 63:685–688

Schmidt U (1978) Vampirfledermäuse. Neue Brehm-Bücherei 515, Ziemsen, Wittenberg Lutherstadt

Stebbings RE (1968) Measurements, composition and behaviour of a large colony of the bat *Pipistrellus pipistrellus*. J Zool Lond 156:15–33

Stutz HP, Haffner M (1985/1986) Activity patterns of non-breeding populations of *Nyctalus noctula* (Mammalia, Chiroptera) in Switzerland. Myotis 23/24:149–155

Swift SM (1980) Activity pattern of pipistrelle bats (*Pipistrellus pipistrellus*) in north-east Scottland. J Zool Lond 190:285–295

Tamsitt JR, Valdivieso D (1961) Notas sobre actividades nocturnas y estados de reproducíón de algunos quirópteros de Costa Rica. Rev Biol Trop 9:219–225

Taylor RJ, O'Neill MG (1988) Summer activity patterns of insectivorous bats and their prey in Tasmania. Aust Wildl Res 15:533–539

Thomas SP (1975) Metabolism during flight in two species of bats, *Phyllostomus hastatus* and *Pteropus gouldii*. J Exp Biol 63:273–293

Thomas SP (1980) The physiology and energetics of bat flight. In: Wilson DE, Gardner AL (eds) Proceedings of the 5th international bat research conference. Texas Tech Press, Lubbock, pp 393–402

Thomas SP, Suthers RA (1972) The physiology and energetics of bat flight. J Exp Biol 57:317–335

Turner DC (1975) The vampire bat. John Hopkins University Press, Baltimore

Tuttle MD (1974) An improved trap for bats. J Mammal 55:475–477

Usman K, Habersetzer J, Subbaraj R, Gopalkrishnaswamy G, Paramanandam K (1980) Behavior of bats during a lunar eclipse. Behav Ecol Sociobiol 7:79–81

Vaughan TA (1976) Nocturnal behavior of the African false vampire bat (*Cardioderma cor*). J Mammal 57:227–248

Voûte AM, Sluiter JW, Grimm MP (1974) The influence of the natural light-dark-cycle on the activity rhythm of pond bats (*Myotis dasycneme* Bioe, 1825) during summer. Oecologia 17: 21–243

Wai-Ping V, Fenton MB (1989) Ecology of spotted bat (*Euderma maculatum*), roosting and foraging behavior. J Mammal 70:617–622

Wimsatt WA (1969) Transient behavior, nocturnal activity patterns and feeding efficiency of vampire bats (*Desmodus rotundus*) under natural conditions. J Mammal 50:233–244

Section IV Conclusion

15 Chronoecology: New Light Through Old Windows - A Conclusion

Stefan Halle and Nils Chr. Stenseth

15.1 A New Term - And a New View

After presenting some introductory material, in the first section of this book, we provided some theoretical considerations that relate individual activity patterns of small mammals to physiological constraints (Chap. 3), and to ecological aspects of temporal behaviour (Chap. 4). The theoretical section also provided a general background to biological rhythms and biological clocks (Chap. 2). The other main section of the book surveyed our current knowledge of activity patterns in various small mammal groups as based on empirical findings. In this concluding chapter we, as the editors, want to provide a synthesis of the information presented in the previous 14 chapters. In doing so we will try to concentrate on the general features in - and behind - the various approaches to activity patterns by a comparison among small mammal species. We would like this synthesis to serve as a baseline to clarify where we are and to point towards future developments - for better understanding, and to address new challenges.

As part of this concluding chapter we are suggesting a new term 'chronoecology' to frame the new field of research outlined in this book. Since chronobiology is a well-established term that represents a fairly active subdiscipline, the reader may suspect that introducing the new term is not more than a compulsory exercise to let things appear more relevant. However, we strongly believe that chronoecology is indeed more than just another word. As a matter of fact, we consider this term to stand for a recently emerging subdiscipline within the domain of behavioural ecology that spans the presently almost insurmountable gap between behavioural ecologists and chronobiologists, who seldom, if ever, are aware of each other. We also strongly believe that this intention is not a convulsive attempt to join things that do not belong together. Rather, we see a promising and rewarding potential for substantial new insights into animal behaviour when the profound knowledge in the two fields is combined in new conceptual and methodological approaches.

Ecological Studies, Vol. 141
Halle/Stenseth (eds.) Activity Patterns in Small Mammals

15.2 Activity Patterns and Evolutionary Ecology

Sibly outlined in his Foreword that the theoretical framework for behavioural ecology is the evolution of life histories. As such, the conceptual approaches of ESS and optimisation in trade-offs among conflicting constraints apply to temporal strategies just like any other behavioural trait with fitness consequences. Fitness is determined by the relationship between behaviour and the environment, and since the environment changes dramatically in the course of the 24-h day, the question of when to do what is indeed within the purview of behavioural ecology. Dealing with activity patterns is different from the examination of time budgets, in which one asks how much time is allocated to what kind of activity. It is also different from the proximately orientated questions in chronobiology and physiology, which focus on how an individual manages to pick a specific time of the day for a specific action. Rather, chronoecology emphasises the ultimate questions of why an individual or species picks a specific time of the day for a specific action, how this timing contributes to individual fitness, and how this trait may have established in the course of evolution. Since ecology is the branch of science which deals with the relationships between organisms and their environment, the fitness consequences of activity timing belong to the field of evolutionary ecology.

Indeed, clocks, activity budgets, and activity patterns are closely related. The daily LD cycle is the most prominent and most predictable short-term change in the environment. The evolution of biological clocks clearly depends on the predictable nature of the environmental fluctuation, but they are such powerful tools because they combine precision with flexibility, as pointed out in Chap. 2 by Bartness and Albers. From the extensive work on chronobiology we know that time assessment mechanisms are very complex and sophisticated, hence, to take up Enright's argument once more (Enright 1970), they can only have evolved because of strong selective pressures to find limited time windows of opportunities. As Weiner stated in Chap. 3 there are four functions of activity that are under natural selection: survival assurance, reproduction, movement, and food acquisition. Locomotion and mechanical work in general cause relatively low energetic costs in small mammals, hence food acquisition and predation risk seem to be the prominent features in this animal group. The decisive factor, however, will obviously depend on the species' ecological life style. Hence, the relationship between environment and activity patterns may best be investigated by taking a comparative ecological approach among species and habitats.

Species not only differ with respect to activity timing, they also differ in the amount of time they have to assign to different activities, which results in a high diversity of activity budgets. Working from very different starting points, several authors in this volume came to the understanding that 'activity', as well as its counterpart 'rest', are vaguely defined terms, and that they are at best generic terms for many different aspects of an individual's daily life. To understand temporal strategies in terms of evolutionary ecology, these terms have to be broken down into their components, asking for each of them separately why a specific time of the day may be more advantageous for performance than others. The constraints for this decision may vary with population density and demography, with sex, age, social status, habitat, and season. Furthermore, for many aspects of behaviour it may not be possible, or helpful, to look at the temporal and spatial dimensions separately; rather, both must be considered together to give a complete picture. Ultimate understanding, however, will seldom result from a mere pattern description, which of course always has to be the first step, but rather will entail concentrating on the underlying processes that generate the observable pattern, as Ziv and Smallwood emphasised in Chap. 9. Thus, a focus on pattern generating processes is characteristic of the field of chronoecology.

15.3 Chronoecology and the Biological Clock

When looking at the various case studies presented in this volume - and we are aware that there are many more, even if only small mammals are considered - we have to recognise that there is no 'general line' of response to the diel rhythm. Rather, we are confronted with a variety of temporal strategies, each seemingly adaptive to its specific circumstances. From this we can, and should, go further by asking 'adaptive to what?', but we will always be constrained to case-related questions and answers. As a conclusion we have to realise that activity patterns are part of the adaptive complex, part of the evolution of the species, and, consequently, part of the species ecological niche.

Two conclusions, however, seem indeed to be universally valid: first, small mammals do reveal systematic activity patterns in the sense of temporal behavioural strategies; and second, to follow such a strategy, they do possess a clock. As Bartness and Albers pointed out in Chap. 2, the basic feature of all clocks is a fluctuation in the melatonin titer with high levels at night-time and a decrease during the day. This hormonal rhythm is in one way or another translated into physiological and behavioural rhythms which basically entrain endogenous processes with the fluctuating environment. To disentangle the sophisticated and often muddled pathways of this translation is the goal of chronobiology. Chronoecology instead accepts that there are clocks that work and asks for the ecological relevance of the clocks. This view is similar to our own experience in everyday life that we are, sometimes, able to get up on time, catch the bus and hold the dates in our appointments diary with total ignorance of how the hands of our clocks are technically driven by the small battery that we have to change once in a while.

In the course of evolution, different species achieved very different solutions to the temporal optimisation problem, which indeed is not surprising because habitats, feeding habits, body size, and social organisation differ considerably. However, despite the enormous variability, there are also common features. One is, as demonstrated by several reviews in this volume, that the start of the active phase is triggered in a very precise and almost schematic manner. Hence, to become active at the right time of the day seems to be under particularly hard selective pressure. The end of the activity phase, on the other hand, is much more flexible and may depend on weather, season, foraging success or whatever. So it is the combination of conservative and flexible elements in the activity triggering mechanism that seems to be the most efficient way to deal with the timing problem. Or, to look at it in another way, the activity triggering mechanisms are internal machineries with environmental antennas. If such sensitive and adjustable machineries exist for different hierarchically nested temporal scales, ranging from ultradian to circadian, seasonal and even long-term components, highly flexible behavioural patterns are the consequence, and it is this flexibility which makes adaptive responses possible (Halle in Chap. 11). We have to realise that the modulation of the daily activity pattern is one of the most direct and immediate responses of animals to environmental challenges (abiotic, intraspecific, and interspecific). This kind of short-term modulation, which bears valuable ecological information, is at best regarded as confounding noise in chronobiology.

15.4 Chronoecology and the Energy Household

The overruling environmental feature in relation to timing problems is the highly predictable daily LD cycle. So, one may wonder, why are there endogenous circadian clocks at all? Wouldn't it be enough with just a simple light-sensitive on/off-switch for activity to trigger the individual's activity phase in a much easier way? As stated in the Introduction, it is the alternation between activity and rest that gives a structure to the continuous flow of time in animals. However, this is only half of the story. In fact, complex internal programs are running to give a fine structure to the interval between 'light on' and 'light off', and activity itself is an orderly sequence of different subroutines. The timing and order of these components is also due to evolutionary optimisation and is closely related to the individual's energy household, as treated in detail by Weiner in Chap. 3 and by Ruf and Heldmaier in Chap. 12.

Metabolism and energetic constraints are crucial features in small mammals that face particular problems from the combination of endothermism, high metabolic rates, lactation, and small body size. Nevertheless, small mammals are a most successful evolutionary model, and particularly so in northern latitudes where energetic constraints are especially severe. The temporal pattern of activity has an important part in solving energetic problems, as demonstrated by a review of the case studies in this volume: they verify that shape and particularly feeding habits are at least as decisive for the activity pattern as body size (in particular see Zielinski in Chap. 5, Palomares and Delibes in Chap. 6 and Erkert in Chap. 14 for illustrative examples). In many chapters there are also strong hints at digestion as an essential factor. This is reasonable because digestion represents the crucial translation from food into available energy which is essential for the next activity bout, necessary to gather new food. The fundamental trade-off in this optimisation process is between energy gain from and the energy costs of activity, and the temporal pattern of when to be active for how long is one important parameter for shaping adaptive strategies.

The theoretical framework for this kind of question is the well-established optimal foraging theory (Charnov 1976; Stephens and Krebs 1986). In a book on the biology of lemmings, Stenseth and Ims (1993) suggested linking optimal foraging theory with optimal life history theory (cf. Engen and Stenseth 1984, 1989). In an extension to this approach we would now like

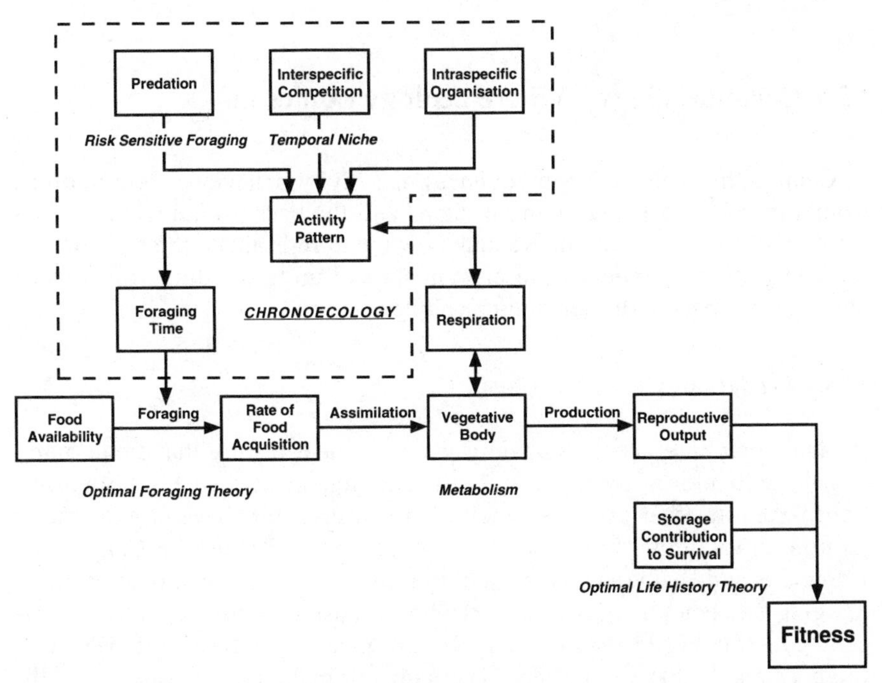

Fig. 15.1. Suggestion of how to link activity patterns with optimal foraging theory, metabolism, optimal life history theory, and fitness measures. Risk sensitive foraging and niche separation may be seen as part of the ecology of activity behaviour when dealing with temporal patterns. The scheme outlines the conceptual framework for the field of chronoecology and its relation to adjacent domains in behavioural ecology

to propose another link between these two aspects of fitness maximisation with activity patterns and their ecological implications (Fig. 15.1). In our view, the resultant scheme outlines the conceptual framework for the field of chronoecology.

Another important aspect is to interpret activity patterns as adaptive strategies of energy saving and thermoregulation. Social synchronisation of activity and non-activity, respectively, is a special issue in this respect because it determines length and times of nest-sharing to save energy. Again, the particular solution a species found in the course of evolution may be very different from optimal solutions in other situations: heat avoidance and water saving in southern latitudes and/or summer may be as important as avoidance of cold and rain in the north and/or winter (e.g. Wauters in Chap. 7 and Daly et al. in Chap. 8). However, we also have to realise that the responses to adverse conditions are graduated, ranging from seasonal phase-shifts (i.e., changes in the activity phase) to food caching and a literal 'switch-off' of activity in daily torpor or even hibernation for months. It is generally not well understood at present under what circumstances the different strategies for overwintering have established in the course of evolution (see Ruf and Heldmaier in Chap. 12 and Merritt and Vessey in Chap. 13).

15.5 Chronoecology – Where Ecology Comes In

In Chap. 4 the ecological consequences of activity behaviour were outlined from a merely conceptual point of view. With the background of the ten review chapters in this volume we may now try to re-evaluate the relevance of these features and to define our current state of understanding, which is the object of this part of the concluding chapter.

15.5.1 Predation

Predator-prey interaction is obviously a prominent feature that small mammals have to address by activity timing. This statement is true in both directions. Like any other predator, small mammal predators have to adjust their hunting times to the activity patterns of their prey (Zielinski in Chap. 5 and Palomares and Delibes in Chap. 6), but in this group the pressure to optimise foraging efficiency is particularly strong because of especially severe energetic constraints (illustrative examples are given by Merritt and Vessey in Chap. 13 and by Erkert in Chap. 14). Again, optimal foraging seems to be the suitable theoretical framework for addressing these questions. As a rule, the availability of prey is patchily distributed, in space as well as in time. Considerable work has been done on the spatial aspect, while the equivalent temporal aspect has as yet received relatively little attention.

Small mammal prey, on the other hand, is often under extremely high predation pressure, so the pressure of finding relatively safe time windows is particularly strong for them. In some cases where there are no relatively safe time windows, distinctive strategies to deal with this problem may have become established (see Halle in Chap. 11). Risk avoidance by activity timing may reasonably be treated within the well-established conceptual framework of risk sensitive foraging (Caraco 1980; Real 1980; Lima et al. 1985; Lima and Dill 1990; Bednekoff 1996; McNamara 1996; Smallwood 1996). A remarkable amount of work on anti-predator strategies has been done on desert rodents (e.g. Daly et al. in Chap. 8), probably because these systems are relatively simple. However, predation risk varies over the 24-h day in most systems, and fitness benefits will always arise if activity timing can lower predation risk. Also, group strategies like synchronous activity to decrease relative individual predation risk by safety-in-numbers belong to this domain (as discussed by Flowerdew in Chap. 10 and Halle in Chap. 11). From a conceptual point of view predation is especially fascinating because it seems reasonable to assume that an evolutionary arms race of temporal strategies might arise. Even more amazing cases occur when small mammals are predators and prey at the same time, like weasels, for instance, as reported in Chap. 5 by Zielinski.

To us, predator-prey interaction is one of the fields where the ecological relevance of diel activity patterns is most obvious. This view is not new, since predation is the one ecological aspect that has been relatively prominent in the literature on activity behaviour up until now. What has changed, however, is the fact that predator-prey interactions are currently dealt with in a more exact and stringent way. While earlier work presents mainly verbal models and plausibility explanations, modern studies try to draw conclusions based on empirical activity data from predators and prey collected at the same time. Hence, two or even more species have to be surveyed simultaneously, which still is demanding, but at least possible today due to new monitoring techniques. This new quality of approach is, in our view, characteristic of studies in the field of chronoecology.

15.5.2 Interspecific Competition

For us, as editors, the greatest surprise in this volume was the clear evidence it gives for the temporal niche concept. In six out of the ten review chapters, the effect of interspecific competition on activity timing is at least mentioned. Chapter 9 by Ziv and Smallwood takes us an important step further: time may not merely be a dimension of the ecological niche that allows for niche separation. Rather, there may even be interspecific strife for the most attractive time windows of the 24-h day, with the dominate species being able to monopolise it. This is indeed much more than just interference avoidance by shifting the activity phase. It is interesting that it is a desert system where this has been demonstrated by experiment (cf. risk sensitive foraging), but there are some hints in this volume that similar situations may also be found in other systems, especially with murid and microtine rodents.

The argument for a temporal niche dimension seems to be a sharp contradiction of Schoener's classical work on relevant niche dimensions (Schoener 1974), but it is not. On the basis of the 81 species reviewed, Schoener did indeed consider activity time as a generally unimportant niche dimension, a statement that was readily adapted as absolute by others in the following decades. However, in the same paper Schoener already confined his assertion by mentioning two important exceptions: (1) when the considered species are predators, and (2) when predation risk is much higher than the ability to process food. The latter especially applies to many small mammals, which are under high predation pressure, and live on tight food resources. So actually, our inference that a temporal niche exists in many small mammals rather supports than questions Schoener's view.

15.5.3 Intraspecific Organisation

While we are quite sure and confident of the empirical findings that predation and competition are important ecological issues for understanding small mammal activity patterns, it is more of a gut feeling that this is also true of the population's social organisation. The reasons for this ambivalence are twofold. From theoretical considerations it is obvious that there are several reasonable traits to suggest an effect of social factors on activity timing (see Halle in Chap. 4). This view is supported by many of the empirical reviews, where indications of social components are found in seven out of the ten chapters. However, it is not much more than just indications, suggestions or, at best, indirect evidence. Interestingly, one of the relatively hard pieces of evidence stems again from a competitive situation (i.e., intraspecific competition between sexes; see Wauters in Chap. 7).

Secondly, we have to realise that obtaining reliable data on the effect of social interactions on activity behaviour is, for various reasons, a most demanding task. First, there is a structural problem, since the social organisation may affect the activity pattern, and, at the same time, the activity pattern may affect the social organisation. This mutual interrelationship is difficult to disentangle, particularly if there are additional effects to address, like the consequences of synchronous or asynchronous behaviour on predation risk for instance. Second, there are severe methodological problems. To survey a social community with respect to activity behaviour implies that several, if not all, individuals have to be monitored at the same time. In order to address questions like age dependence and seasonal changes, monitoring has to be performed on a long-term scale. Both demands can hardly be evaded when dealing with social organisation, but both also push us to the limits of field techniques available today.

In conclusion, we feel that the evidence for social components in activity timing is sufficiently good to let it appear as a promising approach, but that a thorough investigation of this aspect will depend on further methodological developments. Hence, the relationship between social organisation and activity behaviour will probably be a prominent challenge in the field of chronoecology in the decades to come. One inference, however, is clear to us: the understanding of social organisation was until recently almost entirely restricted to spatial behaviour. It is most obvious from this volume that this limited view does not hold. Surveys on the social organisation of animal populations and inter-individual processes have to incorporate temporal behaviour to give a complete picture. Moreover, even spatial behaviour as such may not be understood if the temporal component is ignored.

15.6 Future Challenges in Chronoecology

Chronoecology, and the surveys that stand for this field of research, have opened a new array of questions in behavioural ecology. However, as we have shown in this concluding chapter, it is not a totally novel approach. Rather it matches subdisciplines that belong together, but which have not been integrated for historical reasons. As a second characteristic, chronoecology applies established conceptual frameworks, such as the evolution of life history strategies, evolutionarily stable strategies, metabolic optimisation, optimal foraging theory, risk sensitive foraging, and niche separation, to a new field. For some reason, this has not been done before extensively, but we feel that the time is right to do so now.

So the first and most important task which chronoecology has to do is to provide empirical data in well-designed experimental studies to support and enlighten the arguments framed in the theoretical considerations section of this volume. As the empirical findings section demonstrates, some promising steps have already been taken in this direction, but it has also shown a still considerable lack of hard field data (as clearly pointed out by Flowerdew in Chap. 10 for instance). Progress in this field will also depend on framing new questions, or to be precise, to frame current questions in a slightly different way. At present, for example, we ask what effect predation has *on* activity patterns. The next important step, which more specifically addresses the ultimate approach of evolutionary ecology, is to ask: What are the effects *of* activity patterns on predation? This differentiation may appear as splitting hairs, but in fact it represents a fundamental and qualitative change in the general scientific attitude in this field of research.

What we have coined as chronoecology in this volume is also not new in another sense. Most of the underlying ideas were already presented in the paper by Enright (1970), which for that reason is often cited in this book. So why this gap of more than 20 years, especially since the related conceptual frameworks were developed at that time? It is our strong belief that the reason is methodological. Monitoring animal behaviour, and particularly that of small mammals, is a difficult task and was until recently confined to activity cages. However, the ecological implications of activity behaviour can frankly not be considered at all in a cage set-up. Hence the criticism of ecological conclusions based on cage findings appears almost as a constant thread in this volume, especially in the empirical findings section where it is mentioned in virtually every chapter.

So, the emergence of chronoecology as a new subdiscipline depended heavily on methodological developments which allowed the activity behaviour of small mammals to be studied directly in the field or in the seminatural conditions of large enclosures. Because of this essential dependency on methods, we want to round up this book with an extensive Appendix on the various methods of activity recording and an overview of analytic tools. We would also like to encourage the reader to join us in these new explorations. To do so will be rewarding, since the field of chronoecology is new, promising, and – we believe – relevant.

References

Bednekoff PA (1996) Risk-sensitive foraging, fitness, and life histories: where does reproduction fit into the big picture? Am Zool 36:471–483

Caraco T (1980) On foraging time allocation in a stochastic environment. Ecology 61:119–128

Charnov EL (1976) Optimal foraging: the marginal value theorem. Theor Pop Biol 9:129–136

Engen S, Stenseth NC (1984) A general version of optimal foraging theory: the effect of simultaneous encounters. Theor Pop Biol 26:192–204

Engen S, Stenseth NC (1989) Age-specific optimal diets and optimal foraging tactics: a life historic approach. Theor Pop Biol 36:281–295

Enright JT (1970) Ecological aspects of endogenous rhythmicity. Annu Rev Ecol Syst 1:221–238

Lima SL, Dill LM (1990) Behavioral decisions made under the risk of predation: a review and prospectus. Can J Zool 68:619–640

Lima SL, Valone TJ, Caraco T (1985) Foraging-efficiency – predation-risk trade-off in the grey squirrel. Anim Behav 33:155–165

McNamara JM (1996) Risk-prone behaviour under rules which have evolved in a changing environment. Am Zool 36:484–495

Real L (1980) On uncertainty and the law of diminishing returns in evolution and behavior. In: Staddon JER (ed) Limits to action: the allocation of individual behavior. Academic Press, New York, pp 37–64

Schoener TW (1974) Resource partitioning in ecological communities. Science 185:27–39

Smallwood PD (1996) An introduction to risk sensitivity: the use of Jensen's inequality to clarify evolutionary arguments of adaptation and constraint. Am Zool 36:392–401

Stenseth NC, Ims RA (1993) Food selection, individual growth and reproduction – an introduction. In: Stenseth NC, Ims RA (eds) The biology of lemmings. Academic Press, London, pp 263–280

Stephens DW, Krebs JR (1986) Foraging theory. Princeton University Press, Princeton

Appendix

Measuring and Analysing Activity of Small Mammals in the Laboratory and in the Field

Stefan Halle and Dietmar Weinert

1 Methodology – Reality Constraints to Wishful Thinking

The activity behaviour of an animal species could be fully described if we had complete information about where every individual of a population is at any time and what kind of activity is performed. Unfortunately, this is only a pipe dream, unreachable for various practical reasons. Many small mammal species are nocturnal, cryptic, and reside in a type of habitat that makes direct observation difficult, if not impossible. Even when some methodology is available to overcome these problems, it is commonly impracticable to follow more than one or a few individuals at the same time. Long-term and continuous observations are tedious and extremely labour-consuming when no automatic recording system is available. Finally, even if a more or less reliable method for activity recording is applied, we face the problem that an enormous amount of complex time series data have to be analysed. Only if this last challenge can be surmounted will we have a chance to see the patterns we are looking for. However, like activity recording, several methodological approaches to data analysis are available, each with its specific possibilities and limitations.

The method employed for a particular activity study is, therefore, always a compromise, focusing on one aspect at the costs of other features. Choosing the right method combination for a study objective is indeed essential to obtain the information needed to clarify a specific scientific question. However, one has to be aware of the inevitable limitations and shortcomings of the chosen methods at the same time to avoid obvious or hidden pitfalls in data interpretation. The three crucial decisions to make are (1) where to study the animal, (2) how to record its activity, and (3) how to analyse the activity recordings. For the first decision, cages, enclosures or the open field are the possible options. Depending on the 'environment', a definite set of feasible methods for activity recording is available for the second decision, while the third decision depends on the actual data structure and the aspect of temporal behaviour in focus.

Ecological Studies, Vol. 141
Halle/Stenseth (eds.) Activity Patterns in Small Mammals

2 Recording Activity – Free Choice Among Drawbacks

2.1 Activity Cages

The principal idea of cage experiments is to extract animals from the confusing natural habitat and instead study their behaviour in a simplified environment that encourages surveillance. The advantages of the step from the field into the laboratory are so obvious that this approach, apart from anecdotal reports from observations of free-ranging animals, represents the earliest technique for studying small mammal activity (Szymanski 1920; Johnson 1926; Davis 1933). On some occasions, cages were primarily used to enable direct observations of special behaviours with a distinct time structure, such as reingestion of faeces (Kenagy and Hoyt 1980), social interactions (Bovet 1972), or movements in an artificial tunnel system (Baron and Pottier 1977). The fundamental advantage of experiments in cages, however, is that they facilitate permanent and continuous activity monitoring.

From the earliest studies, so-called 'actographs' (coined by Szymanski 1920) have been frequently employed. In this, a sensitive mechanism transforms the movements of the animal into movements of a cage, which can either be recorded mechanically with smoked or thermograph paper (Johnson 1926; Davis 1933; Benjamini 1988/1989), or, in a somewhat more advanced manner, with electric equipment (e.g. Ostermann 1956; Erkinaro 1969; Loxton et al. 1975). This old method is still used in improved variants, as demonstrated by highly advanced devices for direct computer recording (e.g. Joutsiniemi et al. 1991).

An alternative approach is to house the animal in a fixed cage and to record its movements inside the chamber. In the literature, a variety of different techniques is reported for this purpose, including unique devices such as curtains of moveable metal wires (Crowcroft 1953) and electroacoustical methods (Kleinknecht et al. 1985). More commonly, however, presence in the nest (e.g. Pearson 1962; Lehmann 1976), feeding (e.g. Saarikko and Hanski 1990; Hulsey and Martin 1991), and passages through nest entrances, doorways, or tunnels (e.g. Hatfield 1940; Miller 1955; Myllymäki et al. 1962; Erkinaro 1972; Stebbins 1974; Harland and Millar 1980 and many others) are recorded by microswitches or a variety of other electromechanical installations. More recent technical developments include light beams (e.g. Levitsky 1970; Johst 1973; Gurnell 1975; Lodewijckx et al. 1984; Stubbe et al. 1986; Gaulin et al. 1990; DeCoursey and Menon 1991) and passive infrared motion detectors (Gerkema et al. 1990; Gerkema and van den Leest 1991; Ruf et al. 1991; Weinert and Weiß 1997). Such devices are indeed favourable because they allow monitoring of the animals with minimum disturbance.

The most prominent technique of recording activity patterns in cages, however, is of course the running or exercise wheel. For an innocent observer it may appear bizarre to see a captive animal eagerly working the treadmill, but obviously the method works excellently in a large variety of species (including medium-sized carnivores, Kavanau and Ramos 1975), and particularly with rodents. It normally takes only a few minutes until a naive animal discovers the joy of wheel running, and even wheels placed in the open field are quickly adopted by microtine rodents (Halle, pers. obs.). Running wheels have the additional practical advantage that wheel revolutions are easy to transform into signals that can be continuously recorded by simple or more sophisticated devices, ranging from mechanical marking on paper to computers.

A 'revolving wheel' was employed by Calhoun (1945) for his experiments with *Microtus ochrogaster* and *Sigmodon hispidus*. Considering the practical advantages, it is not surprising that wheel running became the standard measurement of activity in hundreds of laboratory studies during the following decades, and we will not try to review the microcosm of relevant literature here. Recently, new modifications have been developed, like the horizontal activity disk for instance (Borer and Dennis 1991). Nevertheless, the popularity and practical handiness of the method must not divert our attention from the severe problems in interpreting this artificial behaviour with respect to its natural correlate (see De Kock and Rohn 1971; Mather 1981). Furthermore, access to a running wheel does have a feedback effect on behaviour: not only does it induce motor activity, it may also change its daily pattern (Weinert and Weiß 1997).

Since the pioneering methodological work of Calhoun (1945), much effort has been put into the detailed differentiation of discrete cage activities, such as resting in the nest, feeding, drinking, wheel running, and locomotion (Kavanau 1963; Lehmann 1976; Hoogenboom et al. 1984; Gerkema et al. 1990; Reid et al. 1993). The possibility of discriminating between different kinds of activity and to record activity patterns for each behavioural component separately is probably one of the most useful features of activity cages. Cage experiments are also essential for surveys on the biological clock, the physiological mechanisms that trigger activity, and the relationship between activity and metabolism. The implications for ecological aspects of diel activity patterns, however, are limited. Cages represent a most artificial environment, animals are fed on an unnatural diet and, most often, individuals are housed solitarily. The absence of conspecifics must be considered a strange situation for communal living species, and the important role of social interactions must of course be totally ignored in this type of experiment. Moreover, the activity pattern observed may severely depend on the properties of the cage, as shown by Lehmann (1976), for instance.

2.2 Enclosures

Although a familiar term, enclosures are in fact rather difficult to define. There are actually two different kinds of enclosures, indoor and outdoor, which offer considerably different living conditions to the animals. The crucial feature, however, is the enclosure size in relation to the size and activity range of the species being studied: what is a large enclosure for a mouse may be an extremely small enclosure for a rabbit. The borderlines between cages and the open field are shadowy: it is almost impossible to make a reasonable distinction between a large cage and a small indoor enclosure and, consequently, there is no general agreement in the literature for labelling experiments as cage or enclosure studies, respectively. On the other hand, for species with relatively small activity ranges, a large outdoor enclosure may be very much the same as the open field.

With respect to the methodology of activity recording, the technical tasks in small enclosures are very similar to those in cages, while they are almost identical to the field in large enclosures. As a special feature, enclosure experiments are often used as a first step for developing or testing new field techniques. Hence, no method exclusive to enclosures has to be mentioned here. However, experiments with small groups of animals under relatively controlled conditions in medium-sized enclosures represent a distinct class of studies. Rübsamen et al. (1983) performed a radiotelemetry survey with transmitters that allowed discrimination between hopping and other activities of rat-kangaroos (*Aepyprymnus rufescens*). Automatic telemetry systems were developed for different species to gain information about activity and ranging behaviour with a high temporal and spatial resolution (e.g. Chute et al. 1974; Hoek et al. 1985). Lund (1970) used photo-cell equipment to record seasonal changes in the activity patterns of an enclosed population of water voles (*Arvicola terrestris*).

Much more important, however, is the fact that social interactions can occur in the experimental groups, which may be essential for specific studies. Gerkema and Verhulst (1990) employed passive infrared detectors to record the escape response of groups of common voles (*Microtus arvalis*) to predators and warning signals. Rasmuson et al. (1977) registered passages between enclosure compartments as a measure of activity to evaluate behavioural differences between groups from cyclic and non-cyclic populations of field voles (*Microtus agrestis*). Blumenberg (1986) developed a simple but efficient technical device to discriminate individuals when entering hiding places and nest boxes, which allowed correlation of activity with social behaviour in an enclosure population of common voles (*Microtus arvalis*). When using telemetry systems it is possible to monitor the activity behaviour of a single animal living in a group, i.e., in its common social environment. Since similar activities will be recorded with passive infrared detectors, one may combine these two techniques. This provides the possibility of simultaneous activity recording of a group and a single individual (Weinert 1996).

Enclosures seem to be a suitable intermediate arena to investigate certain aspects of animal activity in detail. Some of the advantages of both cage and field studies are combined in enclosures, but unfortunately, surveys are also burdened with the combined disadvantages of the two other approaches. As the most serious drawback, normal social systems cannot be established, at least in relatively small enclosures. In plots of a size multiple to the natural activity range of the investigated species, this effect may not be that severe. However, densities in all types of enclosures are commonly higher than in natural populations, which is most likely to affect the activity pattern. Additional problems are due to 'fence effects', especially in long-term studies, in species with a high natural turnover-rate, and when an exchange of individuals among subpopulations is essential for the demographic machinery. Hence, the utility of enclosure experiments with respect to ecological inquiries is also limited.

2.3 Field Studies

Undoubtedly, field studies are the best way to study ecological aspects of activity behaviour. Animals live in their natural habitat and social environment, feed on their habitual diet, and are exposed to the normal set of extrinsic factors like weather, photoperiod, moonlight, and predation. However, the difficulties in getting reliable information about animal activity are also never as obvious as under field conditions. One has only a limited set of techniques to choose from, and none of them is really satisfying. By focusing on one of the four prominent attributes 'who', 'where', 'when', and 'what', the accuracy of the information about other attributes is inevitably diminished (Table A.1).

It may, for example, be possible to gather fairly detailed data on the activity of one particular individual, but then it is impossible to get the same kind of information for more than one or a few individuals at the same time. Furthermore, one has to be aware that whatever method is employed to survey activity in the field, it is always associated with disturbances. Hence, in every field study it should be carefully considered that the method employed to record activity may possibly affect the natural behaviour which it is intended to gauge.

Direct observations of small mammals in the field are extremely difficult, although not totally impossible, even in cryptic species such as bank voles (*Clethrionomys glareolus*; Mironov 1990). In some instances, camera-aided observation may be possible (e.g. Carthew and Slater 1991), depending on the habitat and the species concerned. The standard methods for gathering quantitative data in the field, however, are trapping, automatic recording, and radiotelemetry.

Table A.1. Advantages and disadvantages of the most important methods of recording animal activity in the field. *TR*: trapping, *AR*: automatic recording with passage counters, *RT*: radiotelemetry

	TR	AR	RT
Advantages			
Natural behaviour	–	+	?
Simultaneous temporal and spatial data	+	–	+
High temporal resolution of data	–	+	+
Long-term and continuous studies	–	+	–
Activity pattern on the group or population level	+	+	–
Intraspecific behavioural variability	+	–	+
Activity data for single individuals	+	–	+
Disadvantages			
Disturbances	+	–	?
Time and work consuming	+	–	+
Free movement of animals restricted	+	–	–
Attraction of animals	+	–	–
Restricted to paths and runways	–	+	–
Only in areas with one dominating species	–	+	–
Experience with method necessary	–	–	+

2.3.1 Trapping

Methodological restrictions are particularly distinct with trapping, which was the earliest field technique for activity studies (Hamilton 1937). The number of captures per time interval is taken as a measurement of activity, based on the reasonable assumption that animals have to be active to enter traps. Not as trivial, however, is the second basic assumption for this method, which is that whenever animals are active, they are trappable. In a strict sense, trapping reveals a temporal pattern of trappability, which need not necessarily reflect the times of activity correctly. As an additional complication it is well known that certain functional groups, juveniles in particular, are often under-represented in trap captures.

Besides this general dilemma, serious practical problems exist with trapping. In order to obtain activity patterns with a sufficient temporal resolution, traps must be checked frequently, which means every 2 h as a minimum (Wójcik and Wołk 1985), but preferably much shorter (Kenagy 1976; Daan and Slopsema 1978; Raptor Group 1982; Hoogenboom et al. 1984). This involves high human activity on the plot, which probably causes disturbances. An alternative approach is to register the exact time of capture by some kind of clockwork (e.g. Bäumler 1975; Requirand 1990).

In any case, traps must of course be checked around the clock in activity studies. To get enough data with short trapping intervals it is also necessary to activate a large number of traps and to use highly attractive bait, which will serve as an additional food resource. Hence, trapping with short time intervals is extremely laborious, causes considerable disturbance, and can thus only be performed over short periods, i.e., some days. Trapping, therefore, gives a snapshot picture of the activity pattern. Long-term changes, as in the course of the season for instance, can only be investigated by repeated trapping sessions with times of recovery (for both animals and observers) in between. There are also, however, important advantages to trapping. The technique is cheap and easy to perform, and, being most important for the social aspect of behaviour, one is able to get information about individual activity schedules, given that animals are individually marked by an easily readable and persistent method.

2.3.2 Automatic Recording with Passage Counters

The main concern with automatic activity recording is to reduce the presence of the observer and all other sources of disturbance to a minimum, but nevertheless obtain data with a high temporal resolution. The second important goal is to allow for long-term and continuous monitoring of activity in the field in order to consider changes in activity behaviour, related to the course of seasons and to population dynamics. The technique basically depends on some device that transforms movements of animals in their natural habitat into electric signals, which can be recorded by means of event recorders or computers (it should be mentioned here, however, that there are good reasons for avoiding computers as a device for direct data-logging in the field: there is more than convincing evidence that computer systems will always fail when nobody is around to fix the problem).

The idea of registering passages is actually quite old and was described for the first time by Calhoun (1945) as part of his ingenious 'population activity apparatus'. Since that time, tunnels, doorways, and similar installations have been frequently used as recording devices in activity cages (see also above). Mossing (1975) presented the prototype of an electromechanical passage counter for recording microtine activity in the field. His idea was to place plastic tubes in the surface runway system of field voles (*Microtus agrestis*). Passing voles operated a coil spring with an attached small magnet, which activated a reed switch, giving a recordable electric signal.

Starting from Mossing's model, Lehmann and Sommersberg (1980) optimised different mechanisms and ended up with a hanging swing gate. This type was employed for the first continuous year-round study of activity in a group of common voles (*Microtus arvalis*) in a small outdoor enclosure. Halle and Lehmann (1987) modified the swing gate model and finally developed a simple and very robust passage counter for long-term use in the field. With this equipment it was even possible to register vole activity continuously under snow cover during the winter in North Sweden (Halle and Lehmann 1992). The method may also be applied in species that do not have easily visible runways, provided that the movements are directed through the passage counters by some kind of barrier (Halle 1988). This, however, presumes that the target species is the only, or at least the largely dominating species of its size in the study area.

Comparable systems can certainly be developed for many other species than microtine rodents. Nevertheless, passage counters are also burdened with drawbacks. They can only register a particular behavioural component, i.e., locomotion within the runway system. This problem may be obviated by combining automatic recording with other monitoring techniques, such as radio-tracking in particular. Most serious, however, is the lack of discrimination between individuals or even classes of animals: with passage counters it is only possible to deal with the overall activity pattern at the group or population level. In response to this problem it is in principle possible to mark some animals with a magnet-collar and to record passages through induction coils, placed like passage counters in the runway system (Halle, unpubl.). This method, however, implies high technical effort and cannot be recommended for use in the field. Much more promising are passive transponder systems, which quite certainly will bring the most important improvement in activity recording techniques in the near future.

2.3.3 Radiotelemetry

The modern technique of radio-tracking seems to solve all methodological problems in a most adequate way. It is known to the observer which animal is wearing which transmitter, hence individual data can be recorded and it is possible to consider intraspecific variations in behaviour. Recording of activity is not restricted to certain 'registration spots' in the habitat, and animals are not affected by any confinements due to the method. Thanks to progressing miniaturisation of radios and particularly batteries, the method is today applicable to rather small animals, although use with very small species, such as shrews for example, may still not be practicable. Most importantly, radiotelemetry is the only method that offers sampling of high resolution data on temporal and spatial behaviour simultaneously.

There are two different localisation procedures, depending on the size of the species concerned and its activity range, respectively. In species with a large activity range, the well-established method of triangulation is the standard technique, which usually needs two observers at a time and a reliable communication system (for application with a small species, see Banks et al. 1975). In smaller species, hand-held antenna systems (Webster and Brooks 1981; Ostfeld 1986) are preferable. As an alternative approach, automatic tracking with a stationary installed antenna system may be employed (e.g. Madison 1981; Wolton 1983; Kolb 1992). As parameters of activity, net activity time, percentage of localisations with activity, and total distance travelled may be evaluated (Palomares and Delibes 1991).

The reader may, however, not be surprised that there are also specific problems with radio-tracking. It is a rather sophisticated and relatively expensive technique that requires some practice before a serious field study can be performed. Manual localisation of animals with probe antennas and particularly triangulation is laborious and time consuming, hence continuous long-term surveys are not possible. Normally it will also be impossible to follow more than a few individuals at the same time. Finally, convincing quantitative studies are still lacking to prove that the behaviour of animals is not affected by wearing a radio-collar, a problem which is particularly crucial in smaller species.

Marking animals with radioactive tags and localisation by means of a Geiger-Müller counting tube may be considered a precursor of radiotelemetry, since the principle is very much the same. The technique has been employed in the past for recording animal activity (Karulin et al. 1976; Airoldi 1979) and is still in use, especially in small fossorial and semifossorial species (e.g. Salvioni 1988). The main advantage of the method is that no battery is needed to generate the signal, since the battery contributes most to the weight of a radiotransmitter and limits the duration of a study. However, one has to be very much aware of the most unpleasant aspects of working with strongly radiating material in the field.

2.4 Activity Recording – A Conclusion

To draw a short conclusion we have to admit that there is no perfect method available today to investigate daily activity patterns with the focus on ecological aspects. Consequently, one must first clearly formulate the question, and as a second step, one has to select the most convenient method for answering it. No general advice can be provided for this decision, because a methodological advantage for one kind of study may be a disadvantage for another. Finally, in interpreting the results, one has to consider carefully the limitations inherent in the method employed. Undoubtedly, much progress in the area of methodology has been made during recent decades. The growing interest in activity behaviour of animals in the field is in fact a direct consequence of new technical developments and possibilities. But still, the solution to methodological problems is a demanding challenge, which again and again should appeal to our fantasy and inventive genius.

3 Analysis of Biological Time Series – Possibilities and Limitations

The analysis of biological time series is aimed at the search for order, which means particularly, isolating components which are driven by regular biological oscillations from those due to random fluctuations. Furthermore, the rhythmic components should be characterised. Depending on the objective of the study, different rhythm characteristics (Halberg et al. 1977) may be of interest. The most definitive, and the most commonly measured value is the period of the rhythm. It is defined as the duration of one complete cycle or as the interval between points at the same phase in consecutive cycles. The phase is a relative quantity, indicating at which time within the cycle a given event (e.g., the onset or the maximum of activity) occurs. It is measured with respect to the entraining rhythm or to some other reference point. Compared with period, phase is a more labile characteristic. The amplitude of a rhythm is not unambiguously defined. Often the difference between peak and trough values in the oscillation is taken as a measure. This would be better called 'the magnitude of oscillation'. In other cases the difference between the mean and the maximum values of the rhythm, as usual in physics, is defined as the amplitude (cf. 'cosinor analysis', Fig. A.3). The most variable of rhythm characteristics is, probably, the wave-form. Changes may be clearly evident in the overt rhythm, but they are difficult to quantify objectively. For a simple oscillation, waveform may be described in terms of period, phase and amplitude; more complex oscillations are best described in terms of harmonic components, derived for example from periodogram analyses.

To discriminate underlying rhythmic processes from random fluctuations and to assess objective estimates of rhythm characteristics directly from the raw data might be complicated due to problems in defining appropriate reference points. Therefore, a number of mathematical algorithms have been developed, some of them labelled with specific names which are established terms in this field of research, but which may be unfamiliar to students coming from different fields. Four of the most common approaches will be described below. Their mathematical algorithm will be explained only as far as it is necessary for critical evaluation of the results. It will be shown what rhythm characteristics may be calculated using these methods, what type of primary data may be analysed, and what the possibilities and limitations are. For more details the following references can be recommended: Enright (1981) for an erudite review of the methods of analysing rhythms, Chatfield (1989), Box et al. (1994) and Fuller (1996) give a good introduction to statistical time series analysis, and Girling (1995) provides an overview of periodogram techniques and their practical use. The paper by Morgan and Minors (1995) is very useful with respect to data collection and pretreatment and provides helpful information concerning the cosinor method. Further references can be found in the sections below.

With respect to software for the analysis of rhythms, many investigators still use their own programs. ANOVA and Fourier analysis may be done with most of the bigger statistical packages. Besides these, a number of data acquisition systems exist which contain software for graphical presentation and mathematical analysis of time series. For activity recording, the following may be recommended: DataCol (Mini-Mitter Co. Inc., P.O. Box 3386, Sunriver, OR 97707, USA), Dataquest (Data Sciences International, 4211 Lexington Avenue, St. Paul, MN 55126, USA), and The Chronobiology Kit (Stanford Software Systems, Stanford University, Stanford, CA 95060, USA). Their disadvantage is that the programs can analyse only data recorded with these systems. In contrast, TAU (Mini-Mitter) may be used for any ASCII data files. The program provides single and double-plotted actograms, mean value chronograms and Chi^2-periodograms.

3.1 Data Collection

The conventional methods of time series analysis require data collected at frequent, regular intervals over a number of cycles. The sampling rate depends on the expected period or on the period being of immediate interest. With respect to the mathematical methods used for the analysis, the sampling interval should be at least half the rhythm period. However, a minimum of 4–6 samples per cycle of the rhythm is highly recommended. With more frequent sampling, the degree of resolution of rhythm characteristics improves.

Recommendation for the length of the time series depends on the longest period expected in the data set. A minimum of three cycles should be covered. This is important not only to achieve statistical significance but also to test the biological relevance. The circadian pattern may comprise a high degree of random fluctuations or any trend. Furthermore, the rhythm characteristics may vary considerably day by day even under highly standardised conditions in the laboratory. Thus, to resolve rhythm characteristics it is often better to make observations at equal intervals over several cycles of the rhythm, rather than making the same number of observations (i.e., a high rate of sampling) within a single cycle of the rhythm.

3.2 Graphic Presentation of Data

The analysis of biological time series should always start with graphic representation of the data. This allows a visual estimate of a trend or noise, but also of rhythm characteristics as shape (wave form), rhythmic components, amplitude, timing of peaks and troughs. On the basis of these estimates appropriate methods for pretreatment and analysis may be chosen. Graphic representation of the data is also extremely important to test the results of computations.

A number of ways exist for depicting time series. They all include a number of cycles, which will be arranged consecutively or be combined over the duration of a single cycle. The simplest method is to plot the measured values depending on time (Fig. A.1a). This permits the visualisation of the

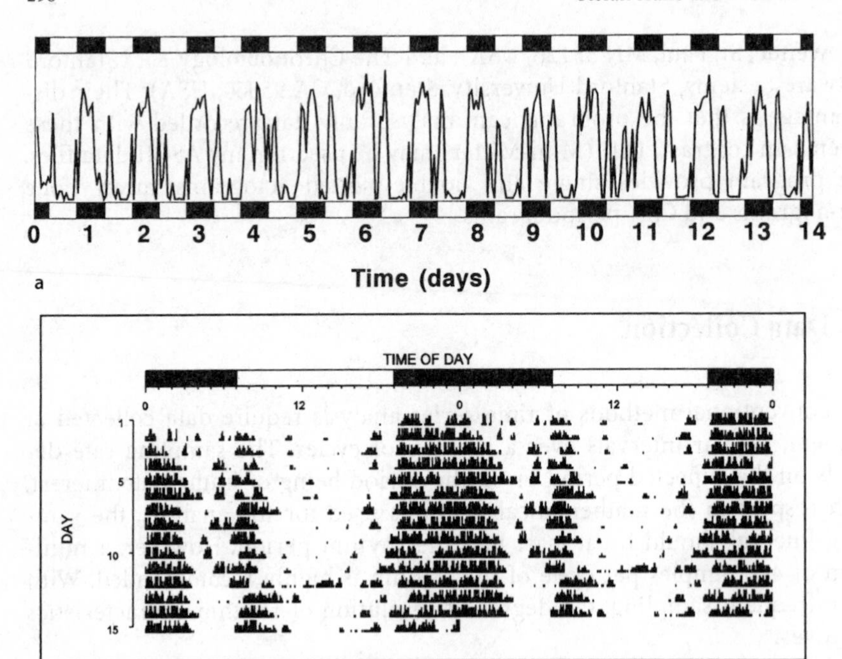

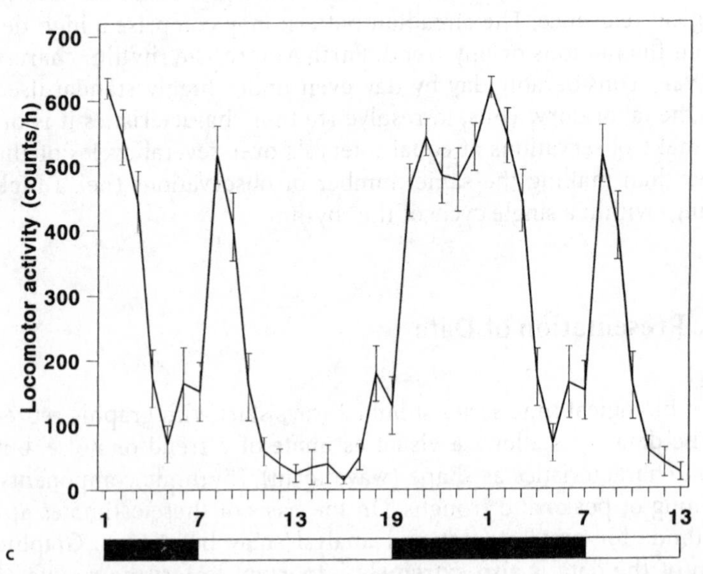

Fig. A.1. Different methods for graphic presentation of locomotor activity data from one female laboratory mouse recorded every minute over 14 days. **a** Time series: the hourly sums of impulses are plotted depending on the time. The *bars below* and *above* indicate the lighting regime. **b** Actogram: double-plotted data from a (5-min values). **c** Mean value chronogram: hourly means over 14 days ± SEM are shown (same data as in a). For better visualisation 1.5 periods are plotted, i.e., the last 12 points are identical with the first ones. Abscissa: time of day in hours. The *black bars below* indicates the lighting regime

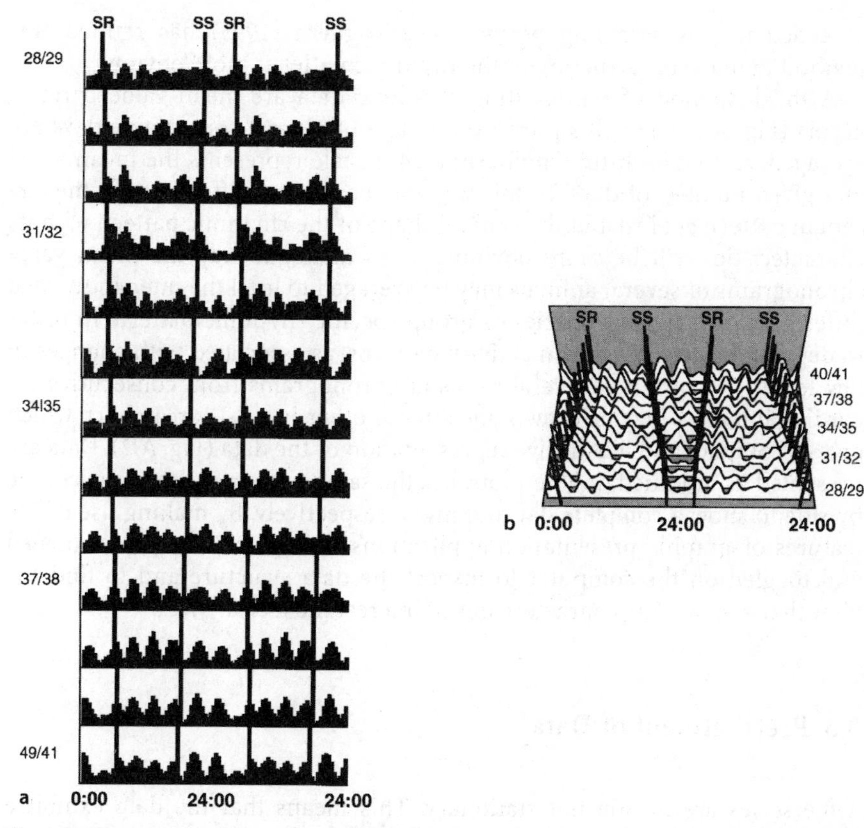

Fig. A.2. Comparison of the convential histogram plot of activity data (a) and the 3-D chronogram (b) of a 3-month record of an enclosed *Microtus oeconomus* population. Each row represents one overlapping 2-week interval (weeks 28 and 29, weeks 29 and 30, and so on). *Vertical lines* (a) and the *rows of sticks* (b), respectively, indicate times of sunrise (*SR*) and sunset (*SS*). The same pattern is drawn twice side by side to show complete days and nights. The 3-D chart gives the additional advantage that the plot may be turned and toggled on the computer for visual inspection and to find the most illustrative perspective

waveform, and rhythm amplitude, period and phase may be estimated. For long data series value/time plots are not practicable. In these cases raster plots (actograms) are preferred showing the cycles (days) one below the other (Fig. A.1b). Conventionally, the data are double-plotted. This means that on the first line the data of the first and the second days were plotted, on the next line the data of the second and the third days and so on. When the active and inactive states are quite distinct, data thus presented permit direct graphic measurement of phase and period. The latter may be measured from the slope of a line joining equivalent points, e.g. the onsets of activity, over a number of days. In some cases it will be useful to convert the data from analogue to digital form by noting only if a mean or threshold value has been exceeded in any sampling period. Diez-Noguera (1995) has refined this method to make the structure of the rhythmic patterns more obvious.

A third method of representing rhythmic data are mean value chronograms (Fig. A.1c). For this purpose the data of several consecutive days are averaged, so that each time point on a 24-h scale represents the mean value of a given number of days. In this way the day-to-day differences in the circadian pattern get lost, but the general shape of the rhythmic pattern with its characteristics will be more obvious. In an analogue way the mean value chronograms of several animals may be averaged to level the inter-individual differences and to get a species or group specific rhythmic pattern. In order to depict long-term trends in activity patterns, as associated with changes in day length for instance, several series of chronograms from consecutive recording intervals may be drawn one after another in a 3-D area-chart, which gives a compact and illustrative representation of the data (Fig. A.2). Data are presented as modified double-plots, i.e., the same pattern is drawn twice side by side to show a complete day and night, respectively. By making use of the features of graphic presentation applications the 3-D charts can be turned and toggled on the computer to inspect the data structure and to find the view that displays the characteristics of interest most clearly.

3.3 Pretreatment of Data

Time series are mostly not stationary. This means that the data exhibit a long-term trend, and/or the variances of subsequent sections are different due to changing amplitude, phase and/or spectral composition. The trend may reflect a superimposed rhythmic process of low frequency. To achieve quasistationarity, which is a prerequisite for time series analyses, long-term trends should be removed and the registration time (length of the time series) should be shortened to a reasonable minimum.

A trend may be reduced by fitting a linear or curvilinear regression line to the data set and subsequently compensating for the slope of the line at each datum point (Peil and Schmerling 1982; Morgan and Minors 1995). The main problem, however, is to define the shape of the trend. Since in most cases no a-priori information exists, this can be done only empirically by visual inspection of the graphic representation of the time series. Sometimes this may lead to an incorrect separation of the trend and the rhythmic signal.

Time series are often noisy due to random spontaneous variations and to oscillations at periods other than that of immediate interest. For example, a data series suspected of having a circadian rhythmicity may also contain ultradian and random fluctuations. A high random noise level may cause a number of problems concerning the visual analysis of graphic representations and the mathematical estimation of accurate rhythm characteristics. Therefore, pretreatment of the data may be required. However, the noise, or better the signal-to-noise ratio itself, may be of interest to characterise the biological time series.

A low-pass filter may be applied to eliminate high frequencies. If the bandwidth of the filter is chosen so as not to obscure periodicities of experimental interest, the corresponding spectral peaks in the periodogram should be enhanced. A smoothing of the time series using a 'moving average' (Morgan and Minors 1995) or empirical regression (Peil and Schmerling 1982) has the effect of attenuating the higher frequency components and the random fluctuations. Autocorrelation of the time series also decreases the level of random noise, while any rhythmic components remain. As a method for periodogram analysis, autocorrelation is less popular in biology (Dutilleul 1995; see, however, below).

Most of the methods of time series analysis require data sampled at equally-spaced intervals of time. If there are any lacks they should be eliminated by interpolation or spline-regression (Schmerling and Peil 1989). The latter may also be used to describe the shape of the rhythm. The strength of this method is that it needs no assumptions concerning the waveform of the rhythmic process.

The right method for testing whether there is a significant rhythmicity or not is the analysis of variance (ANOVA). Commonly the one-factor ANOVA is used. This assesses whether the variation between several time points is significantly greater than the random variation within them, e.g. whether the variation of a given variable is bigger over the 24-h cycle than between consecutive days. As well as tests for parametric data, ANOVA tests are available for non-parametric measurements (the Kruskal-Wallis and Friedman tests). The criteria for use of one or the other are those that are normally applied (for details see any textbook of statistics).

The strength of the ANOVA is that the result does not depend on the shape of the rhythm, which means that no assumptions concerning the waveform have to been made. Furthermore, data with any period can be analysed and times of assessment within one cycle do not have to be evenly spaced. The shortcoming of the ANOVA tests is that, even if they establish that rhythmicity exists, they do not give information as to its characteristics, e.g. waveform, amplitude or phase. Furthermore, the method cannot be used to detect the period which has to be known beforehand.

3.4 Mathematical Methods of Time Series Analyses

3.4.1 Cosinor Analysis

Probably the most frequently used (and misused) method for investigating rhythms is the cosinor analysis. It was first described by Halberg et al. (1967) and is fully documented elsewhere (Nelson et al. 1979; Bingham et al. 1982).

The method models the observed data in terms of a cosine curve (Fig. A.3). That is, the method assumes that the dependence of the variable y_i on the time t_i can be represented by the equation:

$$y_i = C_0 + C_1 \cdot \cos(\varpi t_i + \varphi) + e_i,$$

where C_0 is the mesor or the mean level of the cosine curve, C_1 is the amplitude of the curve, ϖ is $2\pi/\tau$ which is the angular frequency of the curve, τ is the period length, φ is the acrophase or phase angle of the maximum of the curve and e_i is the residual error of the *i*th data point.

When the period length (τ) is known or can be assumed, this equation may be solved by conventional methods of linear least squares regression analysis. The estimated values (C_0, C_1 and φ) are used to characterise the rhythmic process. The mesor (C_0) is the rhythm-determined average, i.e., the mean level of the cosine curve. For equidistant data covering an integral number of cycles, C_0 is equal to the arithmetic mean. The amplitude (C_1) is a measure of one half the extent of rhythmic change in a cycle, or the difference between maximum and mesor. The acrophase (φ) is the lag of the crest time of the best fitting cosine function from a defined reference point. The latter may be an arbitrary clock hour (e.g. 0:00 on day 1 of the study) for circadian rhythms and a date for infradian or circannual rhythms. For example, if the reference point is 0:00, then an acrophase of –7 h means that the crest time (maximum) of the function is at 7:00 a.m. (since 7:00 is later than 0:00 the acrophase has a negative value!). The acrophase is conventionally measured in time units (minutes, hours, days, weeks etc.). Sometimes angular

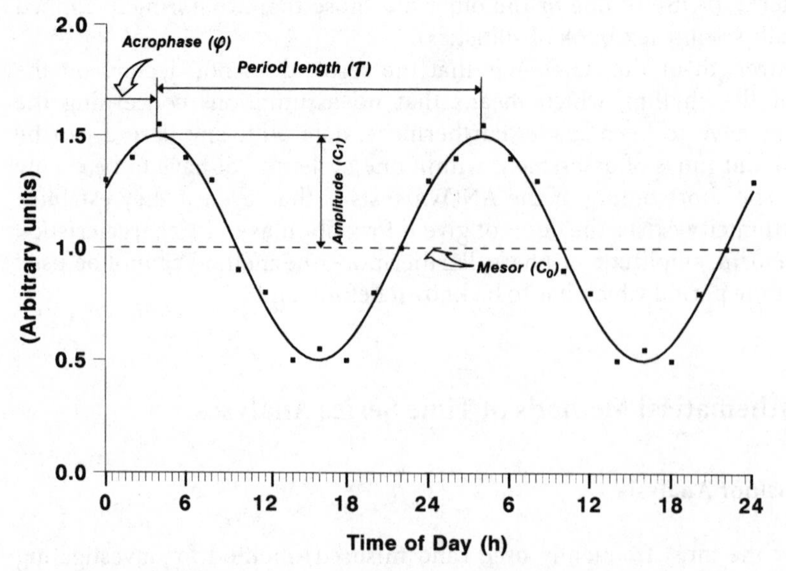

Fig. A.3. Schematic view of the cosine function and its key parameters (mesor, amplitude, period length, and acrophase) used for fitting the curve to a time series (hypothetical data). *Black dots* data points ($y_i = f(t_i)$), *solid line* best fitted cosine curve ($y_i = C_0 + C_1 \cdot \cos(\varpi t_i + \varphi) + e_i$)

measures (radians, degrees) are preferred because they are directly applicable to any period length ($2\pi = 360° =$ period length). The lag of the trough time from a defined reference point is called bathyphase. For some rhythmic processes it may be a more stable characteristic than the acrophase (Halberg et al. 1977).

The significance of the fitted curve may be tested using the F-statistic. If the computed F-value is bigger than that derived from F tables then the amplitude of the fitted curve is different from zero at the chosen level of probability. In other words, it gives a probability of the data being better described by a cosine curve than by a straight line. The test does not indicate that the data or the underlying rhythmic process are cosinusoidal in shape. The F-test can be also used to calculate confidence limits for the amplitude and the acrophase (Bingham et al. 1982). The percentage rhythm (PR) is a measure of how well the cosine function fits the data. It is the percentage of variability accounted for by the fitted curve. By calculating PR for different trial periods one can get some kind of periodogram.

In addition to the single cosinor, a procedure applicable to single biological time series from an individual, methods exist that describe group or population rhythm characteristics. The group mean-cosinor uses the results derived by the single cosinor to determine population amplitude and acrophase for characterising a rhythm in a particular group (Nelson et al. 1979).

Cosinor analysis is a simple procedure giving intelligible results (mesor, amplitude, acrophase) which may be statistically verified. Programs and detailed documentations are easily available. Besides this, the strength of the cosinor is that it can deal with equidistant as well as non-equidistant data and also gives good results with short time series. However, using the cosinor one should be aware that the method does make some assumptions. First of all, the time series should approximately be a cosine curve. If the raw data are far from cosinusoidal, or peaks and troughs are not separated by half the period, the estimates of the rhythmic parameters may be erroneous. To investigate whether the cosine curve describes the shape of the time series adequately a sinusoidality test has been proposed (Nelson et al. 1979). Nevertheless, the method should always be used in conjunction with inspection of the raw data.

The cosinor may also be useful if the shape of the time series cannot be approximated by a cosine curve. In such cases one should ensure that no changes in the shape of the rhythm occur over successive days, or that the shapes of time series (e.g. of different individuals to be compared) are similar. This is important since the estimates of amplitude and acrophase will be different as a consequence of changes of the waveform. Furthermore, one should know how to interpret the estimates. For example, the acrophase of a circadian activity rhythm may be interpreted not as the maximum of the activity but as the 'centre of gravity' of activity in the course of the 24-h day. The amplitudes will be no objective assessment, although relative changes may be taken as a measure if the shapes are similar.

Another feature of the cosinor is that the fitting of a cosine function can be realised by a linear regression procedure only if the period length is known or can be assumed. Otherwise the equation should be solved using non-linear techniques or the period has first to be estimated by methods described later in this chapter. Finally, the residual errors are assumed to be independent random normal deviates with a mean value of zero. This should be tested, otherwise the statistical significance of the fit and the confidence regions for the rhythm parameters must be questioned (De Prins and Waldura 1993; Morgan and Minors 1995). Because the data do not always fit the cosine curve, several attempts have been made using non-sinusoidal functions or by fitting two or more cosines simultaneously (Poesche 1982; Hildebrand et al. 1984; Hardeland 1995).

3.4.2 Chi^2-Periodogram

This method was developed by Sokolove and Bushell (1978) using Enright's formulation of the Whittaker periodogram (Enright 1965). The method is based on the intuitive notion that if one divides a time series into sections of P observations, then the sections will be very similar to each other if the time series has the period P, but not if it has a different period. A measure of the strength of the period P (Q_P) can be obtained by writing the sections one below the other and comparing the variance between the columns with that of the whole data set. If this procedure is repeated with different trial periods and Q_P is plotted against P one can get a periodogram, showing regular oscillations as peaks. The sample distribution of Q_P approximates the Chi^2-distribution, which can be used to test the significance of the obtained periods.

The Chi^2-periodogram may only be computed at periods which are exact multiples of the time interval between successive observations, in other words a period may only be assessed to the nearest sampling interval. To increase the resolution of periodic components it is necessary, therefore, to increase the sampling rate. Components at neighbouring frequencies may be poorly resolved. If periodogram analysis yields a sharp peak then it is likely that a single period is present and its estimate will be sufficiently accurate. A broad maximum cannot be taken to represent a single or a stationary periodic component, and the estimate of period will reflect only an averaged value. In the case of non-stationary time series one can try to split the data into shorter subsets and to analyse them separately or by moving a 3-day window through the data day by day.

The periodogram will show peaks at each multiple of the basic period. For example, if the data contain a period of 12 h the periodogram will show up peaks at 12 h, 24 h, 36 h etc. To decide whether the data contain a real circadian component or whether the peak in the periodogram is only an artifact, one has to rely on visual inspection of the raw data. Statistical tests cannot discriminate harmonics.

If the data are collected such that time-of-occurrence of each event is recorded rather than number of events per unit time the Poisson periodogram may be used which is based on the theory of stochastic point processes (cf. Sokolove and Bushell 1978). Legendre et al. (1981) generalised the Whittaker periodogram to qualitative, non-metric data series. The advantage of Whittaker-type periodogram analysis is that the procedures of calculation are relatively straightforward and easy to apply. It also gives good results with relatively short time series and is tolerant of low signal-to-noise ratios. Furthermore, no assumptions regarding the shape of the rhythmic component are required.

3.4.3 Fourier Analysis

Fourier analysis is a widely used method to reveal rhythmic components in time series (Bloomfield 1976; Kay 1988; Priestley 1992). All algorithms used today refer to the classical Schuster periodogram (Schuster 1906). This method of spectral analysis is based on Fourier's finding that any waveform can be exactly described by a combination of pure sine curves:

$$y_i = a_0 + \sum_{i=1}^{L}\left[a_i \cdot \cos\left(\frac{2\pi it}{N}\right) + b_i \cdot \sin\left(\frac{2\pi it}{N}\right)\right],$$

where L is N / 2 or (N - 1) / 2, depending on whether the number of observations in the series (N) is even or odd. The expression in square brackets represents a single sinusoid with period N/i. The Fourier coefficients (a_i and b_i) are used to calculate the spectral energy.

Fourier analysis produces a periodogram in which each period is associated with its spectral energy or its 'power' (therefore also called a 'power spectrum'). Any regular oscillation will show up as a peak in the periodogram and will not interfere with the detection of other oscillations. The spectral resolution is dependent on the length of the time series and on the interval between measurements. The shortest period is equal to two sampling intervals, the longest is equal to the length of the time series. Formal statistical tests are available for the significance of the peaks. Their usefulness is limited, however, so it is necessary to compare the periodogram with the graphic representation of the raw data.

The period lengths represented were N/1, N/2, N/3, ... N/L (the corresponding frequencies are called Fourier frequencies.). Thus, in the range of shorter periods the resolution will be higher. To achieve a high resolution in the circadian range it is necessary to analyse sufficiently long time series. Moreover, a rhythm with period P will show up in the periodogram only if the length of the time series is an exact multiple of P. A periodicity which is not repeated a whole number of times during the period of observation is subject to the phenomenon known as 'spectral leakage'. The periodogram will exhibit relatively large positive values at those Fourier frequencies closest to the actual frequency, and somewhat smaller positive values at other Fourier frequencies. The 'power' which should be concentrated at a single point of the spectrum is found to 'leak' into neighbouring frequencies.

When the true period is not known in advance, it may easily happen that the duration of the time series is not an exact multiple of the period. In such cases one may drastically increase the length of the data set. Although, a much more practical solution is to repeat the analysis several times with subtle shortening of the data series.

Using Fourier analysis the time series will be represented completely by a sum of sine-curves. This means that any function will be decomposed into sine-curves, even if the waveform is rectangular. Since the equation (see above) does not contain an error term, noise will be decomposed into sine-waves as well. Therefore, the periodogram will comprise the periodogram of the noise term superimposed on the periodogram of the signal. That is the reason why the method is not so tolerant of low signal-to-noise ratios. Moreover, the periodogram will exhibit peaks at some or all harmonics of the basic period (P/1, P/2, P/3 etc.). As a consequence, a periodogram may comprise peaks which are spurious in the sense that they have no biological significance. Nevertheless, it represents a definite characteristic of the analysed time series.

The periodogram described above only covers Fourier frequencies. It is sometimes called 'harmonic analysis' and may be extended to non-Fourier frequencies. This may help to determine the precise location of peaks in the periodogram although peaks at frequencies which are different less than 1/N cannot be discriminated as they will not show distinct peaks. The advantage of extending the periodogram to a continuous set of frequencies is that it may be viewed as an estimate of the spectral density function (Chatfield 1989).

Periodograms are usually smoothed by averaging the ordinates over neighbouring frequencies using a set of weights known as a 'spectral window' (Kay 1988; Priestley 1992). This will reduce contamination of the spectrum by random error. However, the problem of spectral leakage remains, so the resolvability will not improve.

3.4.4 Maximum Entropy Spectral Analysis (MESA)

Autoregressive (maximum entropy) methods have been applied extensively in geophysical research but are becoming increasingly popular for identifying periodic components in biological time series, probably due to their high resolution capability (Burg 1967; Lacoss 1971; Linkens 1979). The MESA method is based on the fact that a periodic time series is predictable (or 'deterministic'). This means that the value at a given time (x_t) is determined by the m previous values (x_{t-1}, x_{t-2}, ... x_{t-m}) and an unpredictable part or error (e_t):

$$x_t = a_1 \cdot x_{t-1} + a_2 \cdot x_{t-2} + \ldots + a_m \cdot x_{t-m} + e_t.$$

The autoregressive coefficients ($a_1, a_2, \ldots a_m$) determine the spectrum of the process. The area under a peak rather than its height, as in Fourier analysis, is a measure of the amplitude of the underlying periodic component. Since it may be difficult to quantify the area, MESA is more useful for determining the frequencies of periodic components than for assessing their relative importance. Statistical tests to confirm the presence of certain periodicities in the data are practically not available.

The strength of MESA lies in its success in identifying regular spectral components and its ability to separate them from one another if their frequencies are close, particularly in short time series and in the case of low signal-to-noise ratios (cf. Fourier analysis). Therefore it is sometimes superior to the methods described above for the resolution of spectral peaks. In long series the results will be not distinguishable from Fourier analysis (spectral window estimates).

When using MESA, problems may occur because time series do not generally admit an exact representation as an autoregressive process. Moreover, it may be difficult to select the right model order. An order which is too low leads to incorrect estimations; in the case of orders which are much too high spurious peaks can occur (Linkens 1979). Although a number of tests have been proposed, neither of the automatic methods gives unambiguous results (Kay 1988). Therefore, it depends mainly on the experience of the investigator to select the right model order. In practice it may be necessary to experiment with different values of m in order to produce the most informative spectral estimate. In any case, m should not exceed N/2.

Both the autoregressive and the classical approaches are based on the idea of Fourier decomposition to reveal the most prominent frequencies in a data series. In contrast to Fourier analysis, where the peaks in the spectra characterise the actual data set even though the periodicity may be biologically non-significant, an autoregressive spectrum can often reveal spurious peaks by choosing an inappropriate value for the order of the process. Therefore, a preliminary calculation of a Fourier spectrum is recommended.

3.5 Time Series Analysis – A Conclusion

A number of methods exists to analyse and present biological time series, but the methods used in one investigation may not be the most appropriate for another. They must be chosen depending on the kind of data and the objective of the study (see overview given in Table A.2). Different characteristics of the rhythm, such as period, phase or amplitude may be of greater or lesser importance.

All methods make assumptions not always corresponding with the biological reality. Particularly, with respect to the shape of the underlying oscillation, sine-functions are mostly used. One should consider, however, that

Table A.2. Advantages and disadvantages of the described methods of time series analysis for estimating period length and/or spectral composition[a]. *Chi²-P.*: Chi²-periodogram, *Fourier*: Fourier-analysis, *MESA*: maximum entropy spectral analysis

	Chi²-P.	Fourier	MESA
Advantages			
Easy to apply	+	+	–
No assumptions concerning waveform	+	–	–
High resolution capability	–	–	+
Good results with:			
short time series	+	–	+
low signal/noise ratios	+	–	+
Spectrum well characterises the data set	–	+	–
'Power' of spectral components	+	+	–
Statistical tests available	+	(+)	–
Disadvantages			
Experience with method necessary	–	–	++
Decomposition into sinusoidal oscillations	–	+	+
For high spectral resolution it is necessary to have:			
long time series	–	+	–
a high sampling rate	+	–	–
Peaks at each multiple of the basic period	+	–	–
Peaks at harmonics of the basic period	–	+	?

[a] In the case of ANOVA and cosinor the period length must be known beforehand.

such waveforms are rather unlikely in biological systems. Moreover, the circadian patterns are the result of interacting oscillations with equal or different period lengths, influencing each other's period, amplitude, and phase. The methods consider, however, only periods with different period length whose contribution to the circadian pattern is simply additive. Furthermore, a statistically significant result indicates not more (and not less) than a sufficient mathematical description of the time series by a particular function or the estimated values. It does not verify that a rhythmic process with this period length is in fact underlying.

Despite the limitations mentioned above, mathematical methods may be very useful for analysing and describing biological rhythms. However, any analysis must be done in conjunction with visual inspection of a graphic representation of the raw data. Furthermore, the biological rather than the statistical value of the significance should be considered.

4 Activity Indices – Special Solutions for Noisy Field Data

Specific problems with the quantitative analysis of activity records arise with field data that are burdened with high variance. This is not only due to the above treated methodological uncertainties associated with field recording, but is also an inevitable consequence of high biological variability and the influence of numerous uncontrollable environmental factors. Hence, field recordings often have to be lumped for several days and/or several individuals in order to derive any idea about the underlying rhythmic pattern. However, with pooling, the basic conditions for time series analyses are corrupted, so they can no longer be applied to the data set. On the other hand, field data are often either long-term data or are related to questions like differences in activity behaviour between treatments or ecological situations. So one has to consider changes of or differences between pattern characteristics. In earlier work the only way to do this was by verbal description, which is qualitative rather than quantitative, and does not allow for statistical testing.

A suggested solution for this kind of problem may be a set of activity indices that value the relative proportion of different aspects of activity patterns. Although very simple from the general data structure point of view, i.e., activity per time interval, activity patterns in fact reveal a very complex information structure. The main idea of activity indices is not to describe behaviour by one super-measure that represents all this information at once, but rather to break the analysis down into digestible bits.

The easiest and most widely applied index is the general activity level, given as number of counts or records (Σ c) per day (DAY) and per number of animals considered (e.g. 'minimum number alive', MNA, in capture-mark-recapture studies):

$$n^* = \frac{\sum c}{\mathrm{DAY} \cdot \mathrm{MNA}}.$$

This simply represents the average activity per animal and day and allows comparisons of activity between categories like sex and age classes, between ecological situations like habitat, predation risk and competition, or between seasons.

Indices that value the relative amount of diurnal against nocturnal activity are also widely used. Several equations have been suggested in the literature of which the I_D-index of diurnality may serve as an example:

$$I_D = \left[\frac{\frac{\sum cL}{hL}}{\frac{\sum cL}{hL} + \frac{\sum cD}{hD}} \right] \cdot 2 - 1,$$

were Σ cL and Σ cD are the number of activity records during the day and night, respectively, and hL and hD are day length and night length. I_D is positive when diurnal activity prevails (maximum: +1 when exclusively active during daytime) and negative when nocturnal activity prevails (minimum: −1 when exclusively active at night). As an important feature, the index value is neither affected by the total number of records nor by a changing photoperiod, which is an essential condition for comparisons.

A similar logic can be applied to specific aspects of activity distribution over the 24-h day. For example, one may ask for differences or changes in the amount of activity at twilight which can be valued by an index I_C of crepuscularity:

$$I_C = \log\left[\frac{\dfrac{\sum c(SR \pm 1) + \sum c(SS \pm 1)}{\sum c}}{\dfrac{4}{48}}\right],$$

where $\Sigma\, c(SR \pm 1)$ and $\Sigma\, c(SS \pm 1)$ are the sums of records during, in this case, the 30-min intervals prior and subsequent to sunrise (SR) and sunset (SS), respectively, and $\Sigma\, c$ is the total number of records. So the activity during four specific 30-min intervals is set in relation to all 48 intervals of 30 min during the 24-h day. I_C is positive when activity around sunrise and sunset is increased compared with the 24-h average activity level, and negative when decreased. The relation is expressed as a logarithm to make increases and decreases directly comparable: an index value of +0.3 indicates twilight activity twice as high as the 24-h average, while the calculation yields –0.3 when it is only half of the 24-h average. This type of index is a most flexible tool and may be applied to any time interval of special interest, such as the first 2 h after sunset, the 3 h around noon, or any other time window.

The autocorrelation function (ACF; Box et al. 1994) in fact belongs to time series analyses, but because it allows only rough approximations of the rhythmic component it is not widely used in experimental biology (see above). However, since the ACF is a very robust and easily applied method it is often appropriate for the analysis of field data. Statistical significance of a rhythm is given when the first positive peak of r in the correlogram exceeds the 95% confidence limit, estimated by

$$\pm 2 \cdot \sqrt{1/N},$$

in which N is the number of time intervals initially included in the ACF. The ratio between r at the first positive peak and the confidence limit can be considered as an index of periodicity I_P, that may be used, for example, to compare the prevalence and distinctness of ultradian rhythmicity in different vole populations or in the course of seasons.

All activity indices may be statistically treated by ANOVAs, and repeated measurement ANOVAs are appropriate when long-term data series are concerned. However, transformation of index data will, in most cases, be necessary to fulfil the condition of normally distributed data. Non-parametric Spearman rank correlation will avoid this problem when activity indices are related to environmental data on an interval scale. Activity indices may be condemned for their extreme statistical crudeness, but one has to be aware that they have been developed for data sets where restrictive methods according to pure statistical doctrine fail. For the time being, the alternatives in this situation are to rely on quick and dirty methods, or to stick with merely qualitative descriptions. However, future developments, such as resampling procedures like bootstrapping and jackknife in particular, may provide promising new approaches.

References

Airoldi JP (1979) Etude du rythme d'activité du campagnol terrestre, *Arvicola terrestris scherman* Shaw. Mammalia 43:25–52

Bäumler W (1975) Activity of some small mammals in the field. Acta Theriol 20:365–377

Banks EM, Brooks RJ, Schnell J (1975) A radiotracking study of home range and activity of the brown lemming (*Lemmus trimucronatus*). J Mammal 56:888–901

Baron G, Pottier J (1977) Determination of activity patterns of *Clethrionomys gapperi* in an artificial tunnel system. Naturaliste Can 104:341–351

Benjamini L (1988/1989) Diel activity rhythms in the levant vole, *Microtus guentheri*. Isr J Zool 35:215–228

Bingham Ch, Arbogast B, Guillaume GC, Lee J-K, Halberg F (1982) Inferential statistical methods for estimating and comparing cosinor parameters. Chronobiologia 9:397–439

Bloomfield P (1976) Fourier analysis of time series: an introduction. Wiley, New York

Blumenberg D (1986) Telemetrische und endoskopische Untersuchungen zur Soziologie, zur Aktivität und zum Massenwechsel der Feldmaus, *Microtus arvalis* (Pall.). Z Angew Zool 20: 301–344

Borer KT, Dennis R (1991) Activity disc and cage for continuous measurement of running activity and core temperature in hamsters. Physiol Behav 50:1057–1061

Bovet J (1972) On the social behavior in a stable group of long-tailed field mice (*Apodemus sylvaticus*). II. Its relations with distribution of daily activity. Behaviour 41:55–67

Box GEP, Jenkin GM, Reinsel GC (1994) Time series analysis: forecasting and control. Prentice Hall, Englewood Cliffs

Burg JP (1967) Maximum entropy spectral analysis. 37th Meeting Soc Explor Geophys, Oklahoma City

Calhoun JB (1945) Diel activity rhythms of the rodents, *Microtus ochrogaster* and *Sigmodon hispidus hispidus*. Ecology 26:251–273

Carthew SM, Slater E (1991) Monitoring animal activity with automated photography. J Wildl Manage 55:689–692

Chatfield C (1989) The analysis of time series, an introduction. Chapman and Hall, London

Chute FS, Fuller WA, Harding PRJ, Herman TB (1974) Radio tracking of small mammals using a grid of overhead wire antennas. Can J Zool 52:1481–1488

Crowcroft P (1953) The daily cycle of activity in British shrews. Proc Zool Soc Lond 123: 715–729

Daan S, Slopsema S (1978) Short-term rhythms in foraging behaviour of the common vole, *Microtus arvalis*. J Comp Physiol A 127:215–227

Davis DHS (1933) Rhythmic activity in the short-tailed vole, *Microtus*. J Anim Ecol 2:232–238

DeCoursey PJ, Menon SA (1991) Circadian photo-entrainment in a nocturnal rodent: quantitative measurement of light-sampling activity. Anim Behav 41:781–785

De Kock LL, Rohn I (1971) Observations on the use of the exercise-wheel in relation to the social rank and hormonal conditions in the bank vole (*Clethrionomys glareolus*), and the Norway lemming (*Lemmus lemmus*). Z Tierpsychol 29:180–195

De Prins J, Waldura J (1993) Sightseeing around the single cosinor. Chronobiol Int 5:395–400

Diez-Noguera A (1995) Non-standard methods of analysis: graphic imaging processing. Biol Rhythm Res 26:202–215

Dutilleul P (1995) Rhythms and autocorrelation analysis. Biol Rhythm Res 26:173–193

Enright JT (1965) The search for rhythmicity in biological time-series. J Theoret Biol 8:426–468

Enright JT (1981) Data analysis. In: Aschoff J (ed) Biological rhythms. Handbook of neurobiology, vol 4. Plenum Press, New York, pp 21–40

Erkinaro E (1969) Der Phasenwechsel der lokomotorischen Aktivität bei *Microtus agrestis* (L.), *M. arvalis* (Pall.) und *M. oeconomus* (Pall.). Aquilo Ser Zool 8:1–31

Erkinaro E (1972) Phase shift of locomotory activity in a birch mouse, *Sicista betulina*, before hibernation. J Zool Lond 168:433–438

Fuller WA (1996) Introduction to statistical time series. Wiley, New York

Gaulin SJC, FitzGerald RW, Wartell MS (1990) Sex differences in spatial ability and activity in two vole species (*Microtus ochrogaster* and *M. pennsylvanicus*). J Comp Psychol 104:88–93

Gerkema MP, van den Leest F (1991) Ongoing ultradian activity rhythms in the common vole, *Microtus arvalis*, during deprivations of food, water and rest. J Comp Physiol A 168:591–597

Gerkema MP, Verhulst S (1990) Warning against an unseen predator: a functional aspect of synchronous feeding in the common vole, *Microtus arvalis*. Anim Behav 40:1169–1178

Gerkema MP, Groos GA, Daan S (1990) Differential elimination of circadian and ultradian rhythmicity by hypothalamic lesions in the common vole, *Microtus arvalis*. J Biol Rhythms 5:81–95

Girling AJ (1995) Periodograms and spectral estimates for rhythm data. Biol Rhythm Res 26: 149–172

Gurnell J (1975) Notes on the activity of wild wood mice, *Apodemus sylvaticus*, in artificial enclosures. J Zool Lond 175:219–229

Halberg F, Tong YL, Johnson EA (1967) Circadian system phase - an aspect of functional morphology; procedures and illustrative examples. In: von Mayersbach H (ed) The cellular aspect of biorhythms. Springer, Berlin Heidelberg New York, pp 20–48

Halberg F, Caradente F, Cornelissen G, Katinas GS (1977) Glossary of chronobiology. Chronobiologia 4 [Suppl 1]:5–189

Halle S (1988) Locomotory activity pattern of wood mice as measured in the field by automatic recording. Acta Theriol 33:305–312

Halle S, Lehmann U (1987) Circadian activity patterns, photoperiodic responses and population cycles in voles. I. Long-term variations in circadian activity patterns. Oecologia 71: 568–572

Halle S, Lehmann U (1992) Cycle-correlated changes in the activity behaviour of field voles, *Microtus agrestis*. Oikos 64:489–497

Hamilton WJ (1937) Activity and home range of the field mouse, *Microtus pennsylvanicus pennsylvanicus* (Ord.). Ecology 18:255–263

Hardeland R (1995) Cosine or non-sinusoidal fitting as a means of detection of rhythms and determination of period lengths in short biological time series. Chronobiol Int 26:194–201

Harland RM, Millar JS (1980) Activity of breeding *Peromyscus leucopus*. Can J Zool 58:313–316

Hatfield DM (1940) Activity and food consumption in *Microtus* and *Peromyscus*. J Mammal 21:29–36

Hildebrand R, Haubitz I, Schultz M (1984) Problems in fitting a cosine curve. Chronobiol Int 1: 93–95

Hoek HN, Heine G, Schwarz G (1985) A device for automatic radio tracking of small mammals using a ground wire antenna in a large enclosure. In: Brooks RP (ed) Nocturnal mammals: techniques for study. School of Forest Resources Research Paper no 48, Pennsylvania State University, pp 39–44

Hoogenboom I, Daan S, Dallinga JH, Schoenmakers M (1984) Seasonal change in the daily timing of behaviour of the common vole, *Microtus arvalis*. Oecologia 61:18–31

Hulsey MG, Martin RJ (1991) A system for automated recording and analysis of feeding behavior. Physiol Behav 50:403–408

Johnson MS (1926) Activity and distribution of certain wild mice in relation to biotic communities. J Mammal 7:245–277

Johst V (1973) Das Aktivitätsmuster der Schermaus *Arvicola terrestris* (L.). Zool Jb (Physiol) 77: 98–106

Joutsiniemi SL, Leinonen L, Laakso ML (1991) Continuous recording of locomotor activity in groups of rats: postweaning maturation. Physiol Behav 50:649–654

Karulin BE, Litvin VY, Nikitina NA, Khlyap LA, Okhotsky YV (1976) A study of activity, mobility and diurnal range in *Microtus oeconomus* on the Yamal peninsula by means of marking with radioactive cobalt. Zool Zh 55:1052–1060

Kavanau JL (1963) Continuous automatic monitoring of the activities of small captive animals. Ecology 44:95–110

Kavanau JL, Ramos J (1975) Influences of light on activity and phasing of carnivores. Amer Natur 109:391–418

Kay SM (1988) Modern spectral estimation: theory and application. Prentice Hall, Englewood Cliffs

Kenagy GJ (1976) The periodicity of daily activity and its seasonal changes in free-ranging and captive kangaroo rats. Oecologia 24:105–140

Kenagy GJ, Hoyt DF (1980) Reingestion of feces in rodents and its daily rhythmicity. Oecologia 44:403–409

Kleinknecht S, Erkert HG, Nelson JE (1985) Circadian and ultradian rhythms of activity and O_2-consumption in three nocturnal Marsupialian species: *Petaurus breviceps*, Phalangeridae; *Dasyuroides byrnei*, Dasyuridae; *Monodelphis domestica*, Didelphidae. Z Säugetierkd 50: 321–329

Kolb HH (1992) The effect of moonlight on activity in the wild rabbit (*Oryctolagus cuniculus*). J Zool Lond 228:661–665

Lacoss RT (1971) Data adaptive spectral analysis methods. Geophysics 36:661–675

Legendre L, Frechette M, Legendre P (1981) The contingency periodogram: a method of identifying rhythms in series of nonmetric ecological data. J Ecol 69:965–979

Lehmann U (1976) Short-term and circadian rhythms in the behaviour of the vole, *Microtus agrestis* (L.). Oecologia 23:185–199

Lehmann U, Sommersberg CW (1980) Activity patterns of the common vole, *Microtus arvalis* - automatic recording of behaviour in an enclosure. Oecologia 47:61–75

Levitsky DA (1970) Feeding patterns of rats in response to fasts and changes in environmental conditions. Physiol Behav 5:291–300

Linkens DA (1979) Maximum entropy analysis of short time-series biomedical rhythms. J Interdiscip Cycle Res 10:145–162

Lodewijckx E, Verhagen R, Verheyen WN (1984) Activity patterns of wild wood mice, *Apodemus sylvaticus* (L.), from the Belgian northern Campine. Annls R Soc Zool Belg 114:291–301

Loxton RG, Raffaelli D, Begon M (1975) Coprophagy and the diurnal cycle of the common shrew, *Sorex araneus*. J Zool Lond 177:449–453

Lund M (1970) Diurnal activity and distribution of *Arvicola terrestris terrestris* L. in an outdoor enclosure. EPPO Publ Ser A 58:147–158

Madison DM (1981) Time patterning of nest visitation by lactating meadow voles. J Mammal 62:389–391

Mather JG (1981) Wheel-running activity: a new interpretation. Mammal Rev 11:41–51

Miller RS (1955) Activity rhythms in the wood mouse, *Apodemus sylvaticus* and the bank vole, *Clethrionomys glareolus*. Proc Zool Soc Lond 125:505–519

Mironov AD (1990) Spatial and temporal organization of populations of the bank vole, *Clethrionomys glareolus*. In: Tamarin RH, Ostfeld RS, Pugh SR, Bujalska G (eds) Social systems and population cycles in voles. Birkhäuser, Basel, pp 181–192

Morgan E, Minors D (1995) The analysis of biological time-series data: some preliminary considerations. Biol Rhythm Res 26:124–148

Mossing T (1975) Measuring small mammal locomotory activity with passage counters. Oikos 26:237–239

Myllymäki A, Aho J, Lind EA, Tast J (1962) Behaviour and daily activity of the Norwegian lemming, *Lemmus lemmus* (L.), during autumn migration. Ann Zool Soc Vanamo 24, 2:1–31

Nelson W, Tong YL, Lee JK, Halberg F (1979) Methods for cosinor-rhythmometry. Chronobiologia 6:305–323

Ostermann K (1956) Zur Aktivität heimischer Muriden und Gliriden. Zool Jb (Physiol) 66: 355–388

Ostfeld RS (1986) Territoriality and mating system of california voles. J Anim Ecol 55:691–706

Palomares F, Delibes M (1991) Assessing three methods to estimate daily activity patterns in radio-tracked mongooses. J Wildl Manage 55:698–700

Pearson AM (1962) Activity patterns, energy metabolism and growth rate of the voles *Clethrionomys rufocanus* and *C. glareolus* in Finland. Ann Zool Soc Vanamo 24:1–58

Peil J, Schmerling S (1982) Empirical regression for trend elimination and smoothing of time series. Gegenbaurs Morph Jahrb 128:324–330

Poesche WH (1982) A non-sinusoidal regression function for last square fitting of rhythmic data. J Interdiscipl Cycle Res 13:1–21

Priestley MB (1992) Spectral analysis and time series. Academic Press, London

Raptor Group (1982) Timing of vole hunting in aerial predators. Mammal Rev 12:169–181

Rasmuson B, Rasmuson M, Nygren J (1977) Genetically controlled differences in behaviour between cycling and non-cycling populations of field vole (*Microtus agrestis*). Hereditas 87: 33–42

Reid AK, Bacha G, Morán C (1993) The temporal organization of behavior on periodic food schedules. J Exp Anal Behav 59:1–27

Requirand C (1990) Pièges à horloge pour micro-mammifères. Mammalia 54:310–312

Rübsamen U, Hume ID, Rübsamen K (1983) Effect of ambient temperature on autonomic thermoregulation and activity patterns in the rufous rat-kangaroo (*Aepyprymnus rufescens*: Marsupialia). J Comp Physiol B 153:175–179

Ruf T, Klingenspor M, Preis H, Heldmaier G (1991) Daily torpor in the Djungarian hamster (*Phodopus sungorus*): interactions with food intake, activity, and social behaviour. J Comp Physiol B 160:609–615

Saarikko J, Hanski I (1990) Timing of rest and sleep in foraging shrews. Anim Behav 40:861–869

Salvioni M (1988) Rythmes d'activité de trois espèces de *Pitymys*: *Pitymys multiplex*, *P. savii*, *P. subterraneus* (Mammalia, Rodentia). Mammalia 52:483–496

Schmerling S, Peil J (1989) Local approximation and its applications in statistics. Gegenbaurs Morph Jahrb 135:255–260

Schuster A (1906) The periodogram and its optical analogy. Proc R Soc Lond A 206:69–100

Sokolove PG, Bushell WN (1978) The chi square periodogram: its utility for analysis of circadian rhythms. J Theor Biol 72:131–160

Stebbins LL (1974) Response of circadian rhythms in *Clethrionomys* mice to a transfer from 60°N to 53°N. Oikos 25:108–113

Stubbe A, Blumenauer V, Schuh J, Dawaa N (1986) Aktivitätsrhythmen von *Microtus brandti* (Radde, 1861) bei unterschiedlicher Gruppengröße. Wiss Beitr Universität Halle-Wittenberg 1985/18 (P22):49–57

Szymanski JS (1920) Aktivität und Ruhe bei Tieren und Menschen. Z Allg Physiol 18:105–162

Webster AB, Brooks RJ (1981) Daily movements and short activity periods of free-ranging meadow voles *Microtus pennsylvanicus*. Oikos 37:80–87

Weinert D (1996) Lower variability in female as compared to male laboratory mice: investigations on circadian rhythms. J Exp Anim Sci 37:121–137

Weinert D, Weiß T (1997) A nonlinear relationship between period length and the amount of activity – age-dependent changes. Biol Rhythm Res 28:105–120

Wójcik JM, Wołk K (1985) The daily activity rhythms of two competitive rodents: *Clethrionomys glareolus* and *Apodemus flavicollis*. Acta Theriol 30:241–258

Wolton RJ (1983) The activity of free-ranging wood mice *Apodemus sylvaticus*. J Anim Ecol 52: 781–794

Subject Index

Ecological Studies

Volumes published since 1993

Volume 101
Effects of Land Use Change on Atmospheric CO_2 Concentrations: South and South East Asia as a Case Study (1993)
V.H. Dale (Ed.)

Volume 102
Coral Reef Ecology (1993)
Y.I. Sorokin

Volume 103
Rocky Shores: Exploitation in Chile and South Africa (1994)
W.R. Siegfried (Ed.)

Volume 104
Long-Term Experiments With Acid Rain in Norwegian Forest Ecosystems (1993)
G. Abrahamsen et al. (Eds.)

Volume 105
Microbial Ecology of Lake Plußsee (1993)
J. Overbeck and R.J. Chrost (Eds.)

Volume 106
Minimum Animal Populations (1994)
H. Remmert (Ed.)

Volume 107
The Role of Fire in Mediterranean-Type Ecosystems (1994)
J.M. Moreno and W.C. Oechel (Eds.)

Volume 108
Ecology and Biogeography of Mediterranean Ecosystems in Chile, California and Australia (1995)
M.T.K. Arroyo, P.H. Zedler, and M.D. Fox (Eds.)

Volume 109
Mediterranean-Type Ecosystems. The Function of Biodiversity (1995)
G.W. Davis and D.M. Richardson (Eds.)

Volume 110
Tropical Montane Cloud Forests (1995)
L.S. Hamilton, J.O. Juvik, and F.N. Scatena (Eds.)

Volume 111
Peatland Forestry. Ecology and Principles (1995)
E. Paavilainen and J. Päivänen

Volume 112
Tropical Forests: Management and Ecology (1995)
A.E. Lugo and C. Lowe (Eds.)

Volume 113
Arctic and Alpine Biodiversity. Patterns, Causes and Ecosystem Consequences (1995)
F.S. Chapin III and C. Körner (Eds.)

Volume 114
Crassulacean Acid Metabolism. Biochemistry, Ecophysiology and Evolution (1996)
K. Winter and J.A.C. Smith (Eds.)

Volume 115
Islands. Biological Diversity and Ecosystem Function (1995)
P.M. Vitousek, L.L. Loope, and H. Adsersen (Eds.)

Volume 116
High Latitude Rainforests and Associated Ecosystems of the West Coast of the Americas: Climate, Hydrology, Ecology and Conservation (1996)
R.G. Lawford, P. Alaback, and E.R. Fuentes (Eds.)

Volume 117
Global Change and Mediterranean-Type Ecosystems (1995)
J. Moreno and W.C. Oechel (Eds.)

Volume 118
Impact of Air Pollutants on Southern Pine Forests (1996)
S. Fox and R.A. Mickler (Eds.)

Volume 119
Freshwater Ecosystems of Alaska. Ecological Syntheses (1997)
A.M. Milner and M.W Oswood (Eds.)

Volume 120
Landscape Function and Disturbance in Arctic Tundra (1996)
J.F. Reynolds and J.D. Tenhunen (Eds.)

Volume 121
Biodiversity and Savanna Ecosystem Processes. A Global Perspective (1996)
O.T. Solbrig, E. Medina, and J.F. Silva (Eds.)

Volume 122
Biodiversity and Ecosystem Processes in Tropical Forests (1996)
G.H. Orians, R. Dirzo, and J.H. Cushman (Eds.)

Volume 123
Marine Benthic Vegetation. Recent Changes and the Effects of Eutrophication (1996)
W. Schramm and P.H. Nienhuis (Eds.)

Volume 124
Global Change and Arctic Terrestrial Ecosystems (1997)
W.C. Oechel et al. (Eds.)

Volume 125
Ecology and Conservation of Great Plains Vertebrates (1997)
F.L. Knopf and F.B. Samson (Eds.)

Volume 126
The Central Amazon Floodplain: Ecology of a Pulsing System (1997)
W.J. Junk (Ed.)

Volume 127
Forest Decline and Ozone: A Comparison of Controlled Chamber and Field Experiments (1997)
H. Sandermann, A.R. Wellburn, and R.L. Heath (Eds.)

Volume 128
Productivity and Sustainability of Southern Forest Ecosystems in a Changing Environment (1998)
R.A. Mickler and S. Fox (Eds.)

Volume 129
Pelagic Nutrient Cycles: Herbivores as Sources and Sinks (1997)
T. Andersen

Volume 130
Vertical Food Web Interactions: Evolutionary Patterns and Driving Forces (1997)
K. Dettner, G. Bauer, and W. Völkl (Eds.)

Volume 131
The Structuring Role of Submerged Macrophytes in Lakes (1998)
E. Jeppesen et al. (Eds.)

Volume 132
Vegetation of the Tropical Pacific Islands (1998)
D. Mueller-Dombois and F.R. Fosberg

Volume 133
Aquatic Humic Substances: Ecology and Biogeochemistry (1998)
D.O. Hessen and L.J. Tranvik (Eds.)

Volume 134
Oxidant Air Pollution Impacts in the Montane Forests of Southern California (1998)
P.R. Miller and J.R. McBride (Eds.)

Volume 135
Predation in Vertebrate Communities: The Bialowieża Primeval Forest as a Case Study (1998)
B. Jedrzejewska and W. Jedrzejewski

Volume 136
Landscape Disturbance and Biodiversity in Mediterranean-Type Ecosystems (1998)
P.W. Rundel, G. Montenegro, and F.M. Jaksic (Eds.)

Volume 137
Ecology of Mediterranean Evergreen Oak Forests (1999)
F. Rodà et al. (Eds.)

Volume 138
Fire, Climate Change and Carbon Cycling in the North American Boreal Forest (2000)
E.S. Kasischke and B. Stocks (Eds.)

Volume 139
Responses of Northern U.S. Forests to Environmental Change (2000)
R. Mickler, R.A. Birdsey, and J. Hom (Eds.)

Volume 140
Rainforest Ecosystems of East Kalimantan: El Niño, Drought, Fire and Human Impacts (2000)
E. Guhardja et al. (Eds.)

Volume 141
Activity Patterns in Small Mammals: An Ecological Approach (2000)
S. Halle and N. C. Stenseth (Eds.)